Engineering Mathematics by Example

Robert Sobot

Engineering Mathematics by Example

 Springer

Robert Sobot
École nationale supérieure de l'électronique et de
ses applications
Paris, France

ISBN 978-3-030-79547-4   ISBN 978-3-030-79545-0   (eBook)
https://doi.org/10.1007/978-3-030-79545-0

This Springer imprint is published by the registered company Springer Nature Switzerland AG
The registered company address is: Gewerbestrasse 11, 6330 Cham, Switzerland

To my math teacher Mr. Miloš Belušević

Preface

This tutorial book resulted from my lecture notes developed for undergraduate engineering courses in mathematics that I have been teaching over the last several years at l'École Nationale Supérieure de l'Électronique et de ses Applications (ENSEA), Cergy in Val d'Oise department, France.

My main inspiration to write this tutorial-type collection of solved problems came from my students who would often ask, "How do I solve this? It is impossible to find the solution", while struggling to logically connect all the little steps and techniques that are required to combine before reaching the solution. In the traditional classical school system, mathematics used to be thought with the help of systematically organized volumes of problems that help us develop "the way of thinking", in other words, to learn how to apply the abstract mathematical concepts to everyday engineering problems. The same is true for music; it is also true for mathematics that in order to reach high level of competence, one must put daily effort in studying typical forms over a long period of time.

In this tutorial book, I choose to give not only the complete solutions to the given problems but also guided hints to techniques being used currently. Therefore, problems presented in this book do not provide review of the rigorous mathematical theory, instead the theoretical background is assumed, while this set of classic problems provides a playground to play and to adopt some of the main problem-solving techniques.

The intended audience of this book are primarily undergraduate students in science and engineering. At the same time, my hope is that students of mathematics at any level will find this book a useful source of practical problems for practice.

Île-de-France, France
November 30, 2020

Robert Sobot

Acknowledgements

I would like to acknowledge all those classic wonderful collections of mathematical problems that I grew up with and used as the source of my knowledge, and to say thank you to their authors for providing me with thousands of problems to work on. Specifically, I would like to acknowledge classic collections written in ex-Yugoslavian, Russian, English and French languages, some are listed in the bibliography. Hence, I do want to acknowledge their contributions, which are clearly visible throughout this book and are now being passed on to my readers.

I would like to thank all of my former and current students who I had opportunity to tutor in mathematics since my student years. Their relentless stream of questions posed with unconstrained curiosity forced me to open my mind, to broaden my horizons and to improve my own understandings.

I want to acknowledge Mr. Allen Sobot for reviewing and verifying some of the problems, for very useful discussions and suggestions, as well as his work on technical redaction of this book.

My sincere gratitude goes to my publisher and editors for their support and making this book possible.

Contents

Acronyms

$f(x), g(x)$	Functions of x
$f'(x), g'(x)$	First derivatives of functions of $f(x), g(x)$
$f(g(x))$	Composite function
$\sin(x)$	Sine of x
$\cos(x)$	Cosine of x
$\tan(x)$	Tangent of x
$\arctan(x)$	Arctangent of x
$\mathrm{sinc}\,(t)$	Sine cardinal of t
$\log(x)$	Logarithm of x, base 10
$\ln(x)$	Natural logarithm of x, base e
$\log_n(x)$	Logarithm of x, base n
$x * y$	Convolution product
i	Imaginary unit, $i^2 = -1$
j	Imaginary unit, $j^2 = -1$
$\Im(z)$	Imaginary part of a complex number z
$\Re(z)$	Real part of a complex number z
$\mathscr{F}(x(t))$	Fourier transform of $x(t)$
$X(\omega)$	Fourier transform of $x(t)$
$\xrightarrow{F}$	Apply property of Fourier transform
$\mathscr{F}^{-1}(x(\omega))$	Inverse Fourier transform of $x(\omega)$
DFT	Discrete Fourier transform
FFT	Fast Fourier transform
$\delta(x)$	Dirac distribution
$\Lambda(x)$	Triangular distribution
$\Pi(x)$	Square (rectangular) distribution
$\mathrm{III}_T(t)$	Dirac comb whose period is T
$\Gamma(t)$	Step function of t
$u(t)$	Step function t
$r(t)$	Ramp function t
$\mathrm{sign}\,(t)$	Sign function t
(a, b)	Pair of numbers
(a, b)	Open interval
$\equiv$	Equivalence
$\{a_1, a_2, a_3, \dots\}$	List of elements, vector, series
$\{a_n\}$	List of elements, vector, series
$a \in \mathbb{C}$	Number a is included in the set of complex numbers
$\forall x$	For all values of x

$\angle$	Angle, argument
$\mathbb{R}$	The set of real numbers
$\mathbb{C}$	The set of complex numbers
$\mathbb{N}$	The set of natural numbers
$\mathbb{Q}$	The set of rational numbers
$\Rightarrow$	It follows that
$\frac{dy}{dx}$	Derivative of y relative to x
u'_x	Partial derivative of $u(x, y, z, \ldots)$ relative to x
$u''_{x,y}$	Second partial derivative of $u(x, y, z, \ldots)$, relative to x then to y
z^*	Complex conjugate of number z
$|x|$	Absolute value of x
$A_{m,n}$	Matrix A whose size is $m \times n$
I_n	The identity square matrix whose size is $n \times n$
$|A|$	Determinant of matrix A
A^T	Transpose of matrix A
Δ	Difference between two variables
Δ	Main determinant of a matrix
Δ_x	Cramer's sub–determinant relative to variable x
$\vec{a}$	Vector a
$\mathbf{a}$	Vector a
$\sum_{i=0}^{\infty} a_i$	Sum of elements $\{a_0, a_1, \ldots, a_\infty\}$
dB	Decibel
dBm	Decibel normalized to "10^{-3}", i.e. "milli"
$H(jx)$	Transfer function

Part I
Algebra and Analysis

Basic Number Theory

1

Important to Know

Assuming $a \in \mathbb{R}$ and $a \neq 0$ basic number operations with exponents and radicals obey the following definitions and rules

$$a^0 \stackrel{\text{def}}{=} 1$$

$$a^1 \stackrel{\text{def}}{=} a$$

$$a^{-n} \stackrel{\text{def}}{=} \frac{1}{a^n}$$

$$a^{\frac{n}{m}} \stackrel{\text{def}}{=} \sqrt[m]{a^n}$$

$$|x| \stackrel{\text{def}}{=} \begin{cases} x & ; \quad (x \geq 0) \\ -x & ; \quad (x < 0) \end{cases}$$

$$a^{n+1} = a^n \cdot a$$

$$a^n \, a^m = a^{m+n}$$

$$\left(a^n\right)^m = a^{n\,m}$$

$$(a\,b)^n = a^n \, b^n$$

$$\frac{a^n}{a^m} = a^n \, a^{-m} = a^{n-m}$$

$$\left(\frac{a}{b}\right)^n = \frac{a^n}{b^n}$$

1.1 Basic Number Operations

1.1 Basic Calculations

1. Calculate the following numbers:

(a) $(-2)^{-3}$

(b) $\left(\dfrac{1}{2}\right)^{-2}$

(c) $\left(\dfrac{1}{2}\right)^{-2} + \left(\dfrac{1}{2}\right)^{-1}$

2. Calculate the following numbers:

(a) $\dfrac{(-2)^{-3} - (-3)^{-2}}{(-4)^{-1}} \left(\dfrac{2}{3}\right)^{-3}$

(b) $\left((-2)^{-1} + (-3)^{-1}\right) \div \left((-3)^{-1} - (-6)^{-1}\right)$

© Springer Nature Switzerland AG 2021
R. Sobot, *Engineering Mathematics by Example*,
https://doi.org/10.1007/978-3-030-79545-0_1

(c) $-4 \times 10^4 + 2.5 \times 10^5$

(d) $0.5^{-1} + 0.25^{-2} + 0.125^{-3} + 0.0625^{-4}$

3. Calculate the following numbers:

(a) 2^0

(b) $\left(-\dfrac{3}{8}\right)^0$

(c) $(\sqrt{3} + \pi - 1)^0$

(d) $(\cos x)^0$, $(|x| < \pi/2)$

(e) $\left(\dfrac{3}{2} + \dfrac{\pi}{6}\right)^0 - \left(\dfrac{4}{27} - \dfrac{7\pi}{9}\right)^0$

4. Simplify the following rational expressions:

(a) $\dfrac{9ab}{18ab}$

(b) $\dfrac{4ab}{6a^2b}$

(c) $\dfrac{7a^2b}{21ac}$

5. Simplify the following:

(a) $2^{3000} \times 3^{2000}$

(b) $3^{200} \times 4^{-300}$

(c) $5^{-2000} \times 2^{3000}$

6. Calculate the following:

(a) $\left(\dfrac{16}{25}\right)^{\frac{1}{2}}$

(b) $25^{-\frac{3}{2}}$

(c) $\left(\dfrac{1}{27}\right)^{-\frac{4}{3}}$

(d) $0.25^{-0.5}$

(e) $\left(\dfrac{1}{256}\right)^{0.375}$

(f) $\left(\dfrac{1}{1024}\right)^{-\frac{2}{5}}$

(g) $\left(\dfrac{1}{4}\right)^{-\frac{3}{2}} + \left(\dfrac{8}{27}\right)^{\frac{2}{3}}$

7. Write the following expressions in the form of radicals ($x, y, z, v > 0$):

(a) $x^{-\frac{2}{5}}$

(b) $x^{-\frac{3}{20}} y^{-\frac{7}{15}}$

(c) $x^{\frac{1}{8}} y^{\frac{2}{25}}$

(d) $\dfrac{x^{-\frac{1}{2}} y^{\frac{3}{4}}}{z^{-\frac{7}{8}} v^{\frac{5}{3}}}$

(e) $\dfrac{x^{-\frac{5}{2}}}{y^{\frac{3}{4}}} z^{-\frac{7}{12}}$

(f) $\left(x^{\frac{2}{3}} y^{\frac{5}{12}}\right)^{-0.75}$

8. Simplify the following expressions ($x > 0$):

(a) $\left(\left(x^{\frac{3}{4}}\right)^2\right)^{-\frac{1}{3}} \div \left(\left(x^{-1}\right)^{\frac{1}{2}}\right)^{-\frac{2}{3}}$

(b) $\left(x^n x^{\frac{1}{n-1}}\right) \div \left(x^{n^2}\right)^{\frac{1}{n-1}}$

9. Simplify the following expressions ($x, y, z > 0$):

(a) $x^{\frac{2}{3}} x^{\frac{3}{4}} x^{-1}$

(b) $x^{\frac{3}{2}} y^{\frac{4}{5}} z^{\frac{5}{6}} \div \left(x^{\frac{5}{4}} y^{\frac{2}{3}} z^{\frac{7}{12}}\right)$

(c) $\left(\left(x^{\frac{3}{4}}\right)^2\right)^{-\frac{1}{3}} \div \left(\left(x^{-1}\right)^{\frac{1}{2}}\right)^{-\frac{2}{3}}$

10. Simplify the following expressions with radicals, $(x, a, b, n > 0)$:

(a) $\left(\sqrt[3]{4a^2b}\right)^2$

(b) $\sqrt[a]{x^n} \div \sqrt[n]{x^a}$

(c) $x\sqrt{x\sqrt{x\sqrt[3]{x}}}\sqrt[3]{x\sqrt{x}}$

(d) $\sqrt[x]{a^{3x-4}} \; \sqrt[x]{a^{1-x}} \; \sqrt[x]{a^{3-x}}$

1.2 **   Absolute Numbers, Equations, and Inequalities

1. Calculate x in the following equations:

(a) $|x - 3| = 7$

(b) $|3 - x| > 4$

(c) $||x| - 5| > 4$

2. Simplify the following expressions:

(a) $\dfrac{|x|}{x}$

(b) $\dfrac{|x| + x}{2}$

(c) $\dfrac{|x| + 2x}{3} + |x| + x$

(d) $\left[\dfrac{x - 2 + |x - 2|}{2}\right]^2 + \left[\dfrac{x - 2 - |x - 2|}{2}\right]^2$

3. Simplify the following expressions $(x, y, z > 0)$:

(a) $x^{\frac{2}{3}} \, x^{\frac{3}{4}} \, x^{-1}$

(b) $x^{\frac{3}{2}} \, y^{\frac{4}{5}} \, z^{\frac{5}{6}} \div \left(x^{\frac{5}{4}} \, y^{\frac{2}{3}} \, z^{\frac{7}{12}}\right)$

(c) $\left(\left(x^{\frac{3}{4}}\right)^2\right)^{-\frac{1}{3}} \div \left(\left(x^{-1}\right)^{\frac{1}{2}}\right)^{-\frac{2}{3}}$

Solutions

Exercise 1.1, page 3

1. By using definition of negative powers:

(a) $2^{-3} = \dfrac{1}{2^3} = \dfrac{1}{8}$

(b) $\left(\dfrac{1}{2}\right)^{-2} = 2^2 = 4$

(c) $\left(\dfrac{1}{2}\right)^{-2} + \left(\dfrac{1}{2}\right)^{-1} = 2^2 + 2 = 6$

2. By obeying the order of operations we write:

(a)   Negative powers of fractions and negative numbers are calculated as

$$\frac{(-2)^{-3} - (-3)^{-2}}{(-4)^{-1}} \times \left(\frac{2}{3}\right)^{-3} = \frac{(-1/2)^3 - (-1/3)^2}{-1/4} \times \left(\frac{3}{2}\right)^3$$

$$= \frac{-1/8 - 1/9}{-1/4} \times \frac{27}{8} = \left(\frac{1}{8} + \frac{1}{9}\right) \times \cancel{4} \times \frac{27}{\cancel{2}_{8}}$$

$$= \frac{17}{72} \times \frac{27}{2} = \frac{17 \times 3 \times \cancel{9}}{\cancel{9} \times 8 \times 2} = \frac{51}{16}$$

(b)   The division symbol is another way to write fractions

$$\left((-2)^{-1} + (-3)^{-1}\right) \div \left((-3)^{-1} - (-6)^{-1}\right) = \left(-\frac{1}{2} - \frac{1}{3}\right) \div \left(-\frac{1}{3} + \frac{1}{6}\right)$$

$$= -\frac{5}{6} \div \left(-\frac{1}{6}\right) = \frac{5}{6} \times 6 = 5$$

(c)   Expressions with decimal numbers may be simplified to the same powers

$$-4 \times 10^4 + 2.5 \times 10^5 = -4 \times 10^4 + 25 \times 10^4 = 21 \times 10^4 = 2.1 \times 10^5$$

(d)   Decimal numbers may be converted into their respective fractional forms. In addition, the powers of two are extensively used in informatics.

$$0.5^{-1} + 0.25^{-2} + 0.125^{-3} + 0.0625^{-4}$$

$$= \left(\frac{1}{2}\right)^{-1} + \left(\frac{1}{4}\right)^{-2} + \left(\frac{1}{8}\right)^{-3} + \left(\frac{1}{16}\right)^{-4}$$

$$= 2 + 4^2 + 8^3 + 16^4 = 2 + (2^2)^2 + (2^3)^3 + (2^4)^4$$

$$= 2 + 2^4 + 2^9 + 2^{16} = 2 + 16 + 512 + 65536 = 66066$$

3. By definition, $x^0 = 1$ for $x \neq 0$, thus

(a)   1   (b)   1   (c)   1   (d)   0

4. Factorization of numbers and variables:

(a) $\dfrac{\cancel{9}1\,\cancel{ab}}{\cancel{18}2\,\cancel{ab}} = \dfrac{1}{2}$ (b) $\dfrac{\cancel{42}\,\cancel{ab}}{\cancel{63}\,a\cancel{b}\cancel{b}} = \dfrac{2}{3a}$ (c) $\dfrac{\cancel{7}1\,a\cancel{b}b}{\cancel{21}3\,\cancel{d}c} = \dfrac{ab}{3c}$

5. Large powers may be combined.

(a) $2^{3000} \times 3^{2000} = (2^3)^{1000} \times (3^2)^{1000} = 8^{1000} \times 9^{1000} = 72^{1000}$

(b) $3^{200} \times 4^{-300} = \dfrac{(3^2)^{100}}{(4^3)^{100}} = \left(\dfrac{9}{64}\right)^{100} = \left(\dfrac{3^2}{8^2}\right)^{100} = \left(\dfrac{3}{8}\right)^{200}$

(c) $5^{-2000} \times 2^{3000} = \dfrac{(2^3)^{1000}}{(5^2)^{1000}} = \left(\dfrac{8}{25}\right)^{1000}$

6. Fractional powers and radicals are interchangeable. Power of power expression is calculated as the product of powers. Number one is neutral for the multiplication and power operations.

(a) $\left(\dfrac{16}{25}\right)^{\frac{1}{2}} = \dfrac{\sqrt{16}}{\sqrt{25}} = \dfrac{4}{5}$

(b) $25^{-\frac{3}{2}} = \left(5^2\right)^{-\frac{3}{2}} = 5^{\left(\cancel{2} \times \frac{-3}{\cancel{2}}\right)} = 5^{-3} = \dfrac{1}{5^3} = \dfrac{1}{125}$

(c) $\left(\dfrac{1}{27}\right)^{-\frac{4}{3}} = 27^{\frac{4}{3}} = \left(3^3\right)^{\frac{4}{3}} = 3^{\left(\cancel{3} \times \frac{4}{\cancel{3}}\right)} = 3^4 = 81$

(d) $0.25^{-0.5} = \left(\dfrac{1}{4}\right)^{-\frac{1}{2}} = 4^{\frac{1}{2}} = \sqrt{4} = 2$

(e) $\left(\dfrac{1}{256}\right)^{0.375} = \left(\dfrac{1}{2^8}\right)^{\frac{3}{8}} = \left(\dfrac{1}{2}\right)^{\left(\cancel{8} \times \frac{3}{\cancel{8}}\right)} = \dfrac{1}{8}$

(f) $\left(\dfrac{1}{1024}\right)^{-\frac{2}{5}} = 1024^{\frac{2}{5}} = \left(2^{10}\right)^{\frac{2}{5}} = 2^{\left(\cancel{10}^{\,2} \times \frac{2}{\cancel{5}^{\,1}}\right)} = 2^4 = 16$

(g)

$\left(\dfrac{1}{4}\right)^{-\frac{3}{2}} + \left(\dfrac{8}{27}\right)^{\frac{2}{3}} = 4^{\frac{3}{2}} + \left(\dfrac{2^3}{3^3}\right)^{\frac{2}{3}} = \left(2^2\right)^{\frac{3}{2}} + \left(\dfrac{2}{3}\right)^{\left(\cancel{3} \times \frac{2}{\cancel{3}}\right)}$

$= 2^{\left(\cancel{2} \times \frac{3}{\cancel{2}}\right)} + \dfrac{4}{9} = 8 + \dfrac{4}{9} = \dfrac{8 \times 9 + 4}{9} = \dfrac{76}{4} = 19$

7. Rational powers are another way of writing radicals, $x^{m/n} = \sqrt[n]{x^m}$.

(a) $x^{-\frac{2}{5}} = \left(\dfrac{1}{x^2}\right)^{\frac{1}{5}} = \dfrac{1}{\sqrt[5]{x^2}}$

(b) $x^{-\frac{3}{20}} y^{-\frac{7}{15}} = \left(\dfrac{1}{x^3}\right)^{\frac{1}{20}} \left(\dfrac{1}{y^7}\right)^{\frac{1}{15}} = \dfrac{1}{\sqrt[20]{x^3} \; \sqrt[15]{y^7}}$

(c) $x^{\frac{1}{8}} y^{\frac{2}{25}} = \sqrt[8]{x} \; \sqrt[25]{y^2}$

(d) $\dfrac{x^{-\frac{1}{2}} y^{\frac{3}{4}}}{z^{-\frac{7}{8}} v^{\frac{5}{3}}} = \dfrac{\dfrac{1}{\sqrt{x}} \sqrt[4]{y^3}}{\dfrac{1}{\sqrt[8]{z^7}} \sqrt[3]{v^5}} = \dfrac{\sqrt[8]{z^7} \; \sqrt[4]{y^3}}{\sqrt{x} \; \sqrt[3]{v^5}}$

(e) $\dfrac{x^{-\frac{5}{2}}}{y^{\frac{3}{4}}} \times z^{-\frac{7}{12}} = \dfrac{1}{\sqrt{x^5} \; \sqrt[4]{x^3} \; \sqrt[12]{z^7}}$

(f)
$$\left(x^{\frac{2}{3}} \, y^{\frac{5}{12}}\right)^{-0.75} = \left(x^{\frac{2}{3}} \, y^{\frac{5}{12}}\right)^{-\frac{3}{4}} = x^{\left(-\frac{2}{3} \cdot \frac{3}{4}\right)} \, y^{\left(-\frac{5}{12} \cdot \frac{3}{4}\right)}$$

$$= x^{-\frac{1}{2}} \, y^{-\frac{5}{16}} = \frac{1}{\sqrt{x} \, \sqrt[16]{y^5}}$$

8. Rational powers are another way of writing radicals, $x^{m/n} = \sqrt[n]{x^m}$.

   (a)   Division of two fractions is transformed into product

$$\left[\left(x^{\frac{3}{4}}\right)^2\right]^{-\frac{1}{3}} \div \left[\left(x^{-1}\right)^{\frac{1}{2}}\right]^{-\frac{2}{3}} = \left\{\frac{a}{b} \div \frac{c}{d} = \frac{a}{b}\frac{d}{c}\right\} = \left(x^{\frac{3}{4}}\right)^{-\frac{2}{3}} \left[\left(x^{-1}\right)^{\frac{1}{2}}\right]^{\frac{2}{3}}$$

$$= x^{-\frac{6}{12}} x^{-\frac{2}{6}} = x^{-\frac{1}{2}-\frac{1}{3}} = x^{-\frac{1}{6}} = \frac{1}{\sqrt[6]{x}}$$

   (b)   Abstract power expressions follow the same rules as "regular" power expressions.

$$\left(x^n \, x^{\frac{1}{n-1}}\right) \div \left(x^{n^2}\right)^{\frac{1}{n-1}} = \left(x^n \, x^{\frac{1}{n-1}}\right)\left(x^{n^2}\right)^{-\frac{1}{n-1}} = x^{n+\frac{1}{n-1}} \, x^{-\frac{n^2}{n-1}}$$

$$= x^{\frac{n^2-n+1}{n-1}} x^{-\frac{n^2}{n-1}} = x^{\frac{n^2-n+1-n^2}{n-1}} = x^{-1} = \frac{1}{x}$$

9. Simplify and convert into the equivalent radicals if there is an opportunity, $(x, y, z > 0)$:

   (a) $x^{\frac{2}{3}} \, x^{\frac{3}{4}} \, x^{-1} = x^{\frac{2}{3}+\frac{3}{4}-1} = x^{\frac{5}{12}} = \sqrt[12]{x^5}$

   (b) $x^{\frac{3}{2}} \, y^{\frac{4}{5}} \, z^{\frac{5}{6}} \div \left(x^{\frac{5}{4}} \, y^{\frac{2}{3}} \, z^{\frac{7}{12}}\right) = x^{\frac{3}{2}-\frac{5}{4}} \, y^{\frac{4}{5}-\frac{2}{3}} \, z^{\frac{5}{6}-\frac{7}{12}} = x^{\frac{1}{4}} \, y^{\frac{2}{15}} \, z^{\frac{1}{4}} = \sqrt[4]{y} \, \sqrt[15]{y^2} \, \sqrt[4]{z}$

   (c)   By definitions of rational division, sum, and negative powers, powers of powers, etc., we write

$$\left[\left(x^{\frac{3}{4}}\right)^2\right]^{-\frac{1}{3}} \div \left[\left(x^{-1}\right)^{\frac{1}{2}}\right]^{-\frac{2}{3}} = \left(x^{\frac{3}{4}}\right)^{-\frac{2}{3}} \left[\left(x^{-1}\right)^{\frac{1}{2}}\right]^{\frac{2}{3}} = x^{-\frac{6}{12}} \, x^{-\frac{2}{6}}$$

$$= x^{-\frac{1}{2}-\frac{1}{3}} = x^{-\frac{5}{6}} = \frac{1}{\sqrt[6]{x^5}}$$

10. By rearranging the expression of radical's argument we find.

   (a) $\left(\sqrt[3]{4a^2b}\right)^2 = \sqrt[3]{2^4 \, a^4 b^2} = \sqrt[3]{2 \cdot 2^3 a^3 a \, b^2} = \sqrt[3]{2(2a)^3 a \, b^2} = 2a\sqrt[3]{2ab^2}$

   (b) $\sqrt[a]{x^n} \div \sqrt[n]{x^a} = x^{\frac{n}{a}} \div x^{\frac{a}{n}} = x^{\frac{n}{a}} x^{-\frac{a}{n}} = x^{\frac{n^2-a^2}{an}}$

(c) Nested radicals are converted to fractional powers and "unfolded" from within.

$$x\sqrt{x\sqrt{x\sqrt[3]{x}}\,\sqrt[3]{x\sqrt{x}}} = x\sqrt{x\sqrt{x\cdot x\,x^{\frac{1}{3}}}\,\sqrt[3]{x\cdot x^{\frac{1}{2}}}} = x\sqrt{x\sqrt{x\cdot x^{\frac{4}{3}}}\,\sqrt[3]{x^{\frac{3}{2}}}}$$

$$= x\sqrt{x\,x^{\frac{2}{3}\cdot\frac{1}{2}}\,x^{\frac{3}{2}\cdot\frac{1}{3}}} = x\sqrt{x^{\frac{5}{3}}\,x^{\frac{1}{2}}} = x\,x^{\frac{5}{3}\cdot\frac{1}{2}}\,x^{\frac{1}{2}}$$

$$= x^{1+\frac{5}{6}+\frac{1}{2}} = x^{\frac{14}{6}} = x^{\frac{7}{3}} = x^{2+\frac{1}{3}} = x^2\sqrt[3]{x}$$

(d) "x-th" radicals obey the same rules as the "ordinary" radicals.

$$\sqrt[x]{a^{3x-4}}\;\sqrt[x]{a^{1-x}}\;\sqrt[x]{a^{3-x}} = a^{\frac{3x-4}{x}}\,a^{\frac{1-x}{x}}\,a^{\frac{3-x}{x}} = a^{\frac{3x-4+1-x+3-x}{x}} = a^{\frac{x}{x}} = a$$

Exercise 1.2, page 5

1. Absolute numbers are defined as $|x| = x$, $(x \geq 0)$ or, $|x| = -x$, $(x < 0)$, that is to say that result of the absolute operation is always a *positive real number*. It is important to realize that an equation with one absolute expression is actually a compact syntax to write two equations. Consequently, the presence of multiple absolute expressions inside a single equation means that we must systematically write separate equations for each combination of positive/negative cases.

(a) One abs function generates two equations, thus two solutions
$$|x - 3| = 7$$

$\underline{x - 3 \geq 0}$ $\therefore$ $|x - 3| = x - 3$ $\therefore$ $x - 3 = 7$ $\therefore$ $\underline{x_1 = 10}$

$\underline{x - 3 < 0}$ $\therefore$ $|x - 3| = -(x - 3)$ $\therefore$ $-(x - 3) = 7$ $\therefore$ $\underline{x_2 = -4}$

Therefore, there are two discrete solutions, $x_1 = 10$ and $x_2 = -4$, see Fig. 1.1.

Fig. 1.1 Example 1.2-1(a)

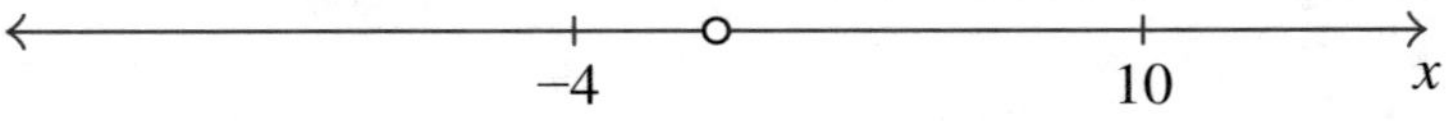

-4 10 x

(b) However, an inequality with abs function generates two inequalities, thus two *intervals* as the solutions. First, we calculate the interval boundaries

$$|3 - x| > 4$$

$\underline{3 - x \geq 0}$ $\therefore$ $|3 - x| = 3 - x$ $\therefore$ $3 - x > 4$ $\therefore$ $\underline{x < -1}$

$\underline{3 - x < 0}$ $\therefore$ $|3 - x| = -(3 - x)$ $\therefore$ $-(3 - x) > 4$ $\therefore$ $\underline{x > 7}$

Therefore, there are two intervals $x \in [-\infty, -1]$ and $x \in [7, +\infty]$, see Fig. 1.2

Fig. 1.2 Example 1.2-1(b)

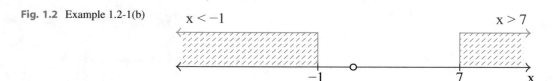

(c)   Due to two abs functions, in total there are four possible inequalities to solve

$$||x| - 5| > 4$$

where, the four possible inequalities are found as,

$\underline{x > 0}$:

$$|x| = x \; \therefore \; |x - 5| > 4 \; \text{if:} \; \underline{x - 5 \geq 0} \; \therefore \; x - 5 > 4 \; \therefore \; \underline{x > 9}$$
$$\text{if:} \; \underline{x - 5 < 0} \; \therefore \; -x + 5 > 4 \; \therefore \; \underline{x < 1}$$

$\underline{x < 0}$:
$$|x| = -x \; \therefore \; |-x - 5| > 4 \; \text{if:} \; \underline{-x - 5 \geq 0} \; \therefore \; -x - 5 > 4 \; \therefore \; \underline{x < -9}$$
$$\text{if:} \; \underline{-x - 5 < 0} \; \therefore \; x + 5 > 4 \; \therefore \; \underline{x > -1}$$

Four inequalities must be satisfied at the same time, we summarize them as (also, see Fig. 1.3)

$$x \in [-\infty, -9] \;\text{ and }\; -1 \leq x \leq 1 \;\text{ and }\; x \in [9, +\infty]$$

Fig. 1.3 Example 1.2-1(c)

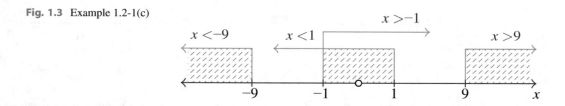

Plot of function $||x| - 5|$ vs. "4" clearly illustrates the intervals where $||x| - 5| > 4$, see Fig. 1.4.

Fig. 1.4 Example 1.2-1(c)

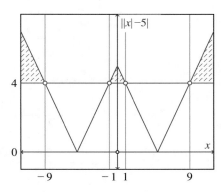

2. Some typical expressions with abs function that are extensively used in discrete mathematics for signal processing.

(a) Signs of abs function's argument are

$\underline{x > 0}:\ |x| = x\ \ \therefore\ \ \dfrac{|x|}{x} = \dfrac{x}{x} = 1$

$\underline{x < 0}:\ |x| = -x\ \ \therefore\ \ \dfrac{|x|}{x} = \dfrac{-x}{x} = -1$

Indeed, these two $|x|/x = x/|x| = \pm 1$ ratios are known as sign function.

(b) Signs of abs function's argument are

$\underline{x > 0}:\ |x| = x\ \ \therefore\ \ \dfrac{|x| + x}{2} = \dfrac{x + x}{2} = \dfrac{2x}{2} = x$

$\underline{x < 0}:\ |x| = -x\ \ \therefore\ \ \dfrac{|x| + x}{2} = \dfrac{-x + x}{2} = \dfrac{0}{2} = 0$

This is one of the possible definitions for the function known as the ramp function.

(c) Similarly,

$\underline{x > 0}:\ |x| = x\ \ \therefore\ \ \dfrac{|x| + 2x}{3} + |x| + x = \dfrac{x + 2x}{3} + x + x = 3x$

$\underline{x < 0}:\ |x| = -x\ \ \therefore\ \ \dfrac{|x| + 2x}{3} + |x| + x = \dfrac{-x + 2x}{3} - x + x = \dfrac{x}{3}$

(d)   Signs of abs function's argument are

$\underline{x - 2 > 0}: \therefore \left[\dfrac{x - 2 + |x - 2|}{2}\right]^2 + \left[\dfrac{x - 2 - |x - 2|}{2}\right]^2$

$= \left[\dfrac{x - 2 + x - 2}{2}\right]^2 + \left[\dfrac{x - 2 - x + 2}{2}\right]^{2 \nearrow 0}$

$= \left[\dfrac{2x - 4}{2}\right]^2 = (x - 2)^2$

$\underline{x - 2 < 0}: \therefore \left[\dfrac{x - 2 + |x - 2|}{2}\right]^2 + \left[\dfrac{x - 2 - |x - 2|}{2}\right]^2$

$= \left[\dfrac{x - 2 - x + 2}{2}\right]^{2 \nearrow 0} + \left[\dfrac{x - 2 + x - 2}{2}\right]^2$

$= \left[\dfrac{2x - 4}{2}\right]^2 = (x - 2)^2$

3. Simplify and convert into the equivalent radicals if there is an opportunity, $(x, y, z > 0)$:

(a) $x^{\frac{2}{3}} \, x^{\frac{3}{4}} \, x^{-1} = x^{\frac{2}{3} + \frac{3}{4} - 1} = x^{\frac{5}{12}} = \sqrt[12]{x^5}$

(b) $x^{\frac{3}{2}} \, y^{\frac{4}{5}} \, z^{\frac{5}{6}} \div \left(x^{\frac{5}{4}} \, y^{\frac{2}{3}} \, z^{\frac{7}{12}}\right) = x^{\frac{3}{2} - \frac{5}{4}} \, y^{\frac{4}{5} - \frac{2}{3}} \, z^{\frac{5}{6} - \frac{7}{12}} = x^{\frac{1}{4}} \, y^{\frac{2}{15}} \, z^{\frac{1}{4}} = \sqrt[4]{y} \, \sqrt[15]{y^2} \, \sqrt[4]{z}$

(c)   By definitions of rational division, sum, and negative powers, powers of powers, etc., we write

$$\left[\left(x^{\frac{3}{4}}\right)^2\right]^{-\frac{1}{3}} \div \left[\left(x^{-1}\right)^{\frac{1}{2}}\right]^{-\frac{2}{3}} = \left(x^{\frac{3}{4}}\right)^{-\frac{2}{3}} \left[\left(x^{-1}\right)^{\frac{1}{2}}\right]^{\frac{2}{3}} = x^{-\frac{6}{12}} \, x^{-\frac{2}{6}}$$

$$= x^{-\frac{1}{2} - \frac{1}{3}} = x^{-\frac{5}{6}} = \dfrac{1}{\sqrt[6]{x^5}}$$

Polynomials

2

Important to Know

Basic polynomial transformations:

$$a(b \pm c) = ab + ac$$
$$(a + b)(c + d) = ac + ad + bc + bd$$
$$(a \pm b)^2 = a^2 \pm 2ab + b^2$$
$$(a^2 - b^2) = (a - b)(a + b)$$
$$(a^3 \pm b^3) = (a \pm b)(a^2 \mp ab + b^2)$$

Factor theorem: reminder r of division $P(x) \div (x - a)$ is equal to $r = P(a)$. Consequently, if $r = P(a) = 0$, then $P(x)$ is divisible by $(x - a)$. That being the case, binomial $(x - a)$ is one of the $P(x)$ factors and $x = a$ one of its roots.

2.1 Exercises

2.1 Multiplication, Identity

1. Given $P(x) = x^2 + x + 1$ and $Q(x) = x^3 - 1$, calculate

   (a) $P(x) + Q(x)$ (b) $P(x) - Q(x)$ (c) $2x P(x) - 2Q(x)$

2. Multiply the following polynomials:

   (a) $(x - 1)(x^2 + x + 1)$ (b) $(x + 3)(x^2 - 3x + 9)$
   (c) $(3x^2 - 5x + 6)(2x - 7)$ (d) $(x^4 + x^3 + 2)(x^2 - 3x + 4)$
   (e) $(3y^2 + 2x - 6xy)(2x^2 + 4xy - 2y^3)$ (f) $(x - y)(y - z)(z - x)$

© Springer Nature Switzerland AG 2021
R. Sobot, *Engineering Mathematics by Example*,
https://doi.org/10.1007/978-3-030-79545-0_2

3. Calculate parameters a, b, c so that $P(x) = Q(x)$, if

    (a) $P(x) = x^3 - 2x^2 + 3$, $Q(x) = (x + 1)(ax^2 + bx + c)$

    (b) $P(x) = 2x^3 - x^2 + x + 4$, $Q(x) = (x - 2)(ax^2 + bx + c)$

4. Given $P(x) = x^3 + ax^2 + bx + c$ calculate parameters a, b, c so that it is divisible by the following binomials: $(x + 1), (x - 1)$, and $(x + 2)$.

5. With the help of Pascal's triangle, calculate

    (a) $(x - 1)^6$ (b) $(x + 2)^5$ (c) $(2x + 1)^4$

2.2   Factor Theorem

1. With the help of the factor theorem, answer the following questions.

    (a)   Calculate the reminder in $(x^4 - 2x^3 + 3x^2 - 4x + 1) \div (x - 1)$

    (b)   Calculate the reminder in $(x^3 + 1) \div (x + 1)$

    (c)   Calculate parameter n so that $P(x) = 4x^5 + nx^4 + 8x^3 + 5x^2 + 3x + 2$ is divisible by $x + 2$.

2.3 ***   Factor Theorem

1. Using the factor theorem, answer the following questions.

    (a)   Calculate parameter n so that
    $P(x) = x^3 - 3nx + 4(n^2 + 1)x - (n^3 + 5)$ is divisible by $(x - 1)$

    (b)   Calculate reminder in $P(x)/Q(x)$, given
    $P(x) = x^{1965} - 256x^{1961} + 1$ and $Q(x) = x^2 - 4x$

    (c)   Calculate reminder in $P(x)/Q(x)$, given that
    $P(x) = x^{2021} - 2x^{2020} - 1$ and $Q(x) = x^2 - 3x + 2$

2.4   Common Factors

1. Factorize the following polynomials by finding the common factors:

    (a) $5a + 5x$ (b) $2a - 2$ (c) $7a - 14$

    (d) $3a^2 + 9$ (e) $3a + 6b + 9$ (f) $6x + ax + bx$

    (g) $9a^2 - 6a + 12$ (h) $a^2 - a^3$ (i) $3a^2 - 6a$

    (j) $x^3a^2 - x^3$ (k) $3a^3 + 2a^2 + a$ (l) $4x^2 - 2x + xy$

2. Factorize the common terms:

    (a) $x^3y^3 - x^3y + x^4y^3$ (b) $x^3y^3 - x^2y^8$

    (c) $6x^2y^2 - 4xy^3$ (d) $5x^3 - 15x^2y^3$

    (e) $6x^3y - 9x^2y^2 + 3x^3y^2$ (f) $x^3 - x^7 - 2x^5$

(g) $a^3b^2 + 2a^4b^2 - 4ab^5$

(h) $3a^3b^3 - 9a^2b^4 + 12a^5b^4$

(i) $a(m+n) + b(m+n)$

(j) $m(a-b) + n(a-b)$

(k) $7q(p-q) + 2p(q-p)$

(l) $3(x+y) + (x+y)^2$

3. Factorize the following polynomials by finding the common terms:

(a) $am - an + bm - bn$

(b) $am - an - bm + bn$

(c) $ab + ay - bx - xy$

(d) $an - ab - mn + mb$

(e) $5ax + 5ay - x - y$

(f) $2x^2 + 2xy - x - y$

(g) $4ym - 4yn - m + n$

(h) $ax^2 - bx^2 - bx + ax - a + b$

(i) $6by - 15bx - 4ay + 10ax$

(j) $5ax^2 - 10ax - bx + 2b - x + 2$

4. Factorize the following polynomials by finding the common terms:

(a) $xyz + x^2y^2 - 3x^4y^5 - 3x^3y^4 - xy - z$

(b) $m^2x^4 - mnx^3 + 2mx^2 - 2nx + n - mx$

5. Factorize the following expressions:

(a) $a^{2n} + a^n$

(b) $a^{3x} - 2a^{2x}b^x$

(c) $2x^{m+n} + 6x^n$

(d) $a^{3x} + 3a^{2x} + 5a^x$

(e) $x^{2n+2} - 2x^{n+1} + 1$

2.5   Polynomial Identities: $a^2 - b^2 = (a+b)(a-b)$

1. Factorize by using the difference of squares identity:

(a) $x^2 - 49$

(b) $a^2 - 36$

(c) $16x^2 - 9$

(d) $9x^2 - 49$

(e) $25 - x^2$

(f) $9 - x^4$

(g) $x^2 - \dfrac{1}{49}$

(h) $\dfrac{x^2}{4} - \dfrac{4}{9}$

(i) $\dfrac{9x^2}{4} - \dfrac{4y^2}{9}$

(j) $\dfrac{49x^2}{25} - 9y^2$

(k) $x^2 - 0.36$

(l) $x^2 - 0.0009$

(m) $0.04x^2 - 0.25$

(n) $0.01x^2 - 0.04y^2$

(o) $x^4y^2 - 0.01$

(p) $0.25x^2y^2 - 0.0001$

2. Factorize the following expressions:

(a) $(x-3)^2 - 4$

(b) $(a+5)^2 - 9$

(c) $y^2 - (x-y)^2$

(d) $x^2 - (x+y)^2$

(e) $(x+2)^2 - 4x^2$

(f) $9x^2 - (x-1)^2$

(g) $(x-y)^2 - 16(x+y)^2$

(h) $(x+2y)^2 - 9(x-2y)^2$

(i) $4(x-y)^2 - 25(x+y)^2$

(j) $36(x-2)^2 - 25(x+1)^2$

(k) $(x+y-z)^2 - (x-y+z)^2$

(l) $(x+y-3)^2 - 25(x+2)^2$

3. Calculate the following *without using a calculator*:

    (a) 98×102 (b) 99×101 (c) 83×77

    (d) 79×81 (e) 18×22 (f) 201×199

    (g) 1.05×0.95 (h) 1.01×0.99 (i) 9.9×10.1

2.6   The Binomial Square $(ax)^2 \pm 2abx + b^2 = (ax \pm b)^2$

1. Factorize the following quadratic and bi-quadratic polynomials:

    (a) $x^2 - 2x + 1$ (b) $x^2 - 6x + 9$ (c) $x^2 - 2\sqrt{2}x + 2$

    (d) $4x^2 - 12x + 9$ (e) $x^4 - 2x^2 + 1$ (f) $x^{2n+2} - 2x^{n+1} + 1$

2.7 *   Viète's Formulas: $x_1 + x_2 = -(b/a), \quad x_1 x_2 = (c/a)$

1. Factorize the following quadratic polynomials:

    (a) $x^2 - 6x + 5$ (b) $x^2 - 9x + 14$

    (c) $x^2 - 6x + 8$ (d) $2x^2 + 3x + 1$

    (e) $3x^2 - 14x + 8$ (f) $-2x^2 + x + 3$

    (g) $-x^2 + 5x - 4$ (h) $-6x^2 + 5x + 4$

    (i) $x^2 + (\sqrt{2} + \sqrt{3})x + \sqrt{6}$

2.8 *   Bi-quadratic Form: $ax^{2n} + bx^n + c$

1. Factorize the following bi-quadratic polynomials:

    (a) $x^4 - 13x^2 + 36$ (b) $x^4 - 10x^2 + 9$ (c) $x^6 - 2x^3 + 1$

    (d) $x^6 + 3x^3 + 2$ (e) $x^6 - 9x^3 + 8$ (f) $2x^{2m+2} - 11x^{m+1} + 9$

2.9 **   Perfect Square $(x + a)^2 = b$

1. Completing the square technique exploits binomial square form.

    (a) $x^2 + 6x - 7 = 0$ (b) $2x^2 - 10x - 3 = 0$

    (c) $-x^2 - 6x - 3 = 0$ (d) $x^2 + 3x$

    (e) $2x^2 + 6x + 2$ (f) $x^2 - x - 6$

2.10 *   Long Division

1. Do the following division:

    (a) $(x^4 - 2x^3 - 7x^2 + 8x + 12) \div (x - 3)$

    (b) $(6x^3 + 10x^2 + 8) \div (2x^2 + 1)$

    (c) $(4x^3 - 7x^2 - 11x + 5) \div (4x + 5)$

    (d) $(x^2 + x - 5) \div (x + 3)$

2. Factorize:

(a) $P(x) = x^3 + x^2 - 4x - 4$ (b) $Q(x) = x^3 - 1$

(c) $Q(x) = x^3 - y^3$ (d) $Q(x) = x^3 + y^3$

2.11 **   Sum of Cubes $a^3 \pm b^3 = (a \pm b)(a^2 \mp ab + b^2)$**

---

1. Factorize:

(a) $Q(x) = x^3 - 1$ (b) $Q(x) = x^3 - y^3$ (c) $Q(x) = x^3 + y^3$

2.12 ***   Multiple Factorization Techniques**

---

1. Factorize the following expressions.

(a) $x^4 + 4$ (b) $x^4 + x^2 + 1$

(c) $x^5 + x + 1$ (d) $(x + y + z)^3 - x^3 - y^3 - z^3$

(e) $(x + 1)(x + 3)(x + 5)(x + 7) + 15$ (f) $(x^2 + x + 1)(x^3 + x^2 + 1) - 1$

2.13 *   Multiple Factorization Techniques**

---

1. Simplify, $(x \neq 0)$:

(a) $\dfrac{x^{n+2}}{2x\,x^n}$ (b) $\dfrac{x^{3n+2} + x^{3n+1}}{x^2\,x^{3n}}$ (c) $\dfrac{x + x^2 + x^3 + \cdots + x^n}{\dfrac{1}{x} + \dfrac{1}{x^2} + \dfrac{1}{x^3} + \cdots + \dfrac{1}{x^n}}$

2.14 *   Partial Fractions**

---

1. Derive the partial fraction form of the following rational functions

(a) $\dfrac{x}{(x + 1)(x - 4)}$ (b) $\dfrac{x + 2}{x^3 - 2x^2}$ (c) $\dfrac{2x^2}{x^4 - 1}$

(d) $\dfrac{x^3 + x^2 - 16x + 16}{x^2 - 4x + 3}$ (e) $\dfrac{1}{x^3 - 1}$ (f) $\dfrac{4}{x^4 + 1}$

(g) $\dfrac{x^3 + x - 1}{(x^2 + 2)^2}$.

Solutions

Exercise 2.1, page 13

1. Given $P(x) = x^2 + x + 1$ and $Q(x) = x^3 - 1$, we write

   (a) $P(x) + Q(x) = (x^2 + x + 1) + (x^3 - 1) = x^3 - \cancel{1} + x^2 + x + \cancel{1}$
   $$= x(x^2 + x + 1)$$

   (b) $P(x) - Q(x) = (x^2 + x + 1) - (x^3 - 1) = -x^3 + 1 + x^2 + x + 1$
   $$= -x^3 + x^2 + x + 2$$

   (c) $2xP(x) - 2Q(x) = 2x(x^2 + x + 1) - 2(x^3 - 1) = \cancel{2x^3} + 2x^2 + 2x - \cancel{2x^3} + 2$
   $$= 2x^2 + 2x + 2 = 2(x^2 + x + 1)$$

2. Multiply the following polynomials:

   (a) $(x - 1)(x^2 + x + 1) = x^3 + \cancel{x^2} + \cancel{x} - \cancel{x^2} - \cancel{x} - 1$
   $$= x^3 - 1$$

   (b) $(x + 3)(x^2 - 3x + 9) = x^3 - \cancel{3x^2} + \cancel{9x} + \cancel{3x^2} - \cancel{9x} + 27$
   $$= x^3 - 3^3$$

   (c) $(3x^2 - 5x + 6)(2x - 7) = 6x^3 - 21x^2 - 10x^2 + 35x + 12x - 42$
   $$= 6x^3 - 31x^2 + 47x - 42$$

   (d) $(x^4 + x^3 + 2)(x^2 - 3x + 4) = x^6 - 3x^5 + 4x^4$
   $$+ x^5 - 3x^4 + 4x^3$$
   $$+ 2x^2 - 6x + 8$$
   $$= x^6 - 2x^5 + x^4 + 4x^3 + 2x^2 - 6x + 8$$

   (e) $(3y^2 + 2x - 6xy)(2x^2 + 4xy - 2y^3)$
   $$= 6x^2y^2 + 12xy^3 - 6y^5 + 4x^3 + 8x^2y - 4xy^3 - 12x^3y - 24x^2y^2 + 12xy^2$$
   $$= -18x^2y^2 + 8xy^3 - 6y^5 + 4x^3 + 8x^2y - 12x^3y + 12xy^4$$

   (f) $(x - y)(y - z)(z - x) = (xy - xz - y^2 + yz)(z - x)$
   $$= \cancel{xyz} - xz^2 - y^2z + yz^2 - x^2y + x^2z + xy^2 - \cancel{xyz}$$
   $$= x^2z + yz^2 + xy^2 - xz^2 - y^2z - x^2y$$

3. Two polynomials are equal if each of their corresponding monomials are equal. The product of $(x + 1)$ binomial and parametric quadratic polynomial is therefore

(a) $Q(x) = (x + 1)(ax^2 + bx + c) = ax^3 + bx^2 + cx + ax^2 + bx + c$
$$= ax^3 + (a + b)x^2 + (b + c)x + c$$

$\therefore$

$Q(x) = P(x)$ $\therefore$ $ax^3 + (a + b)x^2 + (b + c)x + c = x^3 - 2x^2 + 3$

$\underline{x^3 \text{ terms:}}$ $\therefore$ $a = 1$

$\underline{x^0 \text{ terms:}}$ $\therefore$ $c = 3$

$\underline{x^2 \text{ terms:}}$ $a + b = -2$ $\therefore$ $1 + b = -2$ $\therefore$ $b = -3$

$\underline{x^1 \text{ terms:}}$ $b + c = 0$ $\therefore$ $-3 + 3 = 0$ ✓

In conclusion, $P(x) = Q(x)$ $\therefore$ $x^3 - 2x^2 + 3 = (x + 1)(x^2 - 3x + 3)$

(b) $Q(x) = (x - 2)(ax^2 + bx + c) = ax^3 + bx^2 + cx - 2ax^2 - 2bx - 2c$
$$= ax^3 + (b - 2a)x^2 + (c - 2b)x - 2c$$

$\therefore$

$Q(x) = P(x)$ $\therefore$ $ax^3 + (b - 2a)x^2 + (c - 2b)x - 2c = 2x^3 - x^2 + x + 4$

$\therefore$

$\underline{x^3 \text{ terms:}}$ $\therefore$ $a = 2$

$\underline{x^0 \text{ terms:}}$ $-2c = 4$ $\therefore$ $c = -2$

$\underline{x^2 \text{ terms:}}$ $b - 2a = -1$ $\therefore$ $b - 4 = -1$ $\therefore$ $b = 3$

$\underline{x^1 \text{ terms:}}$ $c - 2b = 1$ $\therefore$ $-2 - 6 \neq 1$ ✗

In conclusion, there is no $Q(x)$ that satisfies this $P(x) = Q(x)$ equation.

4. Third order polynomial $Q(x)$ that is divisible by $(x + 1)$, $(x - 1)$, and $(x + 2)$ must have form

$$Q(x) = (x + 1)(x - 1)(x + 2) = (x^2 - 1)(x + 2) = x^3 + 2x^2 - x - 2$$

Therefore, we conclude,

$P(x) = Q(x)$ $\therefore$ $x^3 + ax^2 + bx + c = x^3 + 2x^2 - x - 2$

$\therefore$

$$a = 2, \ b = -1, \ \text{and} \ c = -2$$

5. Formal expression for binomial expansion is

$$(a + b)^n = \sum_{k=0}^{n} \binom{n}{k} a^{n-k} b^k = \sum_{k=0}^{n} \binom{n}{k} a^k b^{n-k} \tag{2.1}$$

and it is used to generate the polynomial development of $(a+b)^n$. Visually, due to their symmetry, the polynomial coefficients are easily arranged in the form of Pascal's triangle, Fig. 2.1. Each row lists polynomial coefficients of the corresponding n-th binomial power. Note that each row coefficient is calculated as the sum of two neighbouring coefficients left and right in the row above.

(a)   Given $n = 6$ and $a = x$ and $b = -1$, in Pascal's triangle we find the following coefficients: $1, 6, 15, 20, 15, 6, 1$. Note how, in accordance to (2.1), the powers of a and b are systematically written in the falling and increasing orders, respectively.

$$(x - 1)^6 = 1\,x^6\,(-1)^0 + 6\,x^5\,(-1)^1 + 15\,x^4\,(-1)^2 + 20\,x^3\,(-1)^3 +$$
$$15\,x^2\,(-1)^4 + 6\,x^1\,(-1)^5 + 1\,x^0\,(-1)^6$$
$$= x^6 - 6x^5 + 15x^4 - 20x^3 + 15x^2 - 6x + 1$$

(b)   Given $n = 5$ and $a = x$ and $b = 2$, in Pascal's triangle we find the following coefficients: $1, 5, 10, 10, 5, 1$, thus we write

$$(x + 2)^5 = 1\,x^5\,2^0 + 5\,x^4\,2^1 + 10\,x^3\,2^2 + 10\,x^2\,2^3 + 5\,x^1\,2^4 + 1\,x^0\,2^5$$
$$= x^5 + 10x^4 + 40x^3 + 80x^2 + 80x + 32$$

(c)   Given $n = 4$ and $a = 2x$ and $b = 1$, in Pascal's triangle we find the following coefficients: $1, 4, 6, 4, 1$, thus we write

$$(2x + 1)^4 = 1\,(2x)^4\,1^0 + 4\,(2x)^3\,1^1 + 6\,(2x)^2\,1^2 + 4\,(2x)^1\,1^3 + 1\,(2x)^0\,1^4$$
$$= 16x^4 + 32x^3 + 24x^2 + 8x + 1$$

Fig. 2.1 Example 2.1-5

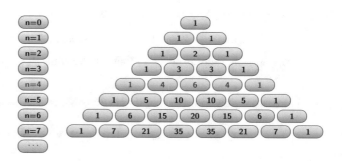

Exercise 2.2, page 14

1. The factor theorem states that a polynomial $P(x)$ has a binomial factor $(x - n)$ if and only if $P(n) = 0$, that is to say n is one of the $P(x)$ roots. Otherwise, if the value $P(n) \neq 0$ we say that $r = P(n)$ is the reminder of division $P(x) \div (x - n)$.

(a)   Given divisor $(x - 1)$, i.e. $n = 1$, we find
$r = P(1) = (1)^4 - 2(1)^3 + 3(1)^2 - 4(1) + 1) = -1$
Therefore $n = 1$ is not the root of $P(x)$.

(b) Given divisor $(x + 1)$, i.e. $n = -1$, we find

$r = P(-1) = (-1)^3 + 1 = 0$

Therefore $n = -1$ is one of the roots of $P(x)$, in other words, $P(x)$ is divisible by $(x+1)$. Indeed,

$$\frac{x^3 + 1}{x + 1} = x^2 - x + 1$$

(c) For $P(x)$ to be divisible by $(x + 2)$ it is necessary that $r = P(-2) = 0$. Therefore, we write,

$$r = P(-2) = 4(-2)^5 + k(-2)^4 + 8(-2)^3 + 5(-2)^2 + 3(-2) + 2$$

$$= -128 + 16k - 64 + 20 - 6 + 2 = 0 \quad \therefore \quad 16k - 176 = 0$$

$$\therefore \quad n = \frac{176}{16} = 11$$

Indeed,

$$\frac{4x^5 + 11x^4 + 8x^3 + 5x^2 + 3x + 2}{x + 2} = 4x^4 + 3x^3 + 2x^2 + x + 1$$

Exercise 2.3, page 14

1. The factor theorem states that a polynomial $P(x)$ has a binomial factor $(x - n)$ if and only if $P(n) = 0$, that is to say n is one of the $P(x)$ roots. Otherwise, if the value $P(n) \neq 0$ we say that $r = P(n)$ is the reminder of division $P(x) \div (x - n)$.

(a) Given binomial $(x - 1)$, the condition of divisibility for $P(x)/(x - 1)$ is that $r = P(1) = 0$, therefore we write,

$$r = P(1) = (1)^3 - 3n(1) + 4(n^2 + 1)(1) - (n^3 + 5)$$

$$= 1 - 3n + 4n^2 + 4 - n^3 - 5$$

$$= -n^3 + 4n^2 - 3n = -n(n^2 - 4n + 3) = -n(n^2 - n - 3n + 3)$$

$$= -n[n(n - 1) - 3(n - 1)] = -n(n - 1)(n - 3) = 0$$

Therefore, there are three values of n that produce $r = 0$, i.e. $n = 0, 1, 3$. Indeed,

$$n = 0 : \quad \frac{P(x)}{x-1} = \frac{x^3 + 4x - 5}{x - 1} = x^2 + x + 5$$

$$n = 1 : \quad \frac{P(x)}{x-1} = \frac{x^3 + 5x - 6}{x - 1} = x^2 + x + 6$$

$$n = 3 : \quad \frac{P(x)}{x-1} = \frac{x^3 + 31x - 32}{x - 1} = x^2 + x + 32$$

(b)   Given $Q(x) = x^2 - 4x = x(x - 4)$, we calculate the division reminders for both $(x - 0)$ and $(x - 4)$ factors. There are multiple x to calculate, thus we must calculate the reminders for each of them $P(0)$ and $P(4)$, i.e. $r(x)$ as

$$r(0) = P(0) = (0)^{1965 \; 0} - 256 (0)^{1961 \; 0} + 1 = 1$$

$$r(4) = P(4) = (4)^{1965} - 256(4)^{1961} + 1 = 4^{1965} - 4^4 (4)^{1961} + 1$$

$$= 4^{1965} - 4^{1965} + 1 = 1$$

In conclusion, reminder is $r = 1$. In other words $P(x) = Q(x) f(x) + 1$, where $f(x)$ is a 1963th order polynomial.

(c)   Given
$$Q(x) = x^2 - 3x + 2 = x^2 - x - 2x + 2 = x(x - 1) - 2(x - 1)$$
$$= (x - 1)(x - 2)$$

we calculate the division reminders for both $(x - 1)$ and $(x - 2)$ factors. There are multiple x to calculate, we must calculate the reminders for each of them $P(1)$ and $P(2)$, i.e. $r(x)$ as

$$r(1) = P(1) = (1)^{2021 \; 1} - 2 (1)^{2020 \; 1} - 1 = -2$$

$$r(2) = P(2) = (2)^{2021} - 2(2)^{2020} - 1 = 2^{2021} - 2^{2021} - 1 = -1$$

Because $r(1) \neq r(2)$ and there are only two points in consideration, $x = 1$ and $x = 2$, we conclude that reminder must be a first order polynomial $r(x) = ax + b$, where its linear coefficients are calculated as

$$\left. \begin{array}{l} \underline{x = 1} : \quad r(1) = -2 = a(1) + b \\ \underline{x = 2} : \quad r(2) = -1 = a(2) + b \end{array} \right\} \quad \therefore \quad \left. \begin{array}{l} a + b = -2 \\ 2a + b = -1 \end{array} \right\} \quad \therefore \quad a = 1, \;\; b = -3$$

In conclusion, the reminder in division $P(x)/Q(x)$ is $r(x) = ax + b = x - 3$.

Exercise 2.4, page 14

1. We find the common factors:

(a) $5a + 5x = 5(a + x)$

(b) $2a - 2 = 2(a - 1)$

(c) $7a - 14 = 7(a - 2)$

(d) $3a^2 + 9 = 3(a^2 + 3)$

(e) $3a + 6b + 9 = 3(a + 2b + 3)$

(f) $6x + ax + bx = x(6 + a + b)$

(g) $9a^2 - 6a + 12 = 3(3a^2 - 2a + 4)$

(h) $a^2 - a^3 = a^2(1 - a)$

(i) $3a^2 - 6a = 3a(a - 2)$

(j) $x^3a^2 - x^3 = x^3(a^2 - 1) = x^3(a + 1)(a - 1)$

(k) $3a^3 + 2a^2 + a = a(3a^2 + 2a + 1)$

(l) $4x^2 - 2x + xy = x(4x - 2 - y)$

2. We search for the highest order common factors:

(a) $x^3y^3 - x^3y + x^4y^3 = \underline{x^3}\,\underline{y}\,y^2 - \underline{x^3}\,\underline{y} + \underline{x^3}\,x\,\underline{y}\,y^2 = \underline{x^3}\,\underline{y}(y^2 - 1 + xy^2)$

(b) $x^3y^3 - x^2y^8 = x^2y^3(x - y^5)$

(c) $6x^2y^2 - 4xy^3 = 2xy^2(3x - 2y)$

(d) $5x^3 - 15x^2y^3 = 5x^2(x - 3y^3)$

(e) $6x^3y - 9x^2y^2 + 3x^3y^2 = 3x^2y(2x - 3y + xy)$

(f) $x^3 - x^7 - 2x^5 = x^3(1 - x^4 - 2x^2)$

(g) $a^3b^2 + 2a^4b^2 - 4ab^5 = ab^2(a^2 + 2a^3 - 4b^3)$

(h) $3a^3b^3 - 9a^2b^4 + 12a^5b^4 = 3a^2b^3(a - 3b + 4a^3b)$

(i) $a(m + n) + b(m + n) = (m + n)(a + b)$

(j) $m(a - b) + n(a - b) = (a - b)(m + n)$

(k) $7q(p - q) + 2p(q - p) = (p - q)(7q - 2p)$

(l) $3(x + y) + (x + y)^2 = (x + y)(3 + x + y)$

3. We factorize in two steps by grouping binomials with the same factors:

(a) $am - an + bm - bn = \underline{a(m - n)} + \underline{b(m - n)} = \underline{(m - n)}(a + b)$

(b) $am - an - bm + bn = a(m - n) - b(m - n) = (m - n)(a - b)$

(c) $ab + ay - bx - xy = a(b + y) - x(b + y) = (b + y)(a - x)$

(d) $an - ab - mn + mb = a(n - b) - m(n - b) = (n - b)(a - m)$

(e) $5ax + 5ay - x - y = 5a(x + y) - (x + y) = (x + y)(5a - 1)$

(f) $2x^2 + 2xy - x - y = 2x(x + y) - (x + y) = (x + y)(2x - 1)$

(g) $4ym - 4yn - m + n = 4y(m - n) - (m - n) = (m - n)(4y - 1)$

(h) $ax^2 - bx^2 - bx + ax - a + b = x^2(a - b) + x(a - b) - (a - b)$
$$= (a - b)(x^2 + x - 1)$$

(i) $6by - 15bx - 4ay + 10ax = 2y(3b - 2a) - 5x(3b - 2a)$
$$= (3b - 2a)(2y - 5x)$$

(j) $5ax^2 - 10ax - bx + 2b - x + 2 = 5ax(x - 2) - b(x - 2) - (x - 2)$
$$= (x - 2)(5ax - b - 1)$$

4. We factorize in two steps by grouping binomials with the same factors:

(a) $xyz + x^2y^2 - 3x^4y^5 - 3x^3y^4z - xy - z$
$$= xy(xy + z) - 3x^3y^4(xy + z) - (xy + z)$$
$$= (xy + z)(xy - 3x^3y^4 - 1)$$

(b) $m^2x^4 - mnx^3 + 2mx^2 - 2nx + n - mx$
$$= mx^3(mx - n) + 2x(mx - n) - (mx - n)$$
$$= (mx - n)(mx^3 + 2x - 1)$$

5. We search for the highest order terms:

(a) $a^{2n} + a^n = \underline{a^n}\, a^n + \underline{a^n} = a^n(a^n + 1)$

(b) $a^{3x} - 2a^{2x}b^x = a^{2x}(a^x - 2b^x)$

(c) $2x^{m+n} + 6x^n = 2x^n(x^m + 3)$

(d) $a^{3x} + 3a^{2x} + 5a^x = a^x(a^{2x} + 3a^x + 5)$

(e) $x^{2n+2} - 2x^{n+1} + 1 = x^{2(n+1)} - 2x^{n+1} + 1 = \left(x^{n+1}\right)^2 - 2\left(x^{n+1}\right) + 1$
$$= \left(x^{n+1} - 1\right)^2$$

Exercise 2.5, page 15

1. Using the difference of squares identity, we simply write

(a) $x^2 - 49 = x^2 - 7^2 = (x + 7)(x - 7)$

(b) $a^2 - 36 = a^2 - 6^2 = (a + 6)(a - 6)$

(c) $16x^2 - 9 = (4x)^2 - 3^2 = (4x + 3)(4x - 3)$

(d) $9x^2 - 49 = (3x)^2 - 7^2 = (3x + 7)(3x - 7)$

(e) $25 - x^2 = 5^2 - x^2 = (5 + x)(5 - x)$

(f) $81 - x^4 = 9^2 - (x^2)^2 = (9 + x^2)(9 - x^2) = (9 + x^2)(3 + x)(3 - x)$

(g) $x^2 - \dfrac{1}{49} = \left(x + \dfrac{1}{7}\right)\left(x - \dfrac{1}{7}\right)$

(h) $\dfrac{x^2}{4} - \dfrac{4}{9} = \left(\dfrac{x}{2}\right)^2 - \left(\dfrac{2}{3}\right)^2 = \left(\dfrac{x}{2} + \dfrac{2}{3}\right)\left(\dfrac{x}{2} - \dfrac{2}{3}\right)$

(i) $\dfrac{9x^2}{4} - \dfrac{4y^2}{9} = \left(\dfrac{3x}{2} + \dfrac{2y}{3}\right)\left(\dfrac{3x}{2} - \dfrac{2y}{3}\right)$

(j) $\dfrac{49x^2}{25} - 9y^2 = \left(\dfrac{7x}{5}\right)^2 - (3y)^2 = \left(\dfrac{7x}{5} + 3\right)\left(\dfrac{7x}{5} - 3\right)$

(k) $x^2 - 0.36 = x^2 - \dfrac{\cancel{36}\ 9}{\cancel{100}\ 25} = x^2 - \left(\dfrac{3}{5}\right)^2 = \left(x + \dfrac{3}{5}\right)\left(x - \dfrac{3}{5}\right)$

(l) $x^2 - 0.0009 = x^2 - 0.03^2 = (x + 0.03)(x - 0.03)$

(m) $0.04x^2 - 0.25 = (0.2x)^2 - 0.5^2 = (0.2x + 0.5)(0.2x - 0.5)$
$$= \dfrac{1}{100}(2x + 5)(2x - 5)$$

(n) $0.01x^2 - 0.04y^2 = (0.1x)^2 - (0.2y)^2$
$$= (0.1x + 0.2y)(0.1x - 0.2y)$$
$$= \frac{1}{100}(x + 2y)(x - 2y)$$

(o) $x^4y^2 - 0.01 = \left(x^2y\right)^2 - 0.1^2 = (x^2y + 0.1)(x^2y - 0.1)$

(p) $0.25x^2y^2 - 0.0001 = (0.5xy + 0.01)(0.5xy - 0.01)$

2. After using the difference of squares identity we simplify if possible:

(a) $(x - 3)^2 - 4 = (x - 3 + 2)(x - 3 - 2) = (x - 1)(x - 5)$

(b) $(a + 5)^2 - 9 = (a + 5 + 3)(a + 5 - 3) = (a + 8)(a + 2)$

(c) $y^2 - (x - y)^2 = (y + x - y)(y - (x - y)) = x(2y - x)$

(d) $x^2 - (x + y)^2 = (x + x + y)(x - x - y) = -y(2x + y)$

(e) $(x + 2)^2 - 4x^2 = (x + 2 + 2x)(x + 2 - 2x) = (3x + 2)(2 - x)$

(f) $9x^2 - (x - 1)^2 = (3x + x - 1)(3x - (x - 1)) = (4x - 1)(2x + 1)$

(g) $(x - y)^2 - 16(x + y)^2 = (x - y + 4x + 4y)(x - y - 4x - 4y)$
$$= (5x + 3y)(-3x - 5y) = -(5x + 3y)(3x + 5y)$$

(h) $(x + 2y)^2 - 9(x - 2y)^2 = (x + 2y + 3x - 6y)(x + 2y - (3x - 6y))$
$$= (4x - 4y)(8y - 2x) = 4(x - y)\, 2(4y - x)$$
$$= 8\,(x - y)(4y - x)$$

(i) $4(x - y)^2 - 25(x + y)^2 = (2x - 2y + 5x + 5y)(2x - 2y - 5x - 5y)$
$$= (7x + 3y)(-3x - 7y) = -(7x + 3y)(3x + 7y)$$

(j) $36(x - 2)^2 - 25(x + 1)^2 = (6x - 12 + 5x + 5)(6x - 12 - 5x - 5)$
$$= (11x - 7)(x - 17)$$

(k) $(x + y - z)^2 - (x - y + z)^2 = (x + y - z + x - y + z)(x + y - z - x + y - z)$
$$= 2x(2y - 2z) = 4x(y - z)$$

(l) $(x + y - 3)^2 - 25(x + 2)^2 = (x + y - 3 + 5x + 10)(x + y - 3 - 5x - 10)$
$$= (6x + y + 7)(y - 4x - 13)$$

3. Practical calculations exploiting on the squares difference identity:

(a) $99 \times 101 = (100 - 1)(100 + 1) = 100^2 - 1 = 9999$

(b) $98 \times 102 = (100 - 2)(100 + 2) = 100^2 - 2^2 = 10{,}000 - 4 = 9996$

(c) $83 \times 77 = (80 + 3)(80 - 3) = 80^2 - 3^2 = 6400 - 9 = 6391$

(d) $79 \times 81 = (80 - 1)(80 + 1) = 80^2 - 1 = 6400 - 1 = 6399$

(e) $18 \times 22 = (20 - 2)(20 + 2) = 400 - 4 = 396$

(f) $201 \times 199 = (200 + 1)(200 - 1) = 40{,}000 - 1 = 39{,}999$

(g) $1.05 \times 0.95 = (1 + 0.05)(1 - 0.05) = 1 - 0.05^2 = 1 - 0.0025 = 0.9975$

(h) $1.01 \times 0.99 = (1 + 0.01)(1 - 0.01) = 1 - 0.01^2 = 1 - 0.0001 = 0.9999$

(i) $9.9 \times 10.1 = (10 - 0.1)(10 + 0.1) = 100 - 0.01 = 99.99$

Exercise 2.6, page 16

1. All trinomials whose form is $(ax)^2 \pm 2abx + b^2$ are factorized into $(ax \pm b)^2$.

   (a) $x^2 - 2x + 1 = (1x)^2 - 2(1)(1)x + 1^2 = (x - 1)^2$

   (b) $x^2 \underline{-6x} + 9 = x^2 \underline{-2 \times 3x} + 3^2 = (x - 3)^2$

   (c) $x^2 - 2\sqrt{2}x + \underline{2} = x^2 - 2\sqrt{2}x + (\sqrt{2})^2 = (x - \sqrt{2})^2$

   (d) $4x^2 - 12x + 9 = (2x)^2 - 2 \times 3 \times 2x + 3^2 = (2x - 3)^2$

   (e) $x^4 - 2x^2 + 1 = (x^2)^2 - 2x^2 + 1 = (x^2 - 1)^2 = (x + 1)^2(x - 1)^2$

   (f) $x^{2n+2} - 2x^{n+1} + 1 = x^{2(n+1)} - 2x^{n+1} + 1 = \left(x^{n+1}\right)^2 - 2\left(x^{n+1}\right) + 1$
   $$= (x^{n+1} - 1)^2$$

Exercise 2.7, page 16

1. The relations between the sum and product of polynomial roots, attributed to François Viète, are rather practical, among other applications in mathematics, for rapid factorization of quadratic polynomials.

   (a)   Given $x^2 - 6x + 5$ we search for factors x_1, x_2 of "+5" so that $x_1 + x_2 = -6$. Being prime number, five is divisible only with $\pm 1, \pm 5$. We find that -1 and -5 factors satisfy both conditions, i.e. their product equals $+5$ and their sum is -6. Thus we write,
   $$x^2 \underline{-6x} + 5 = x^2 \underline{-(1)x - 5x} + 5 = x(x - 1) - 5(x - 1) = (x - 1)(x - 5)$$

   (b)   Two factors of "+14" whose sum equals -9 are -2 and -7, thus we write,

   $$x^2 \underline{-9x} + 14 = x^2 \underline{-2x - 7x} + 14 = x(x - 2) - 7(x - 2) = (x - 2)(x - 7)$$

   (c)   Two factors of "+8" whose sum equals -6 are -2 and -4, thus we write,
   $$x^2 \underline{-6x} + 8 = x^2 \underline{-2x - 4x} + 8 = x(x - 2) - 4(x - 2) = (x - 2)(x - 4)$$

   (d)   In case when the leading coefficient of $ax^2 + bx + c$ does not equal one, i.e. $a = 2 \neq 1$, there are at least two possible techniques to use,

   Technique 1: first, factor the leading coefficient in trinomial as $ax^2 + bx + c = a[x^2 + (b/a)x + (c/a)]$, then search factors of (c/a) whose sum equals (b/a), i.e.

$$2x^2 + 3x + 1 = 2\left(x^2 + \frac{3}{2}x + \frac{1}{2}\right)$$

$$= \left\{1 \times \frac{1}{2} = \frac{1}{2} \quad \text{and} \quad 1 + \frac{1}{2} = \frac{3}{2}\right\}$$

$$= 2\left(x^2 + x + \frac{1}{2}x + \frac{1}{2}\right) = 2\left[x(x+1) + \frac{1}{2}(x+1)\right]$$

$$= 2\left[(x+1)\left(x + \frac{1}{2}\right)\right] = (x+1)(2x+1)$$

Technique 2: search factors of (ac) whose sum equals b, i.e.

$$2x^2 + 3x + 1 = \left\{ac = 2 \times 1 = 2 \quad \therefore \quad 1 \times 2 = 2 \quad \text{and} \quad 1 + 2 = 3\right\}$$

$$= 2x^2 + 3x + 1 = 2x^2 + 2x + x + 1 = 2x(x+1) + x + 1$$

$$= (x+1)(2x+1)$$

(e) $3x^2 - 14x + 8 = \begin{cases} 3 \times 8 = 24, \quad (-12) \times (-2) = 24 \quad \text{and} \\ (-12) + (-2) = -14 \end{cases}$

$$= 3x^2 - 12x - 2x + 8 = 3x(x-4) - 2(x-4)$$

$$= (x-4)(3x-2)$$

(f) $-2x^2 + x + 3 = \left\{-2 \times 3 = -6, \quad (-2) + 3 = 1\right\}$

$$= -2x^2 - 2x + 3x + 3 = -2x(x+1) + 3(x+1)$$

$$= (x+1)(3-2x)$$

(g) $-x^2 + 5x - 4 = -x^2 + x + 4x - 4 = -x(x-1) + 4(x-1) = (x-1)(4-x)$

(h) $-6x^2 + 5x + 4 = -6x^2 - 3x + 8x + 4 = -3x(2x+1) + 4(2x+1)$

$$= (2x+1)(4-3x)$$

(i) $x^2 + (\sqrt{2} + \sqrt{3})x + \sqrt{6} = x^2 + x\sqrt{2} + x\sqrt{3} + \sqrt{2 \times 3}$

$$= x(x + \sqrt{2}) + \sqrt{3}(x + \sqrt{2})$$

$$= (x + \sqrt{2})(x + \sqrt{3})$$

Exercise 2.8, page 16

1. After writing x^{2n} as $(x^n)^2$, we convert the bi-quadratic polynomial form into quadratic relative to (x^n) as:

(a) $x^4 - 13x^2 + 36 = (x^2)^2 - 4(x^2) - 9(x^2) + 36 = x^2(x^2 - 4) - 9(x^2 - 4)$

$$= (x^2 - 4)(x^2 - 9) = (x+2)(x-2)(x+3)(x-3)$$

(b) $x^4 - 10x^2 + 9 = x^4 - x^2 - 9x^2 - 9 = x^2(x^2 - 1) - 9(x^2 - 1)$

$$= (x^2 - 1)(x^2 - 9) = (x+1)(x-1)(x+3)(x-3)$$

(c) $x^6 - 2x^3 + 1 = (x^3)^2 - 2x^3 + 1 = (x^3 - 1)^2$
$$= \{(x^3 - 1) \div (x - 1) = x^2 + x + 1\}$$
$$= [(x^2 + x + 1)(x - 1)]^2 = (x^2 + x + 1)^2(x - 1)^2$$
$$= (x + 1)(x - 1)(x^2 + x + 1)^2$$

(d) $x^6 \underline{+3x^3} + 2 = x^6 + \underline{x^3 + 2x^3} + 2 = x^3(x^3 + 1) + 2(x^3 + 1)$
$$= (x^3 + 1)(x^3 + 2)$$
$$= (x + 1)(x^2 - x + 1)(x + \sqrt[3]{2})(x^2 - x\sqrt[3]{2} + \sqrt[3]{4})$$

(e) $x^6 \underline{+9x^3} + 8 = x^6 \underline{+x^3 + 8x^3} + 8 = x^3(x^3 + 1) + 8(x^3 + 1)$
$$= (x^3 + 1)(x^3 + 8) = (x + 1)(x^2 - x + 1)(x^3 + 8)$$
$$= (x + 1)(x^2 - x + 1)(x + 2)(x^2 - 2x + 4)$$

(f) $2x^{2m+2} - 11x^{m+1} + 9 = 2\left(x^{m+1}\right)^2 - 11\left(x^{m+1}\right) + 9$
$$= 2\left(x^{m+1}\right)^2 - 2\left(x^{m+1}\right) - 9\left(x^{m+1}\right) + 9$$
$$= 2x^{m+1}\left(x^{m+1} - 1\right) - 9\left(x^{m+1} - 1\right)$$
$$= \left(x^{m+1} - 1\right)\left(2x^{m+1} - 9\right)$$

Exercise 2.9, page 16

1. Completing the square technique exploits binomial square form.

(a)   Complete the binomial square form as follows

$$x^2 + 6x - 7 = \underbrace{x^2 + 2(3)x + 3^2}_{(x + 3)^2} \underbrace{-3^2 - 7}_{-16} = (x + 3)^2 - 16 = 0$$

Therefore, the quadratic equation is solved as

$$(x + 3)^2 = 16 \quad \therefore \quad x + 3 = \pm 4 \quad \therefore \quad x_{1,2} = -3 \pm 4 \quad \therefore \quad x_1 = 1, \quad x_2 = -7$$

(b)   The leading coefficient (i.e. "2") is not equal to one so it can be factored. In addition, the linear term coefficient "5" is a prime number, consequently it is not possible to factor "2" that is necessary for the complete square form. The workaround is to multiply and divide "5" by two.

$$2x^2 - 10x - 3 = 2\left(x^2 - 5x - \frac{3}{2}\right) = 2\left(x^2 - 2\left(\frac{5}{2}\right)x - \frac{3}{2}\right)$$

$$= 2\left(\underline{x^2 - 2\left(\frac{5}{2}\right)x + \left(\frac{5}{2}\right)^2} - \left(\frac{5}{2}\right)^2 - \frac{3}{2}\right)$$

$$= 2\left[\left(x - \frac{5}{2}\right)^2 - \frac{25}{4} - \frac{3}{2}\right] = 2\left[\left(x - \frac{5}{2}\right)^2 - \frac{31}{4}\right] = 0$$

$$\therefore$$

$$\left(x - \frac{5}{2}\right)^2 - \frac{31}{4} = 0 \quad \therefore \quad \left(x - \frac{5}{2}\right)^2 = \frac{31}{4}$$

$$\therefore$$

$$x - \frac{5}{2} = \pm\sqrt{\frac{31}{4}} \quad \therefore \quad x_{1,2} = \frac{5}{2} \pm \sqrt{\frac{31}{4}} = \frac{5 \pm \sqrt{31}}{2}$$

(c) Factor the leading coefficient "-1" as

$$-x^2 - 6x + 7 = -[x^2 + 2(3)x + 3^2 - 3^2 - 7] = -[(x+3)^2 - 16] = 0$$

$$\therefore$$

$$(x+3)^2 - 16 = 0 \quad \therefore \quad x + 3 = \pm\sqrt{16} \quad \therefore \quad x_{1,2} = -3 \pm 4$$

$$\therefore$$

$$x_1 = 1, \quad x_2 = -7$$

Note, however, that although the roots of $-x^2 - 6x + 7$ and $x^2 + 6x - 7$ (Example 2.9-1(a)) are equal, the two quadratic polynomials are not the same.

(d) Force the form of complete square as

$$x^2 + 3x = x^2 + 2\left(\frac{3}{2}\right)x + \left(\frac{3}{2}\right)^2 - \left(\frac{3}{2}\right)^2 = \left(x + \frac{3}{2}\right)^2 - \frac{9}{4} = 0$$

$$\therefore$$

$$x + \frac{3}{2} = \pm\frac{3}{2} \quad \therefore \quad x_{1,2} = -\frac{3}{2} \pm \frac{3}{2} \quad \therefore \quad x_1 = 0, \quad x_2 = -3$$

(e) Factorize the leading coefficient, as well as $6 = 2 \times 3$, so that

$$2x^2 + 6x + 2 = 2\left[x^2 + 2\left(\frac{3}{2}\right)x + \left(\frac{3}{2}\right)^2 - \left(\frac{3}{2}\right)^2 + 1\right]$$

$$= 2\left[\left(x + \frac{3}{2}\right)^2 - \frac{9}{4} + 1\right] = 0$$

$$\therefore$$

$$x + \frac{3}{2} = \pm\sqrt{\frac{5}{4}} \quad \therefore \quad x_{1,2} = \frac{-3 \pm \sqrt{5}}{2}$$

(f) $x^2 - x - 6 = 0 \quad \therefore \quad \left(x - \frac{1}{2}\right)^2 - \frac{25}{4} = 0 \quad \therefore \quad x_1 = -2, \quad x_2 = 3$

Exercise 2.10, page 16

1. After rearranging all terms given polynomials in descending powers, the long division is done in the manner similar to division of large numbers.

(a) Given two polynomials, we proceed as follows.

1. Divide only the two highest power terms (underlined) and write the result, that is to say $x^4 \div x = x^3$.

$$(\underline{x^4} - 2x^3 - 7x^2 + 8x + 12) \div (\underline{x} - 3) = \underline{x^3}$$

2. Multiply x^3 with the divisor polynomial and write the product under dividend

$$(x^4 - 2x^3 - 7x^2 + 8x + 12) \div \underline{(x-3)} = x^3$$
$$\underline{x^4 - 3x^3} \longleftarrow \qquad\qquad\qquad \text{Multiply}$$

3. Subtract the vertically aligned terms, i.e. $(x^4 - 2x^2) - (x^4 - 3x^2) = x^3$, then add the next term from dividend, i.e. $-7x^2$. Now, the problem is reduced to division $(x^3 - 7x^2) \div (x - 3)$

$$(x^4 - 2x^3 \underline{-7x^2} + 8x + 12) \div (x - 3) = x^3$$
$$-\underline{(x^4 - 3x^3)} \downarrow$$
$$= 0 \ + x^3 \ -7x^2$$

4 Next term in the solution polynomial is found as $x^3 \div x = x^2$, then repeat the previous steps until the last term in the solution is found as

$$(x^4 - 2x^3 - 7x^2 + 8x + 12) \div (x - 3) = x^3 + x^2 - 4x - 4$$
$$\underline{-(x^4 - 3x^3)}$$
$$x^3 - 7x^2$$
$$\underline{-\ (x^3 - 3x^2)}$$
$$-\ 4x^2 + 8x$$
$$\underline{-\ (-4x^2 + 12x)}$$
$$-\ 4x + 12$$
$$\underline{-\ (-4x + 12)}$$
$$=\ 0$$

Because the last subtraction equals zero means that $r = 0$, that is to say $x = 3$ is zero of $P(x) = x^4 - 2x^3 - 7x^2 + 8x + 12$. It is verified by setting $x = 3$ in $P(x)$ as

$$P(3) = (3)^4 - 2(3)^3 - 7(3)^2 + 8(3) + 12 = 81 - 54 - 63 + 24 + 12 = 0$$

(b) Long division gives,

$$(6x^3 + 10x^2 + 0x + 8) \div (2x^2 + 1) = 3x + 5 + \underbrace{\frac{-3x + 3}{2x^2 + 1}}_{r}$$

$$\underline{6x^3 \quad +0x^2 + 3x}$$
$$10x^2 - 3x + 8$$
$$\underline{10x^2 + 0x + 5}$$
$$\underline{-3x + 3 \neq 0}$$

(c) Long division gives,

$$(4x^3 - 7x^2 - 11x + 5) \div (4x + 5) = x^2 - 3x + 1$$

$$\underline{4x^3 + 5x^2}$$

$$-12x^2 - 11x$$

$$\underline{-12x^2 - 15x}$$

$$4x + 5$$

$$\underline{4x + 5}$$

$$\underline{= 0}$$

(d) Long division gives,

$$(x^2 + x - 5) \div (x + 3) = x - 2 + \underbrace{\frac{1}{x + 3}}_{r}$$

$$\underline{x^2 + 3x}$$

$$-2x - 5$$

$$\underline{-2x - 6}$$

$$\underline{= 1}$$

2. One of the techniques to factorize odd order polynomials exploits two fundamental theorems in algebra (very loosely interpreted as):

1. The total number of polynomial roots (i.e. both real and complex) is equal to the polynomial degree, where the complex roots come in pairs, i.e. each complex root is accompanied by its complex conjugate pair. Consequently, in odd order polynomials there must be at least one real root (i.e. the one that does not have its pair).
2. After multiplying the leading and zero term coefficients, polynomial roots may be found among the factors of that product.

(a) Given $P(x) = x^3 + x^2 - 4x - 4$, the leading coefficient is "$(+1)$" and the zero power term equals "(-4)". Factors of $-4 \times 1 = -4$ are $\{\pm 1, \pm 4, \pm 2\}$. Thus we calculate $P(x)$ for each $x = \pm 1, \pm 4, \pm 2$ as

$$P(-1) = ((-1)^3 + (-1)^2 - 4(-1) - 4) = 0 \quad \therefore \quad \underline{x_1 = -1}$$

$$P(1) = ((1)^3 + (1)^2 - 4(1) - 4) = -6 \quad \therefore \quad P(1) \neq 0$$

$$P(-2) = ((-2)^3 + (-2)^2 - 4(-2) - 4) = 0 \quad \therefore \quad \underline{x_2 = -2}$$

$$P(2) = ((2)^3 + (2)^2 - 4(2) - 4) = 0 \quad \therefore \quad \underline{x_3 = 2}$$

$$P(-4) = ((-4)^3 + (-4)^2 - 4(-4) - 4) = -36 \quad \therefore \quad P(-4) \neq 0$$

$$P(4) = ((4)^3 + (4)^2 - 4(4) - 4) = 60 \quad \therefore \quad P(4) \neq 0$$

Therefore,
$$P(x) = x^3 + x^2 - 4x - 4 = (x - x_1)(x - x_2)(x - x_3) = (x + 1)(x + 2)(x - 2).$$
As it happens, all three roots are real.

(b) Similarly, given $Q(x) = x^3 - 1$, factors of its zero power term are ± 1. We calculate,

$$Q(-1) = (-1)^3 - 1 = -2 \quad \therefore \quad Q(-1) \neq 0$$

$$Q(1) = (1)^3 - 1 = 0 \quad \therefore \quad x_1 = 1$$

That is to say,

$$Q(x) = (x - 1)(ax^2 + bx + c)$$

$$\therefore$$

$$(ax^2 + bx + c) = Q(x) \div (x - x_1) = (x^3) \div (x - 1)$$

By the long division technique, we write

$$(x^3 - 1) \div (x - 1) = x^2 + x + 1$$
$$\underline{-(x^3 - x^2)}$$
$$x^2 - 1$$
$$\underline{-(x^2 - x)}$$
$$x - 1$$
$$\underline{-(x - 1)}$$
$$= 0$$

Therefore, $Q(x) = x^3 - 1 = (x - 1)(x^2 + x + 1)$. Discriminant of the second order polynomial is negative, therefore the last two roots are complex,

$$\Delta = b^2 - 4ac = 1^2 - 4 \times 1 \times 1 = -3 < 0$$

$$\therefore$$

$$x_{2,3} = \frac{-b \pm \sqrt{\Delta}}{2a} = \frac{-1 \pm i \sqrt{3}}{2}$$

$$\therefore$$

$$Q(x) = x^3 - 1 = (x - 1)(x^2 + x + 1)$$

$$= (x - 1)\left(x + \frac{1 + i \sqrt{3}}{2}\right)\left(x + \frac{1 - i \sqrt{3}}{2}\right)$$

(c)   Important polynomial form is known as the *difference of two cubes*. Given $P(x) = x^3 - y^3$ by inspection we conclude that one root of this binomial is $x = y$, i.e. $P(y) = 0$. Therefore, $P(x)$ is divisible by $(x - y)$, we note that Example 2.10-(b) is special case when $y = 1$. Using long division technique we derive this important identity,

$$(x^3 + 0x^2y + 0xy^2 - y^3) \div (x - y) = x^2 + xy + y^2$$
$$\underline{x^3 - x^2y}$$
$$x^2y + 0xy^2$$
$$\underline{x^2y - xy^2}$$
$$xy^2 - y^3$$
$$\underline{xy^2 - y^3}$$
$$= 0$$

Therefore, $(x^3 - y^3) = (x - y)(x^2 + xy + y^2)$.

(d)   Important polynomial form is known as the *sum of two cubes*. Given $P(x) = x^3 + y^3$ by inspection we conclude that one root of this binomial is $x = -y$, i.e. $P(-y) = 0$. Therefore, $P(x)$ is divisible by $(x + y)$. Using long division technique we derive this important identity,

$$(x^3+0x^2y+0xy^2+y^3) \div (x+y) = x^2 - xy + y^2$$

$$\underline{x^3 + x^2y}$$

$$- x^2y+0xy^2$$

$$\underline{-x^2y - xy^2}$$

$$xy^2 + y^3$$

$$\underline{xy^2 + y^3}$$

$$= 0$$

Therefore, $(x^3 + y^3) = (x + y)(x^2 - xy + y^2)$.

Exercise 2.11, page 17

1. Difference/sum of cubes is factorized as follows, see Example 2.10-2.

   (a) $Q(x) = x^3 - 1 = (x - 1)(x^2 + x + 1)$

   (b) $Q(x) = x^3 - y^3 = (x - y)(x^2 + xy + y^2)$

   (c) $Q(x) = x^3 + y^3 = (x + y)(x^2 - xy + y^2)$

Exercise 2.12, page 17

1. Factorization of higher order polynomials usually requires multiple steps and techniques. As a consequence there is no unique path to find the solution, thus among number of possible ways to solve the problem we aim to find ones that are elegant and creative.

   (a) $x^4 + 4 = \underline{x^4+4x^2 + 4}-4x^2 = (x^2 + 2)^2 - (2x)^2$
   $$= (x^2 + 2x + 2)(x^2 - 2x + 2)$$

   (b) $x^4 + x^2 + 1 = \underline{x^4+2x^2 + 1}-x^2 = (x^2 + 1)^2 - (x)^2$
   $$= (x^2 - x + 1)(x^2 + x + 1)$$

   (c) $x^5 + x + 1 = \underline{x^5-x^2}+x^2 + x + 1 = x^2(x^3 - 1) + x^2 + x + 1$
   $$\left\{ x^3 - 1 = (x - 1)(x^2 + x + 1) \text{ see Example 2.10-2(b)} \right\}$$

   $$= x^2\underline{(x - 1)(x^2 + x + 1)} + \underline{(x^2 + x + 1)}$$

   $$= (x^2 + x + 1)[x^2(x - 1) + 1]$$

   $$= (x^2 + x + 1)(x^3 - x^2 + 1)$$

(d) $(x + y + z)^3 - x^3 - y^3 - z^3 = (x + y + z)^3 - x^3 - (y^3 + z^3)$

$$= \begin{cases} a^3 - b^3 & = (a - b)(a^2 + ab + b^2) \\ a^3 + b^3 & = (a + b)(a^2 - ab + b^2) \end{cases}$$

$= (\cancel{x} + y + z - \cancel{x})[(x + y + z)^2 + (x + y + z)\,x + x^2] - (y + z)(y^2 - yz + z^2)$

$= (y + z)\left[(x + y + z)^2 + (x + y + z)\,x + x^2 - y^2 + yz - z^2\right]$

$= (y + z)\left[2xy + 2xz + x^2 \cancel{+ y^2} + 2yz \cancel{+ z^2} + x^2 + xy + zx + x^2 \cancel{- y^2} + yz \cancel{- z^2}\right]$

$= (y + z)\left(3xy + 3xz + 3yz + 3x^2\right) = 3(y + z)\left(\underline{xy} + \underline{xz} + \underline{yz} + \underline{x^2}\right)$

$= 3(y + z)[x(x + y) + z(x + y)] = 3(y + z)(x + y)(x + z)$

(e) $(x + 1)(x + 3)(x + 5)(x + 7) + 15 = (x + 1)(x + 7)\,(x + 3)(x + 5) + 15$
$= (x^2 + 8x + 7)(x^2 + 8x + 15) + 15$

$= (\underline{x^2 + 8x + 7})(\underline{x^2 + 8x + 7} + 8) + 15$

$= \left\{\text{substitution: } x^2 + 8x + 7 = t\right\}$

$= t(t + 8) + 15 = t^2 + 8t + 15$

$= t^2 + 3t + 5t + 15 = t(t + 3) + 5(t + 3) = (t + 3)(t + 5)$

$= \left\{\text{back substitution: } t = x^2 + 8x + 7\right\}$

$= \left(x^2 + 8x + 7 + 3\right)\left(x^2 + 8x + 7 + 5\right)$

$= \left(x^2 + 8x + 10\right)(x^2 + 8x + 12)$

$= \left[x^2 + 2(4)x + 4^2 - 4^2 + 10\right]\left(x^2 + 2x + 6x + 12\right)$

$= \left[(x + 4)^2 - 6\right](x + 2)(x + 6)$

$= \left\{a^2 - b^2 = (a + b)(a - b)\right\}$

$= (x + 4 - \sqrt{6})(x + 4 + \sqrt{6})(x + 2)(x + 6)$

(f) $\left(x^2 + x + 1\right)\left(x^3 + x^2 + 1\right) - 1 = (\underline{x^2 + 1} + x)(x^3 + \underline{x^2 + 1}) - 1$
     $\left\{\text{multiply by } \left(x^2 + 1\right)\right\}$

$= \left(x^2 + 1\right)x^3 + \left(x^2 + 1\right)\left(x^2 + 1\right) + \left(x^2 + 1\right)x + \underline{x^4 - 1}$

$= \left(x^2 + 1\right)x^3 + \left(x^2 + 1\right)^2 + \left(x^2 + 1\right)x + \underline{\left(x^2 - 1\right)\left(x^2 + 1\right)}$

$= \left(x^2 + 1\right)\left[x^3 + \left(x^2 \cancel{+ 1}\right) + x + \left(x^2 \cancel{- 1}\right)\right]$

$= \left(x^2 + 1\right)\left[x^3 + 2x^2 + x\right] = x\left(x^2 + 1\right)\left(x^2 + 2x + 1\right)$

$= x\left(x^2 + 1\right)(x + 1)^2$

Exercise 2.13, page 17

1.

(a) $\dfrac{x^{n+2}}{2x\,x^n} = \dfrac{x\,x^{n+1}}{2x^{n+1}} = \dfrac{x}{2}$

(b) $\dfrac{x^{3n+2} + x^{3n+1}}{x^2\,x^{3n}} = \dfrac{x^{3n+1}\,(x+1)}{x\,x^{3n+1}} = \dfrac{x+1}{x}$

(c) $\dfrac{x + x^2 + x^3 + \cdots + x^n}{\dfrac{1}{x} + \dfrac{1}{x^2} + \dfrac{1}{x^3} + \cdots + \dfrac{1}{x^n}} = \dfrac{x + x^2 + x^3 + \cdots + x^n}{\dfrac{x^{n-1} + x^{n-2} + x^{n-3} + \cdots + 1}{x^n}}$

$\qquad\qquad = \dfrac{\left(x + x^2 + x^3 + \cdots + x^n\right) x^n}{x^{n-1} + x^{n-2} + x^{n-3} + \cdots + 1}$

$\qquad\qquad = \dfrac{\left(1 + x^2 + x^2 + \cdots + x^{n-1}\right) x^{n+1}}{x^{n-1} + x^{n-2} + x^{n-3} + \cdots + 1} = x^{n+1}$

Exercise 2.14, page 17

1. Partial fraction form of a rational function is very useful for calculating its integral. A general case of rational expression of two polynomials, for example, k-th order $P_k(x)$ and m-th order $Q_m(x)$, where $m > k$, may be written as

$$\frac{P_k(x)}{Q_m(x)} = \frac{P(x)}{(x - x_1)(x - x_2)(x - x_3)^n(ax^2 + bx + c)}$$

$$= \frac{\alpha}{x - x_1} + \frac{\beta}{x - x_2} + \frac{\gamma_1}{(x - x_3)^1} + \frac{\gamma_2}{(x - x_3)^2} + \cdots + \frac{\gamma_n}{(x - x_3)^n} + \frac{\delta x + \eta}{(ax^2 + bx + c)}$$

where, $\alpha, \beta, \gamma, \ldots$ are the coefficient constants to be calculated, and partial decomposition rational forms are created as follows:

1. For each unique zero of $Q(x)$, such as x_1, x_2, there is one rational term.
2. For zeros with n multiplicity, such as x_3, there is a series of n rational terms.
3. When quadratic term $(ax^2 + bx + c)$ has complex zeros, then its corresponding numerator must be linear binomial.

(a) In this example, $Q(x)$ has two unique zeros at $x = -1$ and $x = 4$, thus we write

$$\frac{P(x)}{Q(x)} = \frac{x}{(x + 1)(x - 4)} = \frac{A}{x + 1} + \frac{B}{x - 4} = \frac{A(x - 4) + B(x + 1)}{(x + 1)(x - 4)}$$

$$= \frac{Ax - 4A + Bx + B}{(x + 1)(x - 4)} = \frac{(A + B)x + B - 4A}{(x + 1)(x - 4)}$$

Therefore, by equalizing numerators $P(x)$ before and after decomposition, we calculate the unknown constants as

$$x \equiv (A + B)x + B - 4A$$

$$\therefore$$

$$\underline{\text{term } x^1}: \quad 1 = A + B \quad \therefore \quad A = 1 - B$$

$$\underline{\text{term } x^0}: \quad 0 = B - 4A \quad \therefore \quad B = 4A$$

$$\therefore$$

$$B = 4(1 - B), \quad \therefore \quad B = \frac{4}{5} \quad \therefore \quad A = 1 - \frac{4}{5} = \frac{1}{5}$$

Consequently,

$$\frac{P(x)}{Q(x)} = \frac{x}{(x + 1)(x - 4)} = \frac{1}{5(x + 1)} + \frac{4}{5(x - 4)}$$

(b)   In this example, $Q(x)$ has one unique zero at $x = 2$ and one double zero $x = 0$, thus we write

$$\frac{P(x)}{Q(x)} = \frac{x + 2}{x^3 - 2x^2} = \frac{x + 2}{x^2(x - 2)} = \left\{ Q(x) = x^2(x - 2) \quad \therefore \quad x_{1,2} = 0, \quad x_3 = 2 \right\}$$

$$= \frac{A}{x} + \frac{B}{x^2} + \frac{C}{x - 2} = \frac{Ax(x - 2) + B(x - 2) + Cx^2}{x^2(x - 2)}$$

$$= \frac{Ax^2 - 2Ax + Bx - 2B + Cx^2}{x^2(x - 2)} = \frac{(A + C)x^2 + (B - 2A)x - 2B}{x^2(x - 2)}$$

Therefore, by equalizing numerators $P(x)$ before and after decomposition, we calculate the unknown constants as

$$x + 2 \equiv (A + C)x^2 + (B - 2A)x - 2B$$

$$\therefore$$

$$\underline{\text{term } x^2}: \quad 0 = A + C \quad \therefore \quad A = -C$$

$$\underline{\text{term } x^1}: \quad 1 = B - 2A$$

$$\underline{\text{term } x^0}: \quad 2 = -2B \quad \therefore \quad B = -1$$

$$\therefore$$

$$B = -1, \quad \therefore \quad 1 = (-1) - 2A \quad \therefore \quad A = -1 \quad \therefore \quad C = 1$$

Consequently,

$$\frac{P(x)}{Q(x)} = \frac{x + 2}{x^3 - 2x^2} = \frac{x + 2}{x^2(x - 2)} = -\frac{1}{x} - \frac{1}{x^2} + \frac{1}{x - 2}$$

(c) Denominator $Q(x)$ has two unique real zeros at $x = 1$ and $x = -1$, as well as two complex zeros, thus we write

$$\frac{P(x)}{Q(x)} = \frac{2x^2}{x^4 - 1} = \left\{ a^2 - b^2 = (a - b)(a + b) \right\} = \frac{2x^2}{(x^2 - 1)(x^2 + 1)}$$

$$= \frac{2x^2}{(x - 1)(x + 1)(x^2 + 1)} = \frac{A}{x - 1} + \frac{B}{x + 1} + \frac{Cx + D}{x^2 + 1}$$

$$= \frac{A(x + 1)(x^2 + 1) + B(x - 1)(x^2 + 1) + (Cx + D)(x - 1)(x + 1)}{(x - 1)(x + 1)(x^2 + 1)}$$

$$= \frac{(A + B + C)x^3 + (A - B + D)x^2 + (A + B - C)x + A - B - D}{(x - 1)(x + 1)(x^2 + 1)}$$

Therefore, by equalizing numerators $P(x)$ before and after decomposition, we calculate the unknown constants as

$$2x^2 \equiv (A + B + C)x^3 + (A - B + D)x^2 + (A + B - C)x + (A - B - D)$$

$$\therefore$$

$\underline{\text{term } x^3} : \quad 0 = A + B + C$

$\underline{\text{term } x^2} : \quad 2 = A - B + D$

$\underline{\text{term } x^1} : \quad 0 = A + B - C$

$\underline{\text{term } x^0} : \quad 0 = A - B - D$

We solve the above system of equations, for example, by the reduction method. From first equation we have $0 = A + B + C \quad \therefore \quad B = -A - C$, then we write

$$2 = A - B + D \quad \therefore \quad 2 = A - (-A - C) + D = 2A + C + D$$

$$0 = A + B - C \quad \therefore \quad 0 = A + (-A - C) - C = -2C \quad \therefore \quad \underline{C = 0}$$

$$0 = A - B - D \quad \therefore \quad 0 = A - (-A - C) - D = 2A - \cancel{C} - D = 2A - D \quad \therefore \quad 2A = D$$

$$2 = 2A + \cancel{C} + D \quad \therefore \quad D = 2 - 2A$$

$$2A = D \quad \therefore \quad 2A = 2 - 2A \quad \therefore \quad A = \frac{1}{2} \quad \therefore \quad \underline{D = 1}$$

$$B = -A - \cancel{C} \quad \therefore \quad B = -\frac{1}{2}$$

Consequently,

$$\frac{P(x)}{Q(x)} = \frac{2x^2}{x^4 - 1} = \frac{2x^2}{(x - 1)(x + 1)(x^2 + 1)} = \frac{1}{2(x - 1)} - \frac{1}{2(x + 1)} + \frac{1}{x^2 + 1}$$

(d)   In this case numerator polynomial has higher order than the denominator. For that reason, first we use long division then partial fraction decomposition.

$$(x^3 + x^2 - 16x + 16) \div (x^2 - 4x + 3) = x + 5 + \frac{x + 1}{x^2 - 4x + 3}$$

$$(-)\ \underline{x^3 - 4x^2 + 3x}$$

$$5x^2 - 19x + 16$$

$$(-)\quad \underline{5x^2 - 20x + 15}$$

$$x + 1$$

Denominator $Q(x)$ has two unique real zeros of $x = 1$ and $x = 3$, thus we write

$$\frac{P(x)}{Q(x)} = \frac{x + 1}{x^2 - 4x + 3} = \frac{x + 1}{(x - 1)(x - 3)} = \frac{A}{x - 1} + \frac{B}{x - 3}$$

$$= \frac{A(x - 3) + B(x - 1)}{(x - 1)(x - 3)} = \frac{(A + B)x - 3A - B}{(x - 1)(x - 3)}$$

Therefore, by equalizing numerators $P(x)$ before and after decomposition, we calculate the unknown constants as

$$x + 1 \equiv (A + B)x - 3A - B$$

$$\therefore$$

$$\underline{\text{term } x^1}:\ \ 1 = A + B \ \ \therefore \ \ A = 1 - B$$

$$\underline{\text{term } x^0}:\ \ 1 = -3A - B \ \ \therefore \ \ 1 = -3(1 - B) - B$$

$$\therefore$$

$$\underline{B = 2} \ \ \therefore \ \ \underline{A = -1}$$

Consequently,

$$\frac{P(x)}{Q(x)} = \frac{x^3 + x^2 - 16x + 16}{x^2 - 4x + 3} = x + 5 - \frac{1}{x - 1} + \frac{2}{x - 3}$$

(e)   An example of $Q(x)$ with one unique real zero $x = 1$ and two complex zeros.

$$\frac{P(x)}{Q(x)} = \frac{1}{x^3 - 1} = \{\text{see Example 2.2-1}\} = \frac{1}{(x - 1)(x^2 + x + 1)}$$

$$= \frac{A}{x-1} + \frac{Bx+C}{x^2+x+1} = \frac{A(x^2+x+1) + (Bx+C)(x-1)}{(x-1)(x^2+x+1)}$$

$$= \frac{(A+B)x^2 + (A-B+C)x + A-C}{(x-1)(x^2+x+1)}$$

Therefore, by equalizing numerators $P(x)$ before and after decomposition, we calculate the unknown constants as

$$1 \equiv (A+B)x^2 + (A-B+C)x + A - C$$

$$\therefore$$

term x^2 : $0 = A + B$ $\therefore$ $A = -B$

term x^1 : $0 = A - B + C$ $\therefore$ $0 = (-B) - B + C$ $\therefore$ $C = 2B$

term x^0 : $1 = A - C$ $\therefore$ $1 = (-B) - (2B)$ $\therefore$ $B = -\dfrac{1}{3}$

$$C = -\frac{2}{3} \quad \therefore \quad A = \frac{1}{3}$$

Consequently,

$$\frac{P(x)}{Q(x)} = \frac{1}{x^3-1} = \frac{1}{(x-1)(x^2+x+1)} = \frac{1}{3(x-1)} - \frac{x+2}{3(x^2+x+1)}$$

(f)   An example of $Q(x)$ with two complex pairs of zeros. Complete the squares to enforce bi-quadratic form, then use difference of square identities to rearrange given fourth order polynomial.

$$\frac{P(x)}{Q(x)} = \frac{4}{x^4+1} = \frac{4}{\underbrace{x^4+2x^2+1}-2x^2} = \frac{4}{(x^2+1)^2 - \left(\sqrt{2}\,x\right)^2}$$

$$= \frac{4}{(x^2+1-\sqrt{2}\,x)(x^2+1+\sqrt{2}\,x)} = \frac{Ax+B}{x^2-\sqrt{2}\,x+1} + \frac{Cx+D}{x^2+\sqrt{2}\,x+1}$$

$$= \frac{(Ax+B)(x^2+\sqrt{2}\,x+1) + (Cx+D)(x^2-\sqrt{2}\,x+1)}{(x^2-\sqrt{2}\,x+1)(x^2+\sqrt{2}\,x+1)}$$

$$= \frac{(A+C)x^3 + (\sqrt{2}A + B - \sqrt{2}C + D)x^2 + (A+C+\sqrt{2}B - \sqrt{2}D)x + B + D}{(x^2-\sqrt{2}\,x+1)(x^2+\sqrt{2}\,x+1)}$$

Therefore, by equalizing numerators $P(x)$ before and after decomposition, we calculate the unknown constants as

$$4 \equiv (A + C)x^3 + (\sqrt{2}A + B - \sqrt{2}C + D)x^2 + (A + C + \sqrt{2}B - \sqrt{2}D)x + B + D$$

$$\therefore$$

term x^3 : $0 = A + C$ $\therefore$ $A = -C$

term x^2 : $0 = \sqrt{2}A + B - \sqrt{2}C + D$ $\therefore$ $0 = -2\sqrt{2}C + B + D$

term x^1 : $0 = A + C + \sqrt{2}B - \sqrt{2}D$ $\therefore$ $0 = +\sqrt{2}B - \sqrt{2}D$ $\therefore$ $B = D$

term x^0 : $4 = B + D$ $\therefore$ $4 = D + D$ $\therefore$ $\underline{D = 2}$ $\therefore$ $\underline{B = 2}$

$$\therefore \quad 0 = -2\sqrt{2}C + 4 \quad \therefore \quad \underline{C = \sqrt{2}} \quad \therefore \quad \underline{A = -\sqrt{2}}$$

Consequently,

$$\frac{P(x)}{Q(x)} = \frac{4}{x^4 + 1} = \frac{-\sqrt{2}x + 2}{x^2 - \sqrt{2}x + 1} + \frac{\sqrt{2}x + 2}{x^2 + \sqrt{2}x + 1}$$

(g)   An example of $Q(x)$ with multiple complex pair of zeros.

$$\frac{P(x)}{Q(x)} = \frac{x^3 + x - 1}{(x^2 + 2)^2} = \frac{Ax + B}{x^2 + 2} + \frac{Cx + D}{(x^2 + 2)^2} = \frac{(Ax + B)(x^2 + 2) + Cx + D}{(x^2 + 2)^2}$$

$$= \frac{Ax^3 + Bx^2 + (C + 2A)x + 2B + D}{(x^2 + 2)^2}$$

Therefore, by equalizing numerators $P(x)$ before and after decomposition, we calculate the unknown constants as

$$x^3 + x - 1 \equiv Ax^3 + Bx^2 + (C + 2A)x + 2B + D$$

$$\therefore$$

term x^3 : $\underline{A = 1}$

term x^2 : $\underline{B = 0}$

term x^1 : $1 = C + 2A$ $\therefore$ $\underline{C = -1}$

term x^0 : $-1 = 2B^{\,0} + D$ $\therefore$ $\underline{D = -1}$

Consequently,

$$\frac{P(x)}{Q(x)} = \frac{x^3 + x - 1}{(x^2 + 2)^2} = \frac{x}{x^2 + 2} - \frac{x + 1}{(x^2 + 2)^2}$$

Linear Equations and Inequalities

3

Important to Know

Given, for example, system of two linear equations,

$$a_1 x + b_1 y = c_1$$
$$a_2 x + b_2 y = c_2$$

one of the techniques for calculating the solutions is known as **Cramer's** rule

$$x = \frac{\Delta_x}{\Delta}; \quad y = \frac{\Delta_y}{\Delta}; \quad (\Delta \neq 0)$$

where

$$\Delta = \begin{vmatrix} a_1 & b_1 \\ a_2 & b_2 \end{vmatrix} \stackrel{\text{def}}{=} a_1 b_2 - a_2 b_1$$

$$\Delta_x = \begin{vmatrix} c_1 & b_1 \\ c_2 & b_2 \end{vmatrix} \stackrel{\text{def}}{=} c_1 b_2 - c_2 b_1$$

$$\Delta_y = \begin{vmatrix} a_1 & c_1 \\ a_2 & c_2 \end{vmatrix} \stackrel{\text{def}}{=} a_1 c_2 - a_2 c_1$$

3.1 Exercises

3.1 *   Linear Equations

---

1. Comment on the solutions of the following equations

(a) $1 - \dfrac{x-1}{3} = \dfrac{3-x}{3}$

(b) $\dfrac{1+3x}{4} - \dfrac{6x+3}{12} = \dfrac{x}{4}$

(c) $3x + 14 = 5x - 2(x-7)$

(d) $\dfrac{x}{x-2} - \dfrac{2x+3}{x+2} = \dfrac{x^2}{4-x^2}$

© Springer Nature Switzerland AG 2021
R. Sobot, *Engineering Mathematics by Example*,
https://doi.org/10.1007/978-3-030-79545-0_3

2. Solve the following equations

(a) $1 - \dfrac{1}{1 - \dfrac{1}{1 - x}} = 2$

(b) $\dfrac{1}{2 + \dfrac{1}{3 + \dfrac{1}{4 + \dfrac{3}{x + 1}}}} = \dfrac{16}{37}$

3.2 *   Systems of Linear Equations

1. Solve the following systems of equations

(a) $\begin{cases} \dfrac{14}{x} + \dfrac{24}{y} = 10 \\[2ex] \dfrac{7}{x} - \dfrac{18}{y} = -5 \end{cases}$

(b) $\begin{cases} \dfrac{4}{x + y - 1} + \dfrac{1}{x - y + 1} = 1 \\[2ex] \dfrac{18}{x + y - 1} - \dfrac{5}{2(x - y + 1)} = 1 \end{cases}$

3.3 *   Linear Inequalities

1. Solve the following inequalities.

(a) $\left| \dfrac{4x}{2x + 4} \right| < 3$

(b) $\dfrac{(x + 2) |x - 2|}{x^2 + 2} > 1$

(c) $\dfrac{1}{x} + \dfrac{1}{x + 1} < \dfrac{x^2 - 2}{x^2 + x}$

2. Solve the following equation and inequalities:

(a) $-3x^2 + 30x - 75 > 0$

(b) $x^4 - 2x^2 + 1 < 0$

(c) $-2x^2 + 4x - 2 > 0$

(d) $\dfrac{2x^2 - 5x - 3}{-2x + 1} \leq 0$

Solutions

Exercise 3.1, page 43

1.
(a) $1 - \dfrac{x - 1}{3} = \dfrac{3 - x}{3}$ $\therefore$ $\dfrac{3 - x + 1}{3} = \dfrac{3 - x}{3}$ $\therefore$ $4 - \cancel{x} = 3 - \cancel{x}$ $\therefore$ $4 \neq 3$

Conclusion: there are no valid solutions for this equation.

(b) $\dfrac{1 + 3x}{4} - \dfrac{6x + 3}{12} = \dfrac{x}{4}$ $\therefore$ $\dfrac{\cancel{3} + 9x - 6x - \cancel{3}}{12} = \dfrac{3x}{12}$ $\therefore$ $\cancel{3x} = \cancel{3x}$ $\therefore$ $1 = 1$

Conclusion: this equation is true for any x, thus there are infinitely many solutions.

(c) $3x + 14 = 5x - 2(x - 7)$ $\therefore$ $\cancel{3x} + \cancel{14} - \cancel{3x} - \cancel{14} = 0$ $\therefore$ $0 = 0$

Conclusion: this equation is true for any x, thus there are infinitely many solutions.

(d) In this example, the equation is not defined for $x = \pm 2$, because for all three denominators $x = \pm 2$ results in the polynomial division limiting to infinity.

$$\frac{x}{x-2} - \frac{2x+3}{x+2} = \frac{x^2}{4-x^2} \quad \therefore \quad \frac{x(x+2) - (2x+3)(x-2)}{(x-2)(x+2)} = \frac{x^2}{4-x^2} \quad \therefore$$

$$\frac{-x^2 + 3x + 6}{x^2 - 4} = \frac{x^2}{4-x^2} \quad \therefore \quad \frac{x^2 - 3x - 6}{4-x^2} = \frac{x^2}{4-x^2}$$

Which implies that $x^2 - 3x - 6 = x^2$ $\therefore$ $3x + 6 = 0$. However, this equation is valid only if $x = 2$, which is already excluded. Thus we must conclude that the equation in this example does not have valid solutions.

2. "Telescopic" forms of equations are common in iterative processes.

(a) Group numbers, then invert the fractions.

$$1 - \frac{1}{1 - \dfrac{1}{1-x}} = \frac{1}{2} \quad \therefore \quad \frac{1}{1 - \dfrac{1}{1-x}} = 1 - \frac{1}{2} = \frac{1}{2} \quad \therefore \quad \{\text{invert both sides}\}$$

$$\therefore \quad 1 - \frac{1}{1-x} = 2 \quad \therefore \quad \frac{1}{1-x} = -1$$

$$\therefore \quad 1 - x = -1 \quad \therefore \quad x = 2$$

(b) Invert fractions, group the numbers and repeat.

$$\Rightarrow \therefore$$

$$\frac{1}{2 + \dfrac{1}{3 + \dfrac{1}{4 + \dfrac{3}{x+1}}}} = \frac{16}{37} \qquad\qquad \frac{1}{3 + \dfrac{1}{4 + \dfrac{3}{x+1}}} = \frac{5}{16}$$

$$\therefore \qquad\qquad\qquad\qquad\qquad \therefore$$

$$2 + \frac{1}{3 + \dfrac{1}{4 + \dfrac{3}{x+1}}} = \frac{37}{16} \qquad\qquad 3 + \frac{1}{4 + \dfrac{3}{x+1}} = \frac{16}{5}$$

$$\therefore \qquad\qquad\qquad\qquad\qquad \therefore$$

$$\frac{1}{3 + \dfrac{1}{4 + \dfrac{3}{x+1}}} = \frac{37}{16} - 2 \qquad\qquad \frac{1}{4 + \dfrac{3}{x+1}} = \frac{16}{5} - 3 = \frac{1}{5}$$

$$\therefore \Rightarrow \qquad\qquad\qquad\qquad\qquad \therefore$$

$$4 + \frac{3}{x+1} = 5$$

Finally, we find $\quad \dfrac{3}{x+1} = 1 \quad \therefore \quad x+1 = 3 \quad \therefore \quad x = 2$

Exercise 3.2, page 44

1. Use the variable substitution technique

   (a)   Given system, the inverse of variables is substituted by new variables

$$\begin{cases} \dfrac{14}{x} + \dfrac{24}{y} = 10 \\[2mm] \dfrac{7}{x} - \dfrac{18}{y} = -5 \end{cases} \quad \therefore \left\{ t = \dfrac{1}{x}, \; k = \dfrac{1}{y} \right\} \quad \therefore \begin{cases} 14t + 24k = 10 \\[2mm] 7t - 18k = -5 \end{cases}$$

$$\Rightarrow \begin{cases} 14t + 24k = 10 \\ 14t - 36k = -10 \end{cases} \quad \therefore \left\{ (2) - (1) \right\} \quad \therefore \; 60k = 20 \quad \therefore \; k = \dfrac{1}{3}$$

$$\therefore$$

$$\underline{y = 3} \quad \therefore \; 14t + 24\left(\dfrac{1}{3}\right) = 10 \quad \therefore \; 14t + 8 = 10 \quad \therefore \; t = \dfrac{1}{7} \quad \therefore \; \underline{x = 7}$$

   (b)   Two denominators are substituted as

$$\begin{cases} \dfrac{4}{x + y - 1} + \dfrac{1}{x - y + 1} = 1 \\[2mm] \dfrac{16}{x + y - 1} - \dfrac{2}{x - y + 1} = 1 \end{cases} \quad \therefore \left\{ t = \dfrac{1}{x + y - 1}, \; k = \dfrac{1}{x - y + 1} \right\}$$

$$\begin{cases} 4t + k = 1 \\ 16t - 2k = 1 \end{cases} \quad \therefore \; t = \dfrac{1 - k}{4} \quad \therefore \; \cancel{16}4 \; \dfrac{1 - k}{\cancel{4}1} - 2k = 1 \quad \therefore \; k = \dfrac{1}{2}$$

$$\therefore \; 4t + \left(\dfrac{1}{2}\right) = 1 \quad \therefore \; t = \dfrac{1}{8}$$

Putting back the original variables leads into the second system of equations

$$\begin{cases} x + y - 1 = 8 \\ x - y + 1 = 2 \end{cases} \quad \therefore \quad \begin{cases} x + y = 9 \\ x - y = 1 \end{cases} \quad \therefore \; \underline{x = 5, \; y = 4}$$

Exercise 3.3, page 44

1. Comparison of two quantities may be done by comparing, for example, if their ratio is greater than, equal to, or less than one, or if their difference is positive, zero or negative, etc. In addition, absolute values are converted into two equations (inequalities).

(a) $\left| \dfrac{4x}{2x+4} \right| < 3$ $\therefore$ $-3 < \dfrac{4x}{2x+4} < 3$

Therefore, there are two inequalities to solve.

$$\dfrac{4x}{2x+4} < 3 \quad \therefore \quad 4x < 6x + 12 \quad \therefore \quad -2x < 12 \quad \therefore \quad \underline{x < -6}$$

$$\dfrac{4x}{2x+4} > -3 \quad \therefore \quad 4x > -6x - 12 \quad \therefore \quad 10x > -12 \quad \therefore \quad \underline{x > -\dfrac{6}{5}}$$

Conclusion: $x \in [-\infty, -6[$ and $x \in] - \text{⁶/₅}, +\infty]$, where the equalizing points at the interval boundaries, i.e. $x = -6$ and $x = -\text{⁶/₅}$, are not included.

(b) There are two inequalities to solve,

$$\dfrac{(x+2)\,|x-2|}{x^2+2} > 1 \quad \begin{cases} 1: x - 2 \ \geq 0 \ \therefore \ x \geq 2 \ \therefore \ |x-2| = x-2 \\ 2: x - 2 \ < 0 \ \therefore \ x < 2 \ \therefore \ |x-2| = -(x-2) \end{cases}$$

$(x \geq 2):$ $\dfrac{(x+2)(x-2)}{x^2+2} > 1$ $\therefore$ $\dfrac{x^2-4}{x^2+2} > 1$ $\therefore$ $\dfrac{x^{\cancel{2}}\left(1-\frac{4}{x^2}\right)}{x^{\cancel{2}}\left(1+\frac{2}{x^2}\right)} \not> 1$

because numerator is less than one and denominator is greater than one, thus their ratio is always less than one.

$(x < 2):$ $\dfrac{-(x+2)(x-2)}{x^2+2} > 1$ $\therefore$ $\dfrac{4-x^2}{x^2+2} > 1$ $\therefore$ $4 - x^2 > x^2 + 2$

$$\therefore \ 4 - 2 > 2x^2 \ \therefore \ 1 > x^2 \ \therefore \ x < \pm 1$$

Conclusion: $x \in]-1, 1[$, where the boundary points are not included. Indeed, plot of this function shows that its value exceeds one only within the calculated interval, Fig. 3.1

Fig. 3.1 Example 3.1-1(b)

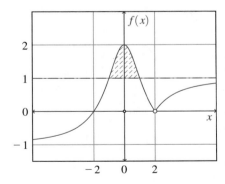

(c)   One possible way to solve this inequality relative to one is to convert it into its equivalent inequality relative to zero. We note that $x \neq 0$ and $x \neq -1$ because of division by zero cases.

$$\frac{1}{x} + \frac{1}{x+1} < \frac{x^2-2}{x^2+x} \quad \therefore \quad \frac{x+1+x}{x(x+1)} < \frac{x^2-2}{x^2+x} \quad \therefore$$

$$(2x+1)(x^2+x) < x(x+1)(x^2-2) \quad \therefore$$

$$x(x+1)(x^2-2) - (2x+1)(x^2+x) > 0 \quad \therefore$$

$$x(x+1)[(x^2-2)-(2x+1)] > 0 \quad \therefore \quad x(x+1)(x^2-2x-3) > 0 \quad \therefore$$

$$\underline{x(x+1)^2(x-3) > 0} \quad \text{the equivalent inequality } f(x)$$

Obviously, there are zeros of this polynomial are $x=0$, $x=-1$, $x=3$ that should be excluded, where $x=-1$ is double zero. We look for the intervals where this inequality $f(x)$ is satisfied. In factored form we look for sign of each factor and the sign of total product, i.e.

x	$[-\infty, -1[$	$]-1, 0[$	$]0; 3[$	$]3; +\infty]$
x	−	−	+	+
$(x+1)^2$	+	+	+	+
$(x-3)$	−	−	−	+
$f(x)$	+	+	−	+

Conclusion, this inequality is valid for $x \in [-\infty, -1[$, $x \in]-1, 0[$, and $x \in]3; +\infty]$, where $x \neq 3$, $x \neq 0$, and $x \neq -1$.

Indeed, zoom-in plot of $f(x)$ function shows that it is positive only within the calculated intervals, Fig. 3.2

Fig. 3.2 Example 3.1-1(c)

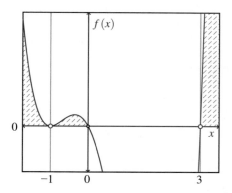

2. Solve the following inequalities:

(a) Knowing that if $AB = 0$, then it must be that either $A = 0$, or $B = 0$, or both A and B equal to zero. More general, if $AB > 0$, then either both $A, B > 0$ or both $A, B < 0$. Similarly, if $AB < 0$, then A and B must have the opposite signs, i.e. one must be positive while the other is negative. With that idea, we use the factorized forms of polynomials, so that there are product terms that help us reach conclusions.

$$-3x^2 + 30x - 75 > 0 \Leftrightarrow -3(x^2 - 10x + 25) > 0$$
$$\Leftrightarrow -3(x^2 - 5x - 5x + 25) > 0$$
$$\Leftrightarrow -3[x(x - 5) - 5(x - 5)] > 0$$
$$\Leftrightarrow \underbrace{-3}_{<0} \ \underbrace{(x - 5)^2}_{>0} > 0 \quad \therefore \quad \text{not possible}$$

because, product of negative and positive terms is always negative. Here, we have a negative number (i.e. "-3") multiplied by square of something (the squared number is always positive), therefore, there is no real solution to this inequality.

(b) $x^4 - 2x^2 + 1 = (x^2)^2 - 2x^2 + 1 = (x^2 - 1)^2 = (x + 1)^2(x - 1)^2 \geq 0 \ \forall x$
Therefore, $x^4 - 2x^2 + 1 < 0$ is never satisfied.

(c) $-2x^2 + 4x - 2 = -2(x^2 - 2x + 1) = -2(x - 1)^2 \leq 0 \ \forall x$
Therefore, $-2x^2 + 4x - 2 > 0$ is never satisfied.

(d) After converting into factorized form, we must verify all combinations that factors of that rational function can have,

$$H(x) = \frac{2x^2 - 5x - 3}{-2x + 1} = \frac{2x^2 - 6x + x - 3}{-2x + 1} \leq 0 \Leftrightarrow \frac{(2x + 1)(x - 3)}{-2x + 1} \leq 0$$

We systematically review all combinations when $H(x) \leq 0$, i.e. the numerator and denominator must have opposite signs, as

	$(2x + 1)$	$(x - 3)$	$(-2x + 1)$	$H(x)$
(1)	$+$	$+$	$-$	$-$
(2)	$-$	$-$	$-$	$-$
(3)	$+$	$-$	$+$	$-$
(4)	$-$	$+$	$+$	$-$

therefore,

(1) : $(2x + 1) > 0$ and $(x - 3) > 0$ and $(-2x + 1) < 0$

$\therefore$ $x > -1/2$ and $x > 3$ and $x > 1/2$ $\therefore$ $\underline{x > 3}$

(2) : $(2x + 1) <$ and $(x - 3) < 0$ and $(-2x + 1) < 0$

$\therefore$ $x < -1/2$ and $x < 3$ and $x > 1/2$ $\therefore$ $\underline{\text{no solution}}$

(3) : $(2x + 1) > 0$ and $(x - 3) < 0$ and $(-2x + 1) > 0$

$\therefore$ $x > -1/2$ and $x < 3$ and $x < 1/2$ $\therefore$ $\underline{-1/2 < x < 1/2}$

(4) : $(2x + 1) < 0$ and $(x - 3) > 0$ and $(-2x + 1) > 0$

$\therefore$ $x < -1/2$ and $x < 3$ and $x < 1/2$ $\therefore$ $\underline{\text{no solution}}$

In conclusion:

$$\frac{(2x + 1)(x - 3)}{-2x + 1} \leq 0 \quad \therefore \quad -\frac{1}{2} \leq x \leq \frac{1}{2} \quad \text{or} \quad x \geq 3$$

Exponential and Logarithmic Functions

4

Important to Know

Basic identities for log and exp functions,

$$\ln e = 1$$

$$\text{if:} \quad a^{f(x)} = a^{g(x)} \Rightarrow f(x) = g(x)$$

$$\text{for:} \quad a > 1 \ \text{ and } \ a^{f(x)} > a^{g(x)} \Rightarrow f(x) > g(x)$$

$$\text{for:} \quad 0 < a < 1 \ \text{ and } \ a^{f(x)} > a^{g(x)} \Rightarrow f(x) < g(x)$$

$$\text{if:} \quad x = \log_a b \Rightarrow a^x = b, \quad (a, b > 0), (a \neq 1)$$

$$\log 1 = 0$$

$$\log_a a = 1$$

$$a^{\log_a b} = b, \quad (a, b > 0), (a \neq 1)$$

$$\log xy = \log x + \log y$$

$$\log x^n = n \, \log x$$

$$\log \frac{x}{y} = \log(x \, y^{-1}) \log x - \log y$$

$$\log_a b = \frac{\ln b}{\ln a}$$

© Springer Nature Switzerland AG 2021
R. Sobot, *Engineering Mathematics by Example*,
https://doi.org/10.1007/978-3-030-79545-0_4

4.1   Exercises

4.1 *   Exponential Function

1. Sketch graphs of the following functions

   (a) $f(x) = 2^x$ (b) $f(x) = 2^x - 2$ (c) $f(x) = \left(\dfrac{1}{2}\right)^x$

   (d) $f(x) = 2^{\sqrt{x^2}}$ (e) $f(x) = \left|5^{|x|} - 2\right|$ (f) $f(x) = 3^x + 3^{-x}$

4.2 *   Exponential Equations and Inequalities

1. Solve the following equations.

   (a) $2^x = 8$ (b) $2^{x-1} = 16$ (c) $2^{x-5} = 3$

   (d) $e^{7-4x} = 6$ (e) $e^{2x} - 3e^x + 2 = 0$ (f) $5^{x+1} - 5^{x-1} = 24$

2. Solve the following equations.

   (a) $\left(x^2 - x - 1\right)^{x^2-1} = 1$

   (b) $2(x + 1)(2x + 1)^x - (x - 1)^x = (2x + 1)^{x+1}$

   (c) $\left(\sqrt{5 + \sqrt{24}}\right)^x + \left(\sqrt{5 - \sqrt{24}}\right)^x = 10$

3. Solve the following inequalities

   (a) $2^{x+2} > \left(\dfrac{1}{4}\right)^{1/x}$ (b) $1 < 3^{|x^2-x|} < 9$ (c) $\sqrt{9^x - 3^{x+2}} > 3^x - 9$

4.3 *   Logarithmic Function

1. Sketch graphs of the following functions

   (a) $f(x) = \log_2 x$ (b) $f(x) = |\log_2 x|$ (c) $f(x) = e^{\ln x}$

2. Without using calculator, calculated values of the following logarithms.

   (a) $\log_2 \dfrac{1}{128}$ (b) $\log_3 \dfrac{1}{81}$ (c) $\log_{\sqrt{2}} 8$

   (d) $\log_{0.1} 1000$ (e) $\log_2 \sqrt[3]{512}$ (f) $\log_a \sqrt[5]{a^2}$

4.4 *   Logarithmic Equations and Inequalities

1. Solve the following equations (without a calculator),

   (a) $x = \log_{10} 10$ (b) $x = \log_{10} 100$ (c) $x = \log_{10} 1000$

   (d) $x = \log_2 16$ (e) $x = \log_{25} \dfrac{1}{625}$ (f) $x = \log_8 \dfrac{1}{4}$

   (g) $x = \log_8 \left(\log_4 \left(\log_2 (16) \right) \right)$

2. Solve the following equations (without a calculator),

   (a) $x = 25^{\log_5 3}$ (b) $x = \log_5 \sqrt{5}$ (c) $x = \log_{2/3} \dfrac{243}{32}$

   (d) $x = \sqrt{\log^2_{1/2}(4)}$ (e) $x = \sqrt{\log_{1/2}(4)}$

3. Solve the following equations.

   (a) $x^{\log_3 x} = 9$ (b) $\log_2 x = 4$ (c) $\ln(3x - 10) = 2$

   (d) $\ln(x^2 - 1) = 3$ (e) $\ln x + \ln(x - 1) = 1$ (f) $\ln(\ln x) = 1$

   (g) $\log_4(x + 1) = 1$

4. Solve the following equations.

   (a) $5^{2(\log_5 2 + x)} - 2 = 5^{x + \log_5 2}$ (b) $3^{(\log_3 x)^2} + x^{\log_3 x} = 162$

   (c) $81^x - 16^x - 2 \times 9^x(9^x - 4^x) + 36^x = 0$

5. Solve the following inequalities.

   (a) $\log \dfrac{x - 1}{x + 2} > 0$ (b) $\log_{2x+3} x^2 < 1$ (c) $\log_{2x^2 - x}(2x + 2) < 1$

Solutions

Exercise 4.1, page 52

1. Regardless of its base, exponential functions alone cross vertical axis $x = 0$ at $y = 1$, then they may be shifted horizontally and/or vertically.

   (a) Basic exponential function $f(x) = 2^x$ whose base is strictly greater than one, Fig. 4.1.

Fig. 4.1
Example 4.1-1(a)(b)(c)

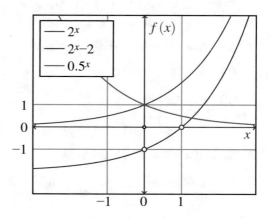

(b)   Basic exponential function shifted along vertical axis $f(x) = 2^x - 2$, Fig. 4.1.

(c)   Basic exponential function $f(x) = 0.5^x$ whose base is positive and strictly inferior to one, Fig. 4.1.

(d)   Simple transformation shows, Fig. 4.2, that

$$f(x) = 2^{\sqrt{x^2}} = 2^{|x|} \quad \therefore \quad \begin{cases} f(x) = 2^x; & x \geq 0 \\ f(x) = 2^{-x}; & x \leq 0 \end{cases}$$

Fig. 4.2 Example 4.1-1(d)

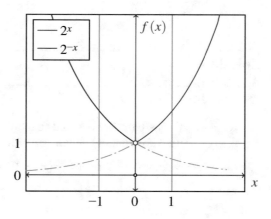

(e) In Example 1-1(d) we found the shape of an absolute exponent. From there we can deduce the following series of transformations

$$5^{|x|} \quad \rightarrow \quad 5^{|x|} - 2 \quad \rightarrow \quad \left|5^{|x|} - 2\right|$$

and derive the function graph in Fig. 4.3, where in the last step the negative portion of the graph is flipped up in the positive region so that the absolute function is satisfied.

Fig. 4.3 Example 4.1-1(e)

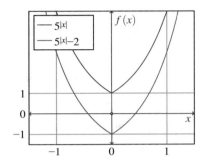 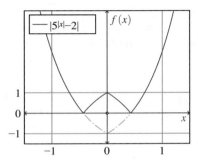

(f) Sum of two exponential forms $f(x) = 3^x + 3^{-x}$ produces graph in Fig. 4.4.

Fig. 4.4 Example 4.1-1(f)

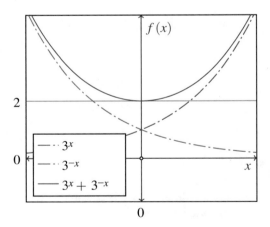

Exercise 4.2, page 52

1. By using the properties of log and exp functions we find

(a) $2^x = 8 \quad \therefore \quad 2^x = 2^3 \quad \therefore \quad x = 3$

(b) $2^{x-1} = 16 \quad \therefore \quad 2^{x-1} = 2^4 \quad \therefore \quad x - 1 = 4 \quad \therefore \quad x = 5$

(c) $2^{x-5} = 3$ $\quad \therefore \quad \log_2\left(2^{x-5}\right) = \log_2 3$ $\quad \therefore \quad x - 5 = \log_2 3 = \dfrac{\ln 3}{\ln 2}$

$\quad \therefore \quad x = \dfrac{5\ln 2 + \ln 3}{\ln 2}$

(d) $e^{7-4x} = 6$ $\quad \therefore \quad \ln\left(e^{7-4x}\right) = \ln 6$ $\quad \therefore \quad 7 - 4x = \ln 6$ $\quad \therefore \quad 7 - \ln 6 = 4x$

$\quad \therefore \quad x = \dfrac{7 - \ln 6}{4}$

(e)   This is quadratic equation in disguise, we use the change of variables technique to convert the original equation into

$$e^{2x} - 3e^x + 2 = 0 \quad \therefore \quad \left(e^x\right)^2 - 3\left(e^x\right) + 2 = 0 \quad \therefore \quad t = e^x$$

$$\therefore \quad t^2 - 3t + 2 = 0$$

$$t^2 - 2t - t + 2 = 0$$

$$t(t - 2) - (t - 2) = 0$$

$$(t - 2)(t - 1) = 0 \quad \therefore \quad t_1 = 2, \quad t_2 = 1$$

Case $t_1 = 2$: $\therefore$ $e^{x_1} = 2$ $\therefore$ $\ln\left(e^{x_1}\right) = \ln 2$ $\therefore$ $\underline{x_1 = \ln 2}$
Case $t_2 = 2$: $\therefore$ $e^{x_2} = 1$ $\therefore$ $\ln\left(e^{x_2}\right) = \ln 1$ $\therefore$ $\underline{x_1 = 0}$

(f)   After factorization, we find

$$5^{x+1} - 5^{x-1} = 24$$

$$5^{x-1}5^2 - 5^{x-1} = 24$$

$$5^{x-1}\left(5^2 - 1\right) = 24$$

$$5^{x-1} \times 24 = 24 \quad \therefore \quad 5^{x-1} = 1 \quad \therefore \quad x - 1 = 0 \quad \therefore \quad x = 1$$

2. (a)   Equation $a^b = 1$ is possible in two cases: (1) if $a > 0$ and $b = 0$, or (2) $a = 1$. Thus for the first case we write

$$\left(x^2 - x - 1\right)^{x^2-1} = 1 \quad \therefore \quad x^2 - x - 1 > 0 \quad \text{and} \quad x^2 - 1 = 0 \quad \therefore \quad x = \pm 1$$

however, only $x = -1$ satisfies both conditions.

The second case leads into the following solutions

$$x^2 - x - 1 = 1 \quad \therefore \quad x^2 - x - 2 = 0 \quad \therefore \quad (x+1)(x-2) = 0$$

$$\therefore$$

$$x = -1, x = 2$$

Thus, $x_1 = -1$ (double root), and $x_2 = 2$.

(b) The equation is converted into the following form.

$$2(x+1)(2x+1)^x - (x-1)^x = (2x+1)^{x+1}$$

$$2(x+1)(2x+1)^x - (2x+1)^{x+1} = (x-1)^x$$

$$(2x+1)^x \left[2(x+1) - (2x+1)\right] = (x-1)^x$$

$$(2x+1)^x \left[2x + 2 1 - 2x - 1\right] = (x-1)^x$$

$$(2x+1)^x = (x-1)^x$$

$$\left(\frac{2x+1}{x-1}\right)^x = 1$$

which is possible if $x = 0$, or

$$\frac{2x+1}{x-1} = 1 \quad \therefore \quad 2x+1 = x-1 \quad \therefore \quad x = -2$$

(c) The use of change of variables technique helps to convert this equation into the form of quadratic equation.

$$\left(\sqrt{5+\sqrt{24}}\right)^x + \left(\sqrt{5-\sqrt{24}}\right)^x = 10$$

$$\left(\sqrt{5+\sqrt{24}} \, \frac{\sqrt{5-\sqrt{24}}}{\sqrt{5-\sqrt{24}}}\right)^x + \left(\sqrt{5-\sqrt{24}}\right)^x = 10$$

$$\left(\frac{\sqrt{(5+\sqrt{24})(5-\sqrt{24})}}{\sqrt{5-\sqrt{24}}}\right)^x + \left(\sqrt{5-\sqrt{24}}\right)^x = 10$$

$$\left(\frac{\sqrt{25-24}}{\sqrt{5-\sqrt{24}}}\right)^x + \left(\sqrt{5-\sqrt{24}}\right)^x = 10$$

$$\frac{1}{\left(\sqrt{5-\sqrt{24}}\right)^x} + \left(\sqrt{5-\sqrt{24}}\right)^x = 10$$

This form is converted into quadratic equation with $\left(\sqrt{5 - \sqrt{24}}\right)^x = t$, which gives

$$\frac{1}{t} + t = 10 \quad \therefore \quad t^2 - 10t + 1 = 0 \quad \therefore \quad t_{1,2} = 5 \pm \sqrt{24}$$

Return to the original variable results in

$$\left(\sqrt{5 - \sqrt{24}}\right)^x = t \quad \therefore \quad \left(\sqrt{5 - \sqrt{24}}\right)^x = 5 - \sqrt{24} \quad \therefore \quad \underline{x = 2}$$

$$\therefore \quad \left(\sqrt{5 - \sqrt{24}}\right)^x = 5 + \sqrt{24} = \frac{1}{5 - \sqrt{24}} \quad \therefore \quad \underline{x = -2}$$

$$\left\{ x^{-n} = \frac{1}{x^n}; \quad (a - b)(a + b) = a^2 - b^2 \right\}$$

3. Inequalities compare two functions within one or multiple intervals.

(a)   The idea is to compare powers of the same base, i.e.

$$2^{x+2} > \left(\frac{1}{4}\right)^{1/x} \quad \therefore \quad 2^{x+2} > 4^{-1/x} \quad \therefore \quad 2^{x+2} > \left(2^2\right)^{-1/x} \quad \therefore \quad 2^{x+2} > 2^{-2/x}$$

Due to the same base, this inequality is equivalent to

$$x + 2 > -\frac{2}{x} \quad \therefore \quad x + 2 + \frac{2}{x} > 0$$

There is discontinuity at $x = 0$ for the inverse function, thus we examine two intervals:

1. $\underline{x > 0}$: Obviously, within interval $x \in]0, +\infty]$ the inequality is always satisfied because all terms on the left side are positive. Therefore, one solution is interval $x > 0$.
2. $\underline{x < 0}$: Within interval $x \in [-\infty, 0[$, terms including x are obviously negative; however, there is one term that is positive (i.e. "+2"). We find that

$$x + 2 + \frac{2}{x} = 0 \quad \therefore \quad x^2 + 2x + 2 = 0 \quad \therefore \quad x_{1,2} \notin \mathbb{R}$$

in other words, there is no change of sign. In conclusion, for $x \in [-\infty, 0[\Rightarrow x + 2 + 2/x < 0$. Therefore, the inequality is not satisfied for $x < 0$.

Relative relationship between $x + 2$ and $-2/x$ is shown in Fig. 4.5.

Fig. 4.5 Example 4.2-3(a)

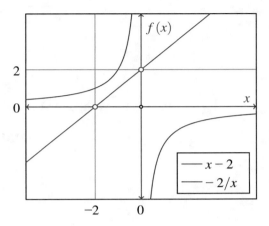

(b)   Double inequality with the absolute expression $|x^2 - x|$ is decomposed in intervals $x^2 - x \geq 0$, and $x^2 - x \leq 0$. In this specific example, however, the equality points are not included (the problem has "strict" inequalities).

1. $x^2 - x < 0$: Within this interval, $x(x - 1) < 0$, that is to say: either $x < 0$ and $x - 1 > 0$ (not possible), or $x > 0$ and $x - 1 < 0$ which is possible if $x \in]0, 1[$ (strict inequality). (Hint: $AB < 0$ means that A and B must have opposite signs.)
2. $x^2 - x > 0$: Within this interval, $x(x - 1) > 0$, that is to say: either $x < 0$ and $x - 1 < 0$ thus $x < 0$, or $x > 0$ and $x - 1 > 0$ thus $x > 1$. (Hint: $AB > 0$ means that A and B must have same signs.)

Given these intervals, inequalities are solved after converting it into the form with same base, i.e.

$$1 < 3^{|x^2 - x|} < 9 \quad \therefore \quad 3^0 < 3^{|x^2 - x|} < 3^2 \quad \therefore \quad 0 < |x^2 - x| \text{ and } |x^2 - x| < 2$$

Which leads into $x^2 - x - 2 < 0$ resulting in $x > -1$ and $x < 2$ interval. Putting everything together, we conclude that $x \in]-1, 2[$ where, $x \neq -1, 0, 2$.

(c)   Following the idea of reducing all terms to the same base, we write

$$\sqrt{9^x - 3^{x+2}} > 3^x - 9$$

$$\sqrt{3^{2x} - 3^x 3^2} > 3^x - 3^2 \quad \Big|^2$$

$$3^{2x} - 3^x 3^2 > 3^{2x} - 2 \times 3^x 3^2 + 3^4$$

$$2 \times 3^x 3^2 - 3^x 3^2 > 3^4$$

$$3^x 3^2 > 3^{4 \, 2}$$

$$\therefore$$

$$x > 2$$

Exercise 4.3, page 52

1. Plots of these three functions are shown in Fig. 4.6. We note that all logarithmic functions have zero at $x = 1$, and that exponent of logarithmic function (and vice versa) is a linear function.

Fig. 4.6
Example 4.3-1(a)(b)(c)

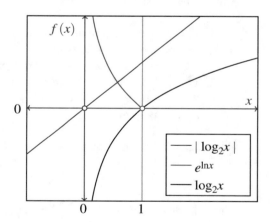

2. By definitions of exp and log functions, we write

(a) $x = \log_2 \dfrac{1}{128} \quad \therefore \quad 2^x = 2^{\log_2 \frac{1}{128}} \quad \therefore \quad 2^x = \dfrac{1}{128} = \dfrac{1}{2^7} = 2^{-7} \quad \therefore \quad x = -7$

(b) $3^x = 3^{\log_3 \frac{1}{81}} \quad \therefore \quad 3^x = \dfrac{1}{81} = \dfrac{1}{3^4} = 3^{-4} \quad \therefore \quad x = -4$

(c) $x = \log_{\sqrt{2}} 8 \quad \therefore \quad \sqrt{2}^x = \sqrt{2}^{\log_{\sqrt{2}} 8} = 8 = \sqrt{2}^6 \quad \therefore \quad x = 6$

(d) $x = \log_{0.1} 1000 \quad \therefore \quad 0.1^x = 1000 = 0.1^{-3} \quad \therefore \quad x = -3$

(e) $x = \log_2 \sqrt[3]{512} \quad \therefore \quad 2^x = \sqrt[3]{512} = 2^{\frac{9}{3}} = 2^3 \quad \therefore \quad x = 3$

(f) $x = \log_a \sqrt[5]{a^2} \quad \therefore \quad a^x = \sqrt[5]{a^2} \quad \therefore \quad x = \dfrac{2}{5}$

1. In examples (a) to (c) note the relationship between the numbers of zeros in 10, 100, 1 000, ... and the respective equation solutions. In this special case of logarithmic functions with the base 10 and decade numbers, the noted shortcut helps us to perform mental calculations of these logarithms.

   (a) $x = \log_{10} 10 \; \therefore \; 10^x = 10 \; \therefore \; x = 1$

   (b) $x = \log_{10} 100 \; \therefore \; 10^x = 100 \; \therefore \; x = 2$

   (c) $x = \log_{10} 1000 \; \therefore \; 10^x = 1000 \; \therefore \; x = 3$

   In addition, writing "log" without specific index number assumes the base 10, while writing "ln" assumes the base e. For example, $x = \log 1{,}000{,}000$ is found by counting the zeros, therefore $x = 6$.

   Examples (d) to (g) illustrate some simple techniques for calculating various logarithms based on the knowledge of relationships between numbers, i.e. squares, roots, the multiplication/division table, and the basic log identities

   (d) $x = \log_2 16 \; \therefore \; 2^x = 16 \; \therefore \; 2^x = 2^4 \; \therefore \; x = 4$

   (e) $x = \log_{25} \dfrac{1}{625} = \log_{25} \dfrac{1}{25^2} = \log_{25} 25^{-2} = -2 \log_{25} 25^{\,1} = -2$

   (f) $x = \log_8 \dfrac{1}{4} = \log_8 \dfrac{1}{\sqrt[3]{64}} = \log_8 \dfrac{1}{\sqrt[3]{8^2}} = \log_8 8^{-2/3} = -\dfrac{2}{3} \log_8 8^{\,1} = -\dfrac{2}{3}$

   (g)   Given $x = \log_8 \left(\log_4 \left(\log_2 (16) \right) \right)$ we start from the right side, as

   $x = \log_8 \left(\log_4 \left(\log_2 (2^4) \right) \right) = \log_8 \left(\log_4 \left(4 \log_2 (2)^{\,1} \right) \right)$

   $= \log_8 \left(\log_4 4^{\,1} \right) \Rightarrow x = \log_8 (1) \Rightarrow 8^x = 8^{\log_8 (1)} \Rightarrow 8^x = 1$

   $\therefore$

   $x = 0$

2. Exercises with problems that include both log and $\sqrt[n]{a}$ functions, as well as absolute values of numbers.

   (a) $x = 25^{\log_5 3} = \left(5^2 \right)^{\log_5 3} = 5^{(2 \log_5 3)} = 5^{\log_5 9} = 9$

   (b) $x = \log_5 \sqrt{5} = \log_5 5^{1/2} = \dfrac{1}{2} \log_5 5^{\,1} = \dfrac{1}{2}$

   (c) $x = \log_{2/3} \dfrac{243}{32} = \log_{2/3} \dfrac{3^5}{2^5} = \log_{2/3} \left(\dfrac{2}{3} \right)^{-5} = -5 \log_{2/3} \dfrac{2}{3}^{\,1} = -5$

   (d) $x = \sqrt{\log_{1/2}^2 (4)} = \left| \log_{1/2} (4) \right| = \left| \log_{1/2} \left(\dfrac{1}{2} \right)^{-2} \right| = \left| -2 \log_{1/2} \dfrac{1}{2}^{\,1} \right| = 2$

   (e) $x = \sqrt{\log_{1/2} (4)} = \sqrt{\log_{1/2} \left(\dfrac{1}{2} \right)^{-2}} = \sqrt{-2 \log_{1/2} \dfrac{1}{2}^{\,1}} = \sqrt{-2} = j \sqrt{2}$

3. By using the properties of log and exp functions we find

(a) $x^{\log_3 x} = 9$ $\therefore$ $\log_3\left(x^{\log_3 x}\right) = \log_3 9$ $\therefore$ $\log_3\left(x^{\log_3 x}\right) = 2$

$\therefore$ $\log_3 x \log_3 x = 2$ $\therefore$ $\left(\log_3 x\right)^2 = 2$ $\therefore$ $\log_3 x = \pm\sqrt{2}$

$\therefore$ $3^{\log_3 x} = 3^{\pm\sqrt{2}}$ $\therefore$ $x = 3^{\pm\sqrt{2}}$

(b) $\log_2 x = 4$ $\therefore$ $2^{\log_2 x} = 2^4$ $\therefore$ $x = 2^4$

(c)   Functions ln and exp are cancelling each other, similar as add/subtract, or multiply/divide, thus we write

$$\ln(3x - 10) = 2 \quad\therefore\quad e^{\ln(3x-10)} = e^2 \quad\therefore\quad 3x - 10 = e^2 \quad\therefore\quad x = \frac{e^2 + 10}{3}$$

(d) $\ln(x^2 - 1) = 3$ $\therefore$ $e^{\ln(x^2-1)} = e^3$ $\therefore$ $x^2 - 1 = e^3$ $\therefore$ $x = \pm\sqrt{e^3 + 1}$

(e) $\ln x + \ln(x - 1) = 1$ $\therefore$ $\ln[x(x - 1)] = 1$ $\therefore$ $e^{\ln[x(x-1)]} = e$

$$x^2 - x - e = 1$$
$$\therefore$$
$$x_{1,2} = \frac{1 \pm \sqrt{1 + 4e}}{2}$$

(f) $\ln(\ln x) = 1$ $\therefore$ $e^{\ln(\ln x)} = e$ $\therefore$ $\ln x = e$ $\therefore$ $e^{\ln x} = e^e$ $\therefore$ $x = e^e$

(g) $\log_4(x + 1) = 1$ $\therefore$ $4^{\log_4(x+1)} = 4^1$ $\therefore$ $x + 1 = 4$ $\therefore$ $x = 3$

4. Convert all terms to the same base.

(a)   A number can be written as exponent of an arbitrary base logarithm, i.e. $2 = 5^{\log_5 2}$, thus

$$5^{2(\log_5 2 + x)} - 2 = 5^{x + \log_5 2}$$

$$\left\{a^{b+c} = a^b \, a^c; \quad a = n^{\log_n a}; \quad x^{ab} = \left(x^a\right)^b\right\}$$

$$5^{2\log_5 2} \, 5^{2x} - 5^{\log_5 2} = 5^x \, 5^{\log_5 2}$$

$$\left(5^{\log_5 2}\right)^2 5^{2x} - 5^{\log_5 2} - 5^x \, 5^{\log_5 2} = 0$$

$$5^{\log_5 2} \left(5^{\log_5 2} \, 5^{2x} - 1 - 5^x\right) = 0$$

$$2 \left(5^{\log_5 2} \, 5^{2x} - 1 - 5^x\right) = 0$$

$$\therefore$$

$$5^{\log_5 2} \, 5^{2x} - 1 - 5^x = 0$$

Which is further converted into a quadratic equation.

$$2 \cdot \left(5^x\right)^2 - 5^x - 1 = 0 \quad \left\{5^x = t\right\}$$

$$2t^2 - t - 1 = 0 \quad \therefore \quad 2t^2 - 2t + t - 1 = 0 \quad \therefore \quad 2t(t-1) + (t-1) = 0$$

$$\therefore$$

$$(t-1)(2t+1) = 0 \quad \therefore \quad t_1 = 1 \quad \text{or} \quad t_2 = -\frac{1}{2}$$

$$\therefore$$

$$5^x = -\frac{1}{2} \quad \text{not possible because} \quad 5^x > 0 \quad \forall x$$

$$5^x = 1 \quad \therefore \quad \underline{x = 0}$$

(b) We recall that $a^{b^2} = a^{b \times b} = (a^b)^b$, thus

$$4^{(\log_4 x)^2} + x^{\log_4 x} = 512 \quad \therefore \quad \left(4^{\log_4 x}\right)^{\log_4 x} + x^{\log_4 x} = 512 \quad \therefore \quad 2\, x^{\log_4 x} = 512$$

$$\therefore$$

$$x^{\log_4 x} = 256 \quad \therefore \quad \left\{\log a^b = b \log a\right\}$$

$$\log_4\left(x^{\log_4 x}\right) = \log_4 256 \quad \therefore \quad \log_4 x \ \log_4 x = \log_4 4^4 = 4 \quad \therefore \quad \log_4 x = \pm 2$$

$$\therefore$$

$$x = 4^{\pm 2} \quad \therefore \quad x_1 = 16, \quad x_2 = \frac{1}{16}$$

(c) After factorization, this equitation is transformed into quadratic equation.

$$81^x - 16^x - 2\, 9^x(9^x - 4^x) + 36^x = 0$$

$$(3^4)^x - (2^4)^x - 2\, 9^{2x} + 2\, 9^x\, 4^x) + (6^2)^x = 0$$

$$3^{4x} - 2^{4x} - 2\, 3^{4x} + 2\, 3^{2x}\, 2^{2x} + 6^{2x} = 0$$

$$3^{4x} - 2\, 3^{4x} - 2^{4x} + 2\, 6^{2x} + 6^{2x} = 0$$

$$-3^{4x} - 2^{4x} + 3\, 6^{2x} = 0$$

$$3^{4x} + 2^{4x} - 3\, 6^{2x} = 0 \quad \Big| \quad /2^{4x}$$

$$\frac{3^{4x}}{2^{4x}} + \frac{2^{4x}}{2^{4x}} - 3\, \frac{3^{2x}\, 2^{2x}}{2^{4x\ 2x}} = 0$$

$$\left[\left(\frac{3}{2}\right)^{2x}\right]^2 - 3\left(\frac{3}{2}\right)^{2x} + 1 = 0$$

Thus, change of variable produces

$$t = \left(\frac{3}{2}\right)^{2x} \quad \therefore \quad t^2 - 3t + 1 = 0 \quad \therefore \quad t_{1,2} = \frac{3 \pm \sqrt{5}}{2}$$

Return to the original variable gives

$$\left(\frac{3}{2}\right)^{2x} = \left(\frac{9}{4}\right)^x = \frac{3 \pm \sqrt{5}}{2}$$

$$\therefore$$

$$x_{1,2} \ln \frac{9}{4} = \ln \frac{3 \pm \sqrt{5}}{2} \quad \therefore \quad x_{1,2} = \frac{\ln(3 \pm \sqrt{5}) - \ln 2}{\ln 9 - \ln 4}$$

5. (a) Logarithmic function is positive when its argument is superior to one, i.e.

$$\log \frac{x-1}{x+2} > 0 \quad \therefore \quad \frac{x-1}{x+2} > 1 \text{ where, } x \neq -2 \text{ and } \lim_{\pm\infty} \frac{x-1}{x+2} = 1$$

We conclude that for $x \in [-\infty, -2[$ it follows that always $(x-1)/(x+2) > 1$, thus the inequality is satisfied. When $x > -2$ the inequality is not satisfied.

(b) We distinguish logarithms whose base is inferior to "1" from those whose base is superior to one. In the case of inverse bases, the two log functions are symmetric and opposite sign.

if $0 < 2x + 3 < 1 \quad \therefore \quad x \in]-3/2, 1[$, then

$$\log_{2x+3} x^2 < 1 = \log_{2x+3}(2x + 3) \quad \therefore \quad x^2 > 2x + 3 \quad \therefore \quad (x < -1) \text{ and } (x > 3)$$

$$\therefore$$

$$x \in]-3/2, -1[$$

if $2x + 3 > 1 \quad \therefore \quad x > -1$, then

$$\log_{2x+3} x^2 < 1 = \log_{2x+3}(2x + 3) \quad \therefore \quad x^2 < 2x + 3 \quad \therefore \quad -1 < x < 3 \text{ and } x \neq 0$$

Therefore, the solution is $x \in]-3/2, 3[\ (x \neq 0)$.

(c)   Logarithm function of an arbitrary base may be evaluated by converting given function into ratio of two logarithmic functions as

$$\log_{2x^2-x}(2x+2) < 1 \quad \left\{ \log_a b = \frac{\ln b}{\ln a} \right\} \quad \therefore \quad \frac{\ln(2x+2)}{\ln(2x^2-x)} < 1$$

In addition, we know that $\ln(1) = 0$, that $\ln x < 0$ if $(0 < x < 1)$, and that $\ln x > 0$ if $(x > 1)$. With that understanding we evaluate both logarithms in this function as follows.

1. Exclude points when denominator equals to zero, when the result of division becomes either infinity or undetermined, thus

$$\ln(2x^2 - x) = 0 \quad \therefore \quad 2x^2 - x = 1 \quad \therefore \quad 2x^2 - x - 1 = 0$$

$$\therefore \quad x = 1, x = -\frac{1}{2}$$

$$\ln(2x^2 - x) < 0 \quad \therefore \quad 0 < x(2x - 1) < 1 \quad \therefore \quad 0 < x < \frac{1}{2}$$

and, this logarithm function is defined for positive argument, i.e.

$$2x^2 - x > 0 \quad \therefore \quad x(2x - 1) > 0 \quad \therefore \quad x < 0 \text{ or, } x > \frac{1}{2}$$

2. Similarly, logarithm in the numerator is defined for

$$2x + 2 > 0 \quad \therefore \quad x > -1 \text{ and,}$$

$$\ln(2x + 2) < 0 \quad \therefore \quad 0 < 2x + 2 < 1 \quad \therefore \quad x < -\frac{1}{2}$$

1. Condition of inequality gives

$$f(x) = \frac{\ln(2x+2)}{\ln(2x^2-x)} < 1 \quad \therefore \quad \ln(2x+2) < \ln(2x^2-x)$$

$$\therefore \quad 2x + 2 < 2x^2 - x$$

$$\therefore \quad 0 < 2x^2 - 3x - 2$$

$$\therefore \quad 0 < 2(x + 1)(x - 2)$$

Which is satisfied when $x > 2$, because values $x < -1$ are already excluded.

(c) *(cont.)* With the above intervals, and knowing signs of numerator and denominator in each interval, we summarize the intervals when inequality is satisfied, see Fig. 4.7, as

$$-1 < x < 0 \ \text{ and } \ \frac{1}{2} < x < 1 \ \text{ and } \ x > 2$$

Fig. 4.7 Example 4.4-5(a)

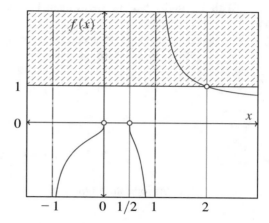

Trigonometry

Important to Know

Among many trigonometric identities, some are more often used than the others,

$$a^2 + b = c^2 \quad \text{(Pythagoras's theorem)}$$

$$\sin^2 \alpha + \cos^2 \beta = 1 \quad \text{(Pythagoras's theorem in disguise, for} \quad c = 1 \text{)}$$

$$|\sin \alpha| = \sqrt{1 - \cos^2 \alpha}$$

$$|\cos \alpha| = \sqrt{1 - \sin^2 \alpha}$$

$$\tan \alpha = \frac{\sin \alpha}{\cos \alpha}$$

Sum to product identities,

$$\sin(\alpha + \beta) = \sin \alpha \cos \beta + \cos \alpha \sin \beta$$

$$\sin(\alpha - \beta) = \sin \alpha \cos \beta - \cos \alpha \sin \beta$$

$$\cos(\alpha + \beta) = \cos \alpha \cos \beta - \sin \alpha \sin \beta$$

$$\cos(\alpha - \beta) = \cos \alpha \cos \beta + \sin \alpha \sin \beta$$

When $\alpha = \beta$, special cases of the sum identities (note: double angle),

$$\sin 2\alpha = 2 \sin \alpha \cos \alpha$$

$$\cos 2\alpha = \cos^2 \alpha - \sin^\alpha$$

Product to sum identities,

$$\sin \alpha \sin \beta = \frac{1}{2} \left(\cos(\alpha - \beta) - \cos(\alpha + \beta) \right)$$

$$\cos \alpha \cos \beta = \frac{1}{2} \left(\cos(\alpha + \beta) + \cos(\alpha - \beta) \right)$$

© Springer Nature Switzerland AG 2021
R. Sobot, *Engineering Mathematics by Example*,
https://doi.org/10.1007/978-3-030-79545-0_5

$$\sin\alpha\cos\beta = \frac{1}{2}\left(\sin(\alpha+\beta)+\sin(\alpha-\beta)\right)$$

Square to double angle identities,

$$\sin^2\alpha = \frac{1-\cos(2\alpha)}{2}$$

$$\cos^2\alpha = \frac{1+\cos(2\alpha)}{2}$$

5.1   Exercises

5.1   Basic Definitions

1. Explain the concept of "similar triangles" (thus, "similar figures" in general) and illustrate the consequent properties that are used in trigonometry.

2. Elementary trigonometric definitions.

   (a) Using right-angled triangles and Pythagoras's theorem illustrate the definitions of $\sin x$, cos, and $\tan x$

   (b) Calculate $\sin x$, cos, and $\tan x$ for each of the following angles in the first quadrant: $0°, 30°, 45°, 60°$, and $90°$.

   (c) Sketch the unit circle and illustrate how angles in II, III, and IV quadrant are related to angles found in the first quadrant.

3. Convert degrees into radians

   (a) $300°$ (b) $330°$ (c) $-18°$

4. Convert radians into degrees

   (a) $\dfrac{3\pi}{4}$ (b) $\dfrac{5\pi}{12}$ (c) 2

5. Calculate exact values

   (a) $\tan\dfrac{\pi}{3}$ (b) $\sin\dfrac{7\pi}{6}$ (c) $\cos\left(-\dfrac{3\pi}{4}\right)$

6. Given right-angled triangle in Fig. 5.1 express lengths of sides a and b as a function of angle θ.

Fig. 5.1 Example 5.1-6

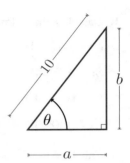

7. Given α, is it possible to have

(a) $\sin \alpha = -\dfrac{1}{5}$ et $\cos \alpha = \dfrac{3}{5}$

(b) $\sin \alpha = \dfrac{4}{\sqrt{17}}$ et $\cos \alpha = \dfrac{1}{\sqrt{17}}$

(c) $\sin \alpha = -\dfrac{\sqrt{15}}{5}$ et $\cos \alpha = \dfrac{\sqrt{5}}{5}$

8. Calculate the exact values of $\cos \alpha$ and $\tan \alpha$ given that

$$\sin \alpha = -\frac{1}{3} \text{ et } \pi < \alpha < \frac{3\pi}{2}$$

9. Calculate the following:

(a) $\cos 150°$ (b) $\sin 120°$ (c) $\tan 300°$

(d) $\tan 225°$ (e) $\tan 480°$

10. Calculate the following:

(a) $\tan 840°$

(b) $\cos 1320°$

(c) $\sin \dfrac{\pi}{6} \sin^2 \dfrac{\pi}{3}$

(d) $\cos \dfrac{\pi}{4} \cos \dfrac{\pi}{6} + \sin \dfrac{\pi}{4} \cos \dfrac{\pi}{6}$

(e) $\sin^2 \dfrac{\pi}{3} - \dfrac{1}{\tan^2 \dfrac{\pi}{3}}$

(f) $\sin \dfrac{\pi}{4} \cos \dfrac{\pi}{4} + \cos \dfrac{\pi}{6} \tan \dfrac{\pi}{3}$

5.2 *   Trigonometry Identities

1. Calculate the following trigonometry functions (without a calculator),

   (a) $\cos\left(\dfrac{\pi}{3}\right)$
         (b) $\tan\left(\dfrac{\pi}{4}\right)$
         (c) $\cos(150°)$

   (d) $\sin(120°)$
         (e) $\tan(-45°)$
         (f) $\tan(-60°)$

   (g) $\tan(300°)$
         (h) $\tan(225°)$
         (i) $\tan(480°)$

   (j) $\tan(840°)$
         (k) $\cos(1320°)$
         (l) $\sin\left(\dfrac{\pi}{6}\right)\sin^2\left(\dfrac{\pi}{3}\right)$

2. Calculate the following expressions (without a calculator),

   (a) $-4\sin\left(\dfrac{7\pi}{12}\right)\sin\left(\dfrac{13\pi}{12}\right)$
         (b) $4\sin\left(\dfrac{5\pi}{8}\right)\cos\left(\dfrac{\pi}{8}\right)$

   (c) $4\cos\left(\dfrac{5\pi}{8}\right)\cos\left(\dfrac{3\pi}{8}\right)$
         (d) $\tan 20°\,\tan 40°\,\tan 80°$

5.3   Sum Identities: $\sin(x \pm y)$, $\cos(x \pm y)$

Identities for trigonometric functions of sum or difference of two angles are

$$\sin(a \pm b) = \sin a \cos b \pm \cos a \sin b$$
$$\cos(a \pm b) = \cos a \cos b \mp \sin a \sin b$$

1. Calculate $\sin\alpha$, $\cos\alpha$, and $\tan\alpha$ of the following angles.

   (a) $\alpha = 15°$
         (b) $\alpha = 75°$
         (c) $\alpha = 105°$

   (d) $\alpha = \pi/12$
         (e) $\alpha = 5\pi/12$
         (f) $\alpha = 7\pi/12$

2. Calculate the following expressions

   (a) $\sin 20° \cos 10° + \cos 20° \sin 10°$
         (b) $\cos 43° \cos 13° + \sin 43° \sin 13°$

3. Calculate $\sin(x+y)$ and $\sin(x-y)$ given that $\cos x = 4/5$ and $\sin y = -3/5$, where x is in IV quadrant and y is in III quadrant.

4. Simplify the following expression

$$\frac{\sqrt{2}\cos\alpha - 2\cos\left(\dfrac{\pi}{4} + \alpha\right)}{2\sin\left(\dfrac{\pi}{4} + \alpha\right) - \sqrt{2}\sin\alpha}$$

5. Derive the double-angle identities for $\cos 2a$, $\sin 2a$, and $\tan 2a$ relative to the single-angle functions.

5.4   Product to Sum Identities

1. Knowing the sum identities, derive the identities for the following products

   (a) $\sin \alpha \sin \beta$ (b) $\cos \alpha \cos \beta$ (c) $\sin \alpha \cos \beta$

2. Knowing the sum identities, derive the identities for the following sums

   (a) $\sin x + \sin y$ (b) $\cos x + \cos y$

3. Calculate the following products.

   (a) $-4 \sin \dfrac{5\pi}{12} \cos \dfrac{7\pi}{12}$ (b) $\dfrac{1}{2} + 2 \sin \left(1 + \dfrac{\pi}{6}\right) \cos \left(1 + \dfrac{\pi}{3}\right)$

   (c) $8 \sin 20° \sin 40° \sin 80°$ (d) $8 \cos 10° \cos 50° \cos 70°$

4. Solve for α in I quadrant if: $\tan \alpha = \sqrt{6} + \sqrt{3} - \sqrt{2} - 2$.

5.5   Trigonometric Equations

1. Solve the following equations for x.

   (a) $\sin x = 0$ (b) $\cos x = 0$

2. Solve the following equations for x.

   (a) $\sin x = \sin \alpha$ (b) $\sin x = \cos x$ (c) $\sin x = \sin 2x$

   (d) $2 \sin^2 x + \sin x = 0$ (e) $\cos (x + \pi/6) = \sin (x - \pi/3)$

3. Solve the following equations for x.

   (a) $\cos \left(\dfrac{\pi}{3} - x\right) = \cos \left(\dfrac{\pi}{6}\right)$; $x \in [\,0; 2\pi\,[$

   (b) $\sin (3x) = \sin \left(x - \dfrac{\pi}{2}\right)$; $x \in \,]-\pi; \pi\,]$

5.6   Trigonometric Inequalities

1. Solve the following inequalities for x.

   (a) $2 \cos x + \sqrt{3} < 0$ (b) $2 \sin x - 1 > 0$

   (c) $\cos x - \sin x < 1$ (d) $\sin 3x - \dfrac{\sqrt{3}}{2} \geq 0$

   (e) $|\sin x| \geq \dfrac{1}{2}$ (f) $|\sin x| \leq \sin x + 2 \cos x$

Solutions

Exercise 5.1, page 68

1. Two figures are said to be similar if one figure can be derived from the other by operations of uniformly scaling, translation, rotation and/or reflection. For example, starting with triangle OMN in Fig. 5.2, triangle OPQ is derived after multiplying each side of OMN by factor of two, similarly any other triangle can be derived by using factor n, for example triangle OKL.

By inspection of Fig. 5.2 it is evident that except for the scaling factors the *form* of all three triangles did not change, thus, they are *similar*. By using Pythagoras's theorem it is not difficult to verify that all three triangles are indeed right-angled whose corresponding sides are simply proportional to each other.

Fig. 5.2 Example 5.1-1

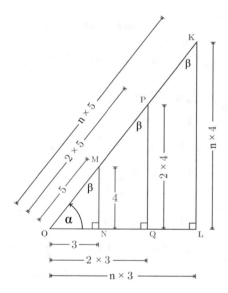

We take a look at angles as well. Obviously, each triangle has one of its angles equal to 90°. In addition, the corresponding sides of α in each triangle are proportional, thus α is shared by all three triangles. Consequently, knowing two angles in a triangle the third angle β is calculated so that their sum adds to $90° + \alpha + \beta = 180°$. In conclusion, direct consequence of similarity is that not only the figure sides are proportional (including the factor $n = 1$) but also their corresponding angles are *equal*. The similarity of right-angled triangles is among most important and exploited properties in mathematics.

2. By inspection of Fig. 5.2 we write ratios among triangle sides.

(a) There are three basic ratios,

1. The cathetus forming α (i.e. adjacent) and hypotenuse (the side opposite the right angle) are related as

$$\frac{ON}{OM} = \frac{OQ}{OP} = \frac{OL}{OK} = \frac{3}{5} = \frac{\cancel{6}3}{\cancel{10}\,5} = \cdots = \frac{\cancel{n} \times 3}{\cancel{n} \times 5} = \text{const.}$$

2. The cathetus opposite to α and hypotenuse are related as

$$\frac{MN}{OM} = \frac{PQ}{OP} = \frac{KL}{OK} = \frac{4}{5} = \frac{\cancel{8}4}{\cancel{10}\,5} = \cdots = \frac{\cancel{n} \times 4}{\cancel{n} \times 5} = \text{const.}$$

3. The catheti opposite and adjacent to α are related as

$$\frac{MN}{ON} = \frac{PQ}{OQ} = \frac{KL}{OL} = \frac{4}{3} = \frac{\cancel{8}4}{\cancel{6}3} = \cdots = \frac{\cancel{n} \times 4}{\cancel{n} \times 3} = \text{const.}$$

The fact that these three ratios are constant for all corresponding similar triangles is very useful because, given a right-angled triangle, by knowing the side ratios we can calculate various unknown sides and angles.

Thus, relative to α we name these three ratios as

$$\frac{\text{adjacent}}{\text{hypotenuse}} = \cos\alpha, \qquad \frac{\text{opposite}}{\text{hypotenuse}} = \sin\alpha, \qquad \frac{\text{opposite}}{\text{adjacent}} = \tan\alpha$$

Similarly, the equivalent ratios exist relative to β,

$$\frac{\text{adjacent}}{\text{hypotenuse}} = \cos\beta, \qquad \frac{\text{opposite}}{\text{hypotenuse}} = \sin\beta, \qquad \frac{\text{opposite}}{\text{adjacent}} = \tan\beta$$

We search these ratios related to α et β when triangles are rotated, scaled, translated and/or reflected by looking for "adjacent" and "opposite" sides relative to the angle under consideration. In addition, it is useful to notice that by knowing $\sin x$ and $\cos x$, the third ratio, i.e. $\tan x$ is also known because

$$\frac{\sin x}{\cos x} = \frac{\dfrac{\text{opposite}}{\cancel{\text{hypotenuse}}}}{\dfrac{\text{adjacent}}{\cancel{\text{hypotenuse}}}} = \frac{\text{opposite}}{\text{adjacent}} = \tan x$$

(b)   Angles between $0°$ and $90°$ are said to be "in the first quadrant" of the unit circle. They are reused to calculate trigonometric functions of angles in other three quadrants by rotating their respective right-angled triangles into the first quadrant.

Case: $x = 0°$ In this special case, right-angled triangle whose hypothenuse equals one degenerates into a horizontal line segment, as a consequence the hypotenuse is found "on top" of adjacent side, thus the two are equal. At the same time, the opposite side of the triangle is reduced to zero. By definitions we write

$$\cos 0° = \frac{\text{adjacent}}{\text{hypotenuse}} = 1,$$

$$\sin 0° = \frac{0}{\text{hypotenuse}} = 0,$$

$$\tan 0° = \frac{\sin 0°}{\cos 0°} = \frac{0}{1} = 0$$

Case: $x = 30° = \pi/6$ We calculate trigonometric functions with the help of triangle in Fig. 5.3. Here, hypotenuse is two times longer than a one of the cathetus, thus the two associated angles must be in the same proportion. We conclude the following:

$$90° + \alpha + \beta = 180°$$

$$\alpha = 2\beta$$

$$\therefore$$

$$90° + 3\beta = 180°$$

$$\beta = 30° \quad \therefore \quad \alpha = 60°$$

In addition, Pythagoras's theorem gives

$$2 = \sqrt{1^2 + b}$$

$$\therefore$$

$$a = \sqrt{3}$$

By definitions we write

$$\cos 30° = \frac{\text{adjacent}}{\text{hypotenuse}} = \frac{\sqrt{3}}{2}$$

$$\sin 30° = \frac{\text{opposite}}{\text{hypotenuse}} = \frac{1}{2},$$

$$\tan 30° = \frac{\text{opposite}}{\text{adjacent}} = \frac{1}{\sqrt{3}}$$

(b) *(cont.)*

Case: $x = 60° = \pi/3$ We calculate trigonometric functions of this angle also with the help of triangle in Fig. 5.3, it is also found in the same triangle. Again, knowing the three sides by definitions we write

$$\cos 60° = \frac{\text{adjacent}}{\text{hypotenuse}} = \frac{1}{2}$$

$$\sin 60° = \frac{\text{opposite}}{\text{hypotenuse}} = \frac{\sqrt{3}}{2},$$

$$\tan 60° = \frac{\text{opposite}}{\text{adjacent}} = \frac{\sqrt{3}}{1}$$

Fig. 5.3 Example 5.1-2

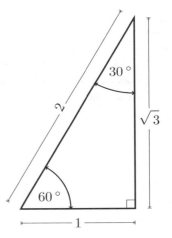

Case: $x = 45° = \pi/4$ Here, the two catheti are equal, thus the two associated angles must be in the same proportion. By Pythagoras's theorem, we calculate hypotenuse

$$c = \sqrt{1^2 + 1^2} = \sqrt{2}$$

Therefore, by definition we conclude the following (Fig. 5.4):

$$\cos 45° = \frac{\text{adjacent}}{\text{hypotenuse}} = \frac{1}{\sqrt{2}} = \frac{\sqrt{2}}{2}$$

$$\sin 45° = \frac{\text{opposite}}{\text{hypotenuse}} = \frac{1}{\sqrt{2}} = \frac{\sqrt{2}}{2}$$

$$\tan 45° = \frac{\text{opposite}}{\text{adjacent}} = \frac{1}{1} = 1$$

(b)   *(cont.)*

Fig. 5.4 Example 5.1-2

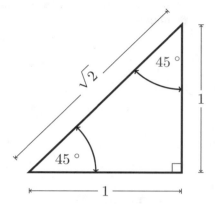

Case: $x = 90°$ In this special case, right-angled triangle whose hypothenuse equals one degenerates into a vertical line segment, as a consequence the hypotenuse is found "on top" of opposite side, thus the two are equal. At the same time, the adjacent side of the triangle is reduced to zero. By definitions we write

$$\cos 90° = \frac{0}{\text{hypotenuse}} = 0,$$

$$\sin 90° = \frac{\text{opposite}}{\text{hypotenuse}} = 1,$$

$$\tan 90° = \frac{\text{opposite}}{\text{adjacent}} = \frac{1}{0} = \infty$$

(c)   Trigonometric functions of sin and cos are commonly summarized in a graph known as "unit circle", see Fig. 5.5. The circle radius, i.e hypotenuse, is normalized to one. Consequently, the length of adjacent side becomes numerically equal $\cos \alpha$ and the length of opposite side becomes numerically equal to $\sin \alpha$, because

$$\cos \alpha = \frac{\text{adjacent}}{\text{hypotenuse}} = \frac{\text{adjacent}}{1} \quad \therefore \quad \text{adjacent} = \cos \alpha$$

$$\sin \alpha = \frac{\text{opposite}}{\text{hypotenuse}} = \frac{\text{opposite}}{1} \quad \therefore \quad \text{opposite} = \sin \alpha$$

(c) *(cont.)*

In the first quadrant we find angles $0°$, $30°$, $45°$, $60°$, and $90°$, see Fig. 5.5. Relative to the first quadrant, angles in the second quadrant are shifted by $90°$, thus we find angles:

$$30° + 90° = 120°$$

$$45° + 90° = 135°$$

$$60° + 90° = 150°$$

$$90° + 90° = 180°$$

Angles in the third quadrant are shifted again by another $90°$, thus we find angles:

$$30° + 180° = 210°$$

$$45° + 180° = 225°$$

$$60° + 180° = 240°$$

$$90° + 180° = 270°$$

Angles in the third quadrant are shifted again by the total of $270°$, thus we find angles:

$$30° + 270° = 300°$$

$$45° + 270° = 315°$$

$$60° + 270° = 330°$$

$$90° + 270° = 360° = 0°$$

The symmetry among these "key" angles is clearly visible in the graph, where we can visualize the associated right-angled triangle for each angle and deduce its corresponding values of $\sin x$ and $\cos x$, as listed in Fig. 5.5.

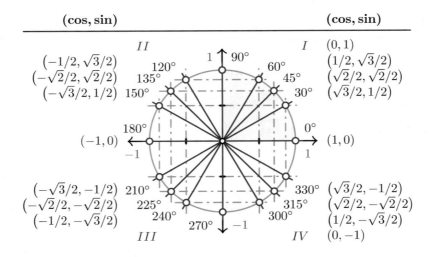

Fig. 5.5 Example 5.1-2

(c)   *(cont.)*

For example, an angle $\beta = 90° + \alpha$ in the second quadrant and its corresponding α angle in the first quadrant create similar right-angled triangles, see Fig. 5.6. By definitions we read its $\sin \beta$ value as the segment length at the vertical axis, and its $\cos \beta$ value as the segment length at the horizontal axis. After comparing side lengths of similar triangles in second and first quadrant (marked in solid red lines), while accounting for the positive/negative signs, we find the following identities

$$\sin\left(\alpha + \frac{\pi}{2}\right) = \sin\beta = \cos\alpha$$

$$-\cos\left(\alpha + \frac{\pi}{2}\right) = -\cos\beta = \sin\alpha$$

where the negative sign of $\cos \beta$ is direct consequence of its position in the second quadrant. In conclusion, by knowing $\sin\alpha$ and $\cos\alpha$ in the first quadrant, we easily conclude that $\cos\alpha = \sin(\alpha + 90°)$ and $\sin\alpha = -\cos(\alpha + 90°)$.

For example,

$$\left.\begin{array}{ll}\cos 120° & = -\sin 30° = -\dfrac{1}{2} \\[2mm] \sin 120° & = \cos 30° = \dfrac{\sqrt{3}}{2}\end{array}\right\} \quad \therefore \quad \tan 120° = \frac{\sqrt{3}/2}{-1/2} = -\sqrt{3}$$

Fig. 5.6 Example 5.1-2

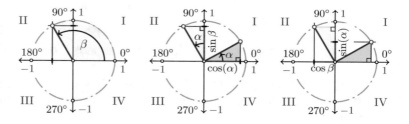

Similarly, for example, we can read angles in fourth quadrant as negative values of the equivalent angles in the first quadrant. After rotation of 180°, see Fig. 5.7, we can deduce the following

$$\cos(\pi \pm \alpha) = -\cos \pm\alpha$$

$$\sin(\pi - \alpha) = -\sin(-\alpha) = \sin\alpha$$

Fig. 5.7 Example 5.1-2

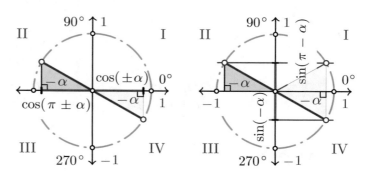

In conclusion, knowing "key" angles in the first quadrant, unit circle is used to visually calculate their correspond angles in the other three quadrants.

3. Knowing that $180° = \pi$ rad, we use a simple proportion equation as follows.

(a) $\dfrac{300°}{180°} = \dfrac{\alpha}{\pi}$ $\therefore$ $\alpha = \pi \dfrac{\cancel{300°}\,5}{\cancel{180°}\,3} = \dfrac{5\pi}{3}$

(b) $\dfrac{330°}{180°} = \dfrac{\alpha}{\pi}$ $\therefore$ $\alpha = \pi \dfrac{\cancel{330°}\,11}{\cancel{180°}\,6} = \dfrac{11\pi}{6}$

(c) $\dfrac{-18°}{180°} = \dfrac{\alpha}{\pi}$ $\therefore$ $\alpha = \pi \dfrac{\cancel{-18°}\,-1}{\cancel{180°}\,10} = -\dfrac{\pi}{10}$

4. Knowing that $180° = \pi$ rad, we use a simple proportion equation as follows.

(a) $\dfrac{\alpha}{180°} = \dfrac{3\pi/4}{\pi}$ $\therefore$ $\alpha = \dfrac{3\cancel{\pi}/4}{\cancel{\pi}} \overset{45°}{\cancel{180°}} = 135°$

(b) $\dfrac{\alpha}{180°} = \dfrac{5\pi/12}{\pi}$ $\therefore$ $\alpha = \dfrac{5\cancel{\pi}/12}{\cancel{\pi}} \overset{15°}{\cancel{180°}} = 75°$

(c) $\dfrac{\alpha}{180°} = \dfrac{2}{\pi}$ $\therefore$ $\alpha = \dfrac{2}{\pi} 180° = \dfrac{360°}{\pi}$

5. With the help of unit circle, after accounting for the positive negative angle directions, we calculate

(a) $\tan \dfrac{\pi}{3} = \dfrac{\sin(\pi/3)}{\cos(\pi/3)} = \dfrac{\sqrt{3}/\cancel{2}}{1/\cancel{2}} = \sqrt{3}$

(b) $\sin \dfrac{7\pi}{6} = \sin\left(\dfrac{\cancel{6}\pi}{\cancel{6}} + \dfrac{\pi}{6}\right) = \sin\left(6\pi + \dfrac{\pi}{6}\right) = -\sin\left(\dfrac{\pi}{6}\right) = -\dfrac{1}{2}$

(c) $\cos\left(-\dfrac{3\pi}{4}\right) = \cos\left(\dfrac{\pi}{4} \pm \pi\right) = -\cos \dfrac{\pi}{4} = -\dfrac{\sqrt{2}}{2}$

6. By definition of $\sin x$ and $\cos x$ functions, assuming a, b to be the two catheti and c hypotenuse, we write,

$$\cos\theta = \dfrac{\text{adjacent}}{\text{hypotenuse}} = \dfrac{a}{10} \quad \therefore \quad a = 10\cos\theta$$

$$\sin\alpha = \dfrac{\text{opposite}}{\text{hypotenuse}} = \dfrac{b}{10} \quad \therefore \quad b = 10\sin\theta$$

7. Knowing that hypotenuse of a right-angled triangle equals the unit circle radius, i.e. equals one, then by Pythagoras's theorem, assuming a, b to be the two catheti and c hypotenuse, we write,

$$a = \cos\alpha, \quad b = \sin\alpha$$

$$\therefore$$

$$a^2 + b^2 = 1^2 \quad \therefore \quad \cos^2\alpha + \sin^2\alpha = 1^2$$

which is direct consequence of hypotenuse $c = 1$ and definitions for $\sin x$ and $\cos x$ functions. Therefore we verify each case.

(a) $\sin^2 \alpha + \cos^2 \alpha = \left(-\dfrac{1}{5}\right)^2 + \left(\dfrac{3}{5}\right)^2 = \dfrac{1}{25} + \dfrac{9}{25} = \dfrac{10}{25} \neq 1$ ✗

(b) $\sin^2 \alpha + \cos^2 \alpha = \left(\dfrac{4}{\sqrt{17}}\right)^2 + \left(\dfrac{1}{\sqrt{17}}\right)^2 = \dfrac{16}{17} + \dfrac{1}{17} = \dfrac{17}{17} = 1$ ✓

(c) $\sin^2 \alpha + \cos^2 \alpha = \left(-\dfrac{\sqrt{15}}{5}\right)^2 + \left(\dfrac{\sqrt{5}}{5}\right)^2 = \dfrac{15}{25} + \dfrac{5}{25} = \dfrac{20}{25} \neq 1$ ✗

8. Knowing the relationship

$$\cos^2 \alpha + \sin^2 \alpha = 1^2$$

and given

$$\sin \alpha = -\frac{1}{3} \quad \text{et} \ \pi < \alpha < \frac{3\pi}{2}$$

we write

$$\cos^2 \alpha + \sin^2 \alpha = 1 \quad \therefore \quad \cos^2 \alpha = 1 - \sin^2 \alpha = 1 - \left(-\frac{1}{3}\right)^2$$

$$\therefore$$

$$\cos \alpha = \sqrt{1 - \frac{1}{9}} \quad \therefore \quad \cos \alpha = \frac{\sqrt{8}}{3} = \frac{2\sqrt{2}}{3}$$

However, α is in the <u>third quadrant</u>, thus its $\cos x$ is *negative*, see Fig. 5.5, i.e. $\cos \alpha = -2\sqrt{2}/3$. Therefore, we write

$$\tan \alpha = \frac{\sin \alpha}{\cos \alpha} = \frac{-1/\cancel{3}}{-2\sqrt{2}/\cancel{3}} = \frac{1}{2\sqrt{2}} = \frac{\sqrt{2}}{4}$$

9. With the help of unit circle, we write,

(a) $\cos 150° = \cos(180° - 30°) = -\cos 30° = -\dfrac{\sqrt{3}}{2}$

(b) $\sin 120° = \cos(90° + 30°) = \cos 30° = \dfrac{\sqrt{3}}{2}$

(c)

$$\tan 300° = \tan(360° - 60°) = \tan(-60°) = \frac{\sin(-60°)}{\cos(-60°)} = \frac{-\sin(60°)}{\cos(60°)}$$

$$= \frac{-\sqrt{3}/\cancel{2}}{1/\cancel{2}} = -\sqrt{3}$$

(d) $\tan 225° = \tan(180° + 45°) = \tan 45° = 1$

$\tan 480° = \tan(360° + 120°) = \tan 120° = \tan(180° - 60°)$

(e)
$$= \frac{\sin(180° - 60°)}{\cos(180° - 60°)} = \frac{\sin 60°}{-\cos 60°} = \frac{\sqrt{3}/2}{-1/2} = -\sqrt{3}$$

10. With the help of unit circle, we calculate

(a) $\tan 840° = \tan(2 \times 360° + 120°) = \tan 120° = -\sqrt{3}$

$\cos 1320° = \cos(3 \times 360° + 240°) = \cos 240° = \cos(180° + 60°)$

(b)
$$= -\cos 60° = -\frac{1}{2}$$

(c) $\sin \dfrac{\pi}{6} \sin^2 \dfrac{\pi}{3} = \dfrac{1}{2}\left(\dfrac{\sqrt{3}}{2}\right)^2 = \dfrac{1}{2}\dfrac{3}{4} = \dfrac{3}{8}$

(d) $\cos \dfrac{\pi}{4} \cos \dfrac{\pi}{6} + \sin \dfrac{\pi}{4} \cos \dfrac{\pi}{6} = \dfrac{1}{\sqrt{2}}\dfrac{\sqrt{3}}{2} + \dfrac{1}{\sqrt{2}}\dfrac{\sqrt{3}}{2} = \sqrt{\dfrac{3}{2}}$

(e) $\sin^2 \dfrac{\pi}{3} - \dfrac{1}{\tan^2 \dfrac{\pi}{3}} = \left(\dfrac{\sqrt{3}}{2}\right)^2 - \dfrac{1}{\left(\sqrt{3}\right)^2} = \dfrac{3}{4} - \dfrac{1}{3} = \dfrac{9-4}{12} = \dfrac{5}{12}$

(f) $\sin \dfrac{\pi}{4} \cos \dfrac{\pi}{4} + \cos \dfrac{\pi}{6} \tan \dfrac{\pi}{3} = \dfrac{1}{\sqrt{2}}\dfrac{1}{\sqrt{2}} + \dfrac{\sqrt{3}}{2}\sqrt{3} = \dfrac{1}{2} + \dfrac{3}{2} = 2$

Exercise 5.2, page 70

1. Main idea is to decompose given angles into the basic angles found in the unity circle, whose sin and cos (therefore, also tan) values are already known. In addition, understanding of relations among similar angles, for example, $180° \pm \alpha$ and $90° \pm \alpha$ is very important, (Fig. 5.8).

Fig. 5.8 Example 5.2-1

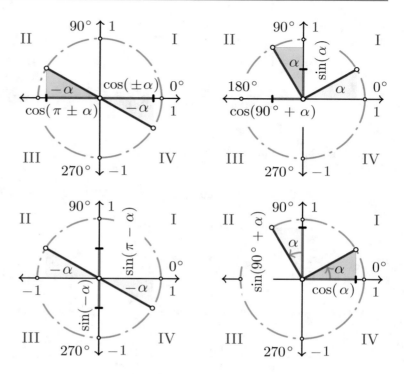

(a)   Important to know values of sin and cos are found on the unity circle,

$$\cos\left(\frac{\pi}{3}\right) = \frac{1}{2}$$

(b)   Function tan is found by sin and cos ratio

$$\tan\left(\frac{\pi}{4}\right) = \frac{\sin(\pi/4)}{\cos(\pi/4)} = \frac{\sqrt{2}/2}{\sqrt{2}/2} = 1$$

(c)   Difference of 180 and 30 gives

$$\cos(150°) = \cos(180° - 30°) = -\cos(30°) = -\frac{\sqrt{3}}{2}$$

(d)   Sum of 90 and 30 gives

$$\sin(120°) = \sin(90° + 30°) = \cos(30°) = \frac{\sqrt{3}}{2}$$

(e)   Functions sin and cos of negative angles result in

$$\tan(-45°) = \frac{\sin(-45°)}{\cos(-45°)} = \frac{-\sin(45°)}{\cos(45°)} = -\frac{\sin(45°)}{\cos(45°)} = -\tan(45°) = -1$$

(f)   Functions sin and cos of negative angles result in

$$\tan(-60°) = \frac{\sin(-60°)}{\cos(-60°)} = \frac{-\sin(60°)}{\cos(60°)} = -\frac{\sin(60°)}{\cos(60°)} = -\frac{\sqrt{3}/2}{1/2} = -\sqrt{3}$$

(g)   Difference of 360 and 60 gives

$$\tan(300°) = \tan(360° - 60°) = \tan(-60°) = \frac{\sin(-60°)}{\cos(-60°)} = \frac{-\sin(60°)}{\cos(60°)}$$

$$= -\frac{\sqrt{3}/2}{1/2} = -\sqrt{3}$$

(h)   Sum of 180 and 45 gives

$$\tan(225°) = \tan(180° + 45°) = \frac{\sin(180° + 45°)}{\cos(180° + 45°)} = \frac{-\sin(+45°)}{-\cos(+45°)}$$

$$= \tan 45° = 1$$

(i)   Difference of 360, 120, and 60 gives

$$\tan(480°) = \tan(360° + 120°) = \tan(180° - 60°) = \tan(-60°)$$

$$= -\tan(60°) = -\sqrt{3}$$

(j)   Multiple revolutions around the 2π circle result in

$$\tan(840°) = \tan(2 \times 360° + 120°) = \tan(180° - 60°) = \tan(-60°)$$

$$= -\tan(60°) = -\sqrt{3}$$

(k)   Multiple revolutions around the 2π circle result in

$$\cos(1320°) = \cos(3 \times 360° + 240°) = \cos(180° + 60°) = -\cos(60°) = -1/2$$

(l)   More complicated problems may be also reduced as, for example,

$$\sin\left(\frac{\pi}{6}\right) \sin^2\left(\frac{\pi}{3}\right) = \frac{1}{2}\left(\frac{\sqrt{3}}{2}\right)^2 = \frac{1}{2}\frac{3}{4} = \frac{3}{8}$$

2.   The idea is to use trigonometric identities for products of sin and cos functions,

(a)   Product of two sin functions is identical to

$$-4\sin\left(\frac{7\pi}{12}\right)\sin\left(\frac{13\pi}{12}\right) = -4\frac{1}{2}\left[\cos\left(\frac{7\pi}{12} - \frac{13\pi}{12}\right) - \cos\left(\frac{7\pi}{12} + \frac{13\pi}{12}\right)\right]$$

$$= -2\left[\cos\left(-\frac{6\pi}{12}\right) - \cos\left(\frac{20\pi}{12}\right)\right]$$

$$= -2\left[\cos\left(\frac{\pi}{2}\right)^{0} - \cos\left(\frac{5\pi}{3}\right)\right]$$

$$= -2\left[-\cos\left(-\frac{\pi}{3}\right)\right] = 1$$

(b)   Inter-product of sin and cos functions is identical to

$$4 \sin \left(\frac{5\pi}{8} \right) \cos \left(\frac{\pi}{8} \right) = 4 \frac{1}{2} \left[\sin \left(\frac{5\pi}{8} + \frac{\pi}{8} \right) + \sin \left(\frac{5\pi}{8} - \frac{\pi}{8} \right) \right]$$

$$= 4 \frac{1}{2} \left[\sin \left(\frac{3\pi}{4} \right) + \sin \left(\frac{\pi}{2} \right) \right] = 4 \frac{1}{2} \left[\frac{\sqrt{2}}{2} + 1 \right]$$

$$= \sqrt{2} + 2$$

(c)   Product of cos functions results in

$$4 \cos \left(\frac{5\pi}{8} \right) \cos \left(\frac{3\pi}{8} \right) = 4 \frac{1}{2} \left[\cos \left(\frac{5\pi}{8} + \frac{3\pi}{8} \right) + \cos \left(\frac{5\pi}{8} - \frac{3\pi}{8} \right) \right]$$

$$= 4 \frac{1}{2} \left[\cos (\pi) + \cos \left(\frac{\pi}{4} \right) \right] = 4 \frac{1}{2} \left[-1 + \frac{\sqrt{2}}{2} \right]$$

$$= \sqrt{2} - 2$$

(d)   This product is found first, the products of numerators, then the product of denominators,

$$\tan 20° \ \tan 40° \ \tan 80° = \frac{\sin 20° \ \sin 40° \ \sin 80°}{\cos 20° \ \cos 40° \ \cos 80°}$$

$\therefore$

$$\sin 20° \sin 40° \sin 80° = \frac{1}{2} \left[\cos(20° - 40°) - \cos(20° + 40°) \right] \sin 80°$$

$$= \frac{1}{2} \left[\underset{\cos 20°}{\cancel{\cos(-20°)}} - \cos(60°) \right] \sin 80°$$

$$= \frac{1}{2} \left[\frac{1}{2} (\sin 100° + \sin 60°) - \underset{1/2}{\cancel{\cos 60°}} \sin 80° \right]$$

$$= \frac{1}{2} \frac{1}{2} \left(\underset{}{\cancel{\sin 100°}} + \underset{\sqrt{3}/2}{\cancel{\sin 60°}} - \cancel{\sin 80°} \right)$$

$$= \frac{\sqrt{3}}{8}$$

here, we note that $\sin 100° = \sin(90° + 10°)$ and $\sin(80°) \sin(90° - 10°)$, in other words, these two values are equal thus cancel. Similarly, we find that $\cos 100° = -\cos 80°$, and the product of denominators is calculated as

$$\therefore$$

$$\cos 20° \cos 40° \cos 80° = \frac{1}{2}[\cos(20° + 40°) + \cos(20° - 40°)]\cos 80°$$

$$= \frac{1}{2}\left[\underset{\nearrow^{1/2}}{\cos 60°} + \underset{\nearrow^{\cos 20°}}{\cos(-20°)}\right]\cos 80°$$

$$= \frac{1}{2}\left[\frac{1}{2}\cos 80° + \cos 20° \cos 80°\right]$$

$$= \frac{1}{2}\left[\underbrace{\frac{1}{2}\cos 80°}_{\cos(90-10)°} + \underbrace{\frac{1}{2}\cos 100°}_{\cos(90+10)°} + \underset{\nearrow^{1/2}}{\frac{1}{2}\cos 60°}\right]$$

$$= \frac{1}{8}$$

Therefore, we conclude that

$$\tan 20° \tan 40° \tan 80° = \frac{\sqrt{3}/8}{1/8} = \sqrt{3}$$

Exercise 5.3, page 70

1. With the help of trigonometric functions for angle sum identities, we find

   (a) Angle of 15° can be seen as (45° − 30°), so that

   $$\sin 15° = \sin(45° - 30°) = \sin 45° \cos 30° - \cos 45° \sin 30°$$

   $$= \frac{\sqrt{2}}{2}\frac{\sqrt{3}}{2} - \frac{\sqrt{2}}{2}\frac{1}{2} = \frac{\sqrt{2}}{4}(\sqrt{3} - 1)$$

   $$\cos 15° = \cos(45° - 30°) = \cos 45° \cos 30° + \sin 45° \sin 30°$$

   $$= \frac{\sqrt{2}}{2}\frac{\sqrt{3}}{2} + \frac{\sqrt{2}}{2}\frac{1}{2} = \frac{\sqrt{2}}{4}(\sqrt{3} + 1)$$

   $$\tan 15° = \frac{\sin 15°}{\cos 15°} = \frac{\frac{\sqrt{2}}{4}(\sqrt{3} - 1)}{\frac{\sqrt{2}}{4}(\sqrt{3} + 1)}\frac{\sqrt{3} - 1}{\sqrt{3} - 1}$$

   $$= \frac{(\sqrt{3} - 1)(\sqrt{3} - 1)}{(\sqrt{3} + 1)(\sqrt{3} - 1)} = \frac{3 - 2\sqrt{3} + 1}{3 - 1} = \frac{2(2 - \sqrt{3})}{2} = 2 - \sqrt{3}$$

(b)   Angle of 75° can be seen as $(45° + 30°)$, so that

$$\sin 75° = \sin(45° + 30°) = \sin 45° \cos 30° + \cos 45° \sin 30°$$

$$= \frac{\sqrt{2}}{2}\frac{\sqrt{3}}{2} + \frac{\sqrt{2}}{2}\frac{1}{2} = \frac{\sqrt{2}}{4}(\sqrt{3} + 1)$$

$$\cos 75° = \cos(45° + 30°) = \cos 45° \cos 30° - \sin 45° \sin 30°$$

$$= \frac{\sqrt{2}}{2}\frac{\sqrt{3}}{2} - \frac{\sqrt{2}}{2}\frac{1}{2} = \frac{\sqrt{2}}{4}(\sqrt{3} - 1)$$

$$\tan 75° = \frac{\sin 75°}{\cos 75°} = \frac{\frac{\sqrt{2}}{4}(\sqrt{3} + 1)}{\frac{\sqrt{2}}{4}(\sqrt{3} - 1)} \frac{\sqrt{3} + 1}{\sqrt{3} + 1}$$

$$= \frac{(\sqrt{3} + 1)(\sqrt{3} + 1)}{(\sqrt{3} + 1)(\sqrt{3} - 1)} = \frac{3 + 2\sqrt{3} + 1}{3 - 1} = \frac{2(2 + \sqrt{3})}{2} = 2 + \sqrt{3}$$

(c)   Angle of 105° can be seen as $(60° + 45°)$, so that

$$\sin 105° = \sin(60° + 45°) = \sin 60° \cos 45° + \cos 60° \sin 45°$$

$$= \frac{\sqrt{3}}{2}\frac{\sqrt{2}}{2} + \frac{1}{2}\frac{\sqrt{2}}{2} = \frac{\sqrt{2}}{4}(\sqrt{3} + 1)$$

$$\cos 105° = \cos(60° + 45°) = \cos 60° \cos 45° - \sin 60° \sin 45°$$

$$= \frac{1}{2}\frac{\sqrt{2}}{2} - \frac{\sqrt{3}}{2}\frac{\sqrt{2}}{2} = \frac{\sqrt{2}}{4}(1 - \sqrt{3})$$

$$\tan 105° = \frac{\sin 105°}{\cos 105°} = \frac{\frac{\sqrt{2}}{4}(\sqrt{3} + 1)}{\frac{\sqrt{2}}{4}(1 - \sqrt{3})} \frac{1 + \sqrt{3}}{1 + \sqrt{3}}$$

$$= \frac{(\sqrt{3} + 1)(\sqrt{3} + 1)}{(1 - \sqrt{3})(1 + \sqrt{3})} = \frac{3 + 2\sqrt{3} + 1}{1 - 3} = \frac{2(2 + \sqrt{3})}{-2 - 1} = -(2 + \sqrt{3})$$

(d)   Similarly, angle of $\pi/12$ can be seen as $(\pi/3 - \pi/4)$, so that

$$\sin\frac{\pi}{12} = \sin\left(\frac{\pi}{3} - \frac{\pi}{4}\right) = \sin\frac{\pi}{3}\cos\frac{\pi}{4} - \cos\frac{\pi}{3}\sin\frac{\pi}{4}$$

$$= \frac{\sqrt{3}}{2}\frac{\sqrt{2}}{2} - \frac{1}{2}\frac{\sqrt{2}}{2} = \frac{\sqrt{2}}{4}(\sqrt{3} - 1)$$

$$\cos\frac{\pi}{12} = \cos\left(\frac{\pi}{3} - \frac{\pi}{4}\right) = \cos\frac{\pi}{3}\cos\frac{\pi}{4} + \sin\frac{\pi}{3}\sin\frac{\pi}{4}$$

$$= \frac{1}{2}\frac{\sqrt{2}}{2} + \frac{\sqrt{3}}{2}\frac{\sqrt{2}}{2} = \frac{\sqrt{2}}{4}(\sqrt{3} + 1)$$

$$\tan\frac{\pi}{12} = \frac{\sin\dfrac{\pi}{12}}{\cos\dfrac{\pi}{12}} = \frac{\dfrac{\sqrt{2}}{4}(\sqrt{3} - 1)}{\dfrac{\sqrt{2}}{4}(\sqrt{3} + 1)} \quad \frac{\sqrt{3} - 1}{\sqrt{3} - 1}$$

$$= \frac{(\sqrt{3} - 1)(\sqrt{3} - 1)}{(\sqrt{3} + 1)(\sqrt{3} - 1)} = \cdots = 2 - \sqrt{3}$$

(e)   Angle of $5\pi/12$ can be seen as $\left(\frac{\pi}{4} + \frac{\pi}{6}\right)$, so that

$$\sin\left(\frac{5\pi}{12}\right) = \sin\left(\frac{\pi}{4} + \frac{\pi}{6}\right) = \cdots = \frac{\sqrt{2}}{4}(\sqrt{3} + 1)$$

$$\cos\left(\frac{5\pi}{12}\right) = \cos\left(\frac{\pi}{4} + \frac{\pi}{6}\right) = \cdots = \frac{\sqrt{2}}{4}(\sqrt{3} - 1)$$

$$\tan\left(\frac{5\pi}{12}\right) = \frac{\sin\left(\dfrac{5\pi}{12}\right)}{\cos\left(\dfrac{5\pi}{12}\right)} = \frac{\dfrac{\sqrt{2}}{4}(\sqrt{3} + 1)}{\dfrac{\sqrt{2}}{4}(\sqrt{3} - 1)} = \cdots = 2 + \sqrt{3}$$

(f)   Angle of $7\pi/12$ can be seen as $\left(\frac{\pi}{3} + \frac{\pi}{4}\right)$, so that

$$\sin\left(\frac{7\pi}{12}\right) = \sin\left(\frac{\pi}{3} + \frac{\pi}{4}\right) = \cdots = \frac{\sqrt{2}}{4}(\sqrt{3} + 1)$$

$$\cos\left(\frac{7\pi}{12}\right) = \cos\left(\frac{\pi}{3} + \frac{\pi}{4}\right) = \cdots = \frac{\sqrt{2}}{4}(1 - \sqrt{3})$$

$$\tan\left(\frac{7\pi}{12}\right) = \frac{\sin\left(\dfrac{7\pi}{12}\right)}{\cos\left(\dfrac{7\pi}{12}\right)} = \cdots = -(2 + \sqrt{3})$$

2. Knowing the sum identities, we write

(a) $\sin 20° \cos 10° + \cos 20° \sin 10° = \sin(20° + 10°) = \sin 30° = \dfrac{1}{2}$

(b) $\cos 43° \cos 13° + \sin 43° \sin 13° = \cos(43° - 13°) = \cos 30° = \dfrac{\sqrt{3}}{2}$

3. Knowing Pythagoras's theorem, unit circle and the sum identities, we write

$$\left. \begin{array}{l} \cos x \;\; = {}^4\!/_5 \;\; \therefore \;\; \sin x = \sqrt{1 - (4/5)^2} = 3/5 \\[2mm] \sin y \;\; = -{}^3\!/_5 \;\; \therefore \;\; \cos y = \sqrt{1 - (-3/5)^2} = 4/5 \end{array} \right\} \quad \therefore \quad \left\{ \begin{array}{l} \sin x \;\; = -\dfrac{3}{5} \;\; \text{(IV)} \\[3mm] \cos y \;\; = -\dfrac{4}{5} \;\; \text{(III)} \end{array} \right.$$

$$\sin(x + y) = \sin x \cos y + \cos x \sin y = \left(-\dfrac{3}{5}\right)\left(-\dfrac{4}{5}\right) + \left(\dfrac{4}{5}\right)\left(-\dfrac{3}{5}\right) = \dfrac{12}{25} - \dfrac{12}{25} = 0$$

$$\sin(x - y) = \sin x \cos y - \cos x \sin y = \left(-\dfrac{3}{5}\right)\left(-\dfrac{4}{5}\right) - \left(\dfrac{4}{5}\right)\left(-\dfrac{3}{5}\right) = \dfrac{12}{25} + \dfrac{12}{25} = \dfrac{24}{25}$$

4. With the help of sum identities we write

$$\frac{\sqrt{2}\cos\alpha - 2\cos\left(\dfrac{\pi}{4} + \alpha\right)}{2\sin\left(\dfrac{\pi}{4} + \alpha\right) - \sqrt{2}\sin\alpha} = \frac{2\left(\dfrac{\sqrt{2}}{2}\cos\alpha - \cos\left(\dfrac{\pi}{4} + \alpha\right)\right)}{2\left(\sin\left(\dfrac{\pi}{4} + \alpha\right) - \dfrac{\sqrt{2}}{2}\sin\alpha\right)}$$

$$= \frac{\cancel{2}\left(\cos\dfrac{\pi}{4}\cancel{\cos\alpha} - \cos\dfrac{\pi}{4}\cancel{\cos\alpha} + \sin\dfrac{\pi}{4}\sin\alpha\right)}{\cancel{2}\left(\sin\dfrac{\pi}{4}\cos\alpha + \cos\dfrac{\pi}{4}\cancel{\sin\alpha} - \cos\dfrac{\pi}{4}\cancel{\sin\alpha}\right)}$$

$$= \frac{\cancel{\sin\dfrac{\pi}{4}}\sin\alpha}{\cancel{\sin\dfrac{\pi}{4}}\cos\alpha} = \tan\alpha$$

5. After setting $a = b$ in the sum identities we have

$$\sin(a + a) = \underline{\sin 2a} = \sin a \cos a + \cos a \sin a = \underline{2\sin a \, \cos a}$$

$$\cos(a + a) = \underline{\cos 2a} = \cos a \cos a - \sin a \sin a = \underline{\cos^2 a - \sin^2 a}$$

$$\tan 2a = \frac{\sin 2a}{\cos 2a} = \frac{\sin a \cos a + \cos a \sin a}{\cos a \cos a - \sin a \sin a} = \frac{\sin a \cos a + \cos a \sin a}{\cos a \cos a \left(1 - \dfrac{\sin a \, \sin a}{\cos a \, \cos a}\right)}$$

$$= \frac{\dfrac{\sin a \cancel{\cos a}}{\cos a \cancel{\cos a}} + \dfrac{\cancel{\cos a} \sin a}{\cancel{\cos a} \cos a}}{1 - \tan^2 a}$$

$$= \frac{2 \tan a}{1 - \tan^2 a}$$

Further, we can write

$$\left.\begin{array}{ll} \cos^2 a + \sin^2 a & = 1 \\ \cos^2 a - \sin^2 a & = \cos 2a \end{array}\right\} \therefore \begin{cases} 2 \sin^2 a & = 1 - \cos 2a \therefore \sin^2 a = \dfrac{1 - \cos 2a}{2} \\ 2 \cos^2 a & = 1 + \cos 2a \therefore \cos^2 a = \dfrac{1 + \cos 2a}{2} \end{cases}$$

In addition,

$$\sin 2a = 2 \sin a \cos a \frac{\cos a}{\cos a} = 2 \frac{\sin a}{\cos a} \frac{\cos^2 a}{1} = 2 \tan a \frac{\cos^2 a}{\cos^2 a + \sin^2 a}$$

$$= 2 \tan a \frac{1}{\dfrac{\cos^2 a}{\cos^2 a} + \dfrac{\sin^2 a}{\cos^2 a}}$$

$$= \frac{2 \tan a}{1 + \tan^2 a}$$

$$\cos 2a = \frac{\cos^2 a - \sin^2 a}{1} = \frac{\cos^2 a - \sin^2 a}{\cos^2 a + \sin^2 a} = \frac{\cancel{\cos^2 a}\left(1 - \dfrac{\sin^2 a}{\cos^2 a}\right)}{\cancel{\cos^2 a}\left(1 + \dfrac{\sin^2 a}{\cos^2 a}\right)}$$

$$= \frac{1 - \tan^2 a}{1 + \tan^2 a}.$$

These identities are used to integrate rational functions by trigonometric substitutions.

Exercise 5.4, page 71

1. Given the sum identities

$$\sin(\alpha + \beta) = \sin \alpha \cos \beta + \cos \alpha \sin \beta \tag{5.1}$$

$$\sin(\alpha - \beta) = \sin \alpha \cos \beta - \cos \alpha \sin \beta \tag{5.2}$$

$$\cos(\alpha + \beta) = \cos \alpha \cos \beta - \sin \alpha \sin \beta \tag{5.3}$$

$$\cos(\alpha - \beta) = \cos \alpha \cos \beta + \sin \alpha \sin \beta \tag{5.4}$$

(a) Difference between (5.4) and (5.3) gives

$$\cos(\alpha - \beta) - \cos(\alpha + \beta) = \underline{\cos\alpha\cos\beta} + \sin\alpha\sin\beta - \underline{\cos\alpha\cos\beta} + \sin\alpha\sin\beta$$
$$= 2\sin\alpha\sin\beta$$

$$\therefore$$

$$\sin\alpha\sin\beta = \frac{1}{2}\left(\cos(\alpha - \beta) - \cos(\alpha + \beta)\right) \tag{5.5}$$

(b) Sum of (5.3) and (5.4) gives

$$\cos(\alpha + \beta) + \cos(\alpha - \beta) = \cos\alpha\cos\beta - \underline{\sin\alpha\sin\beta} + \cos\alpha\cos\beta + \underline{\sin\alpha\sin\beta}$$
$$= 2\cos\alpha\cos\beta$$

$$\therefore$$

$$\cos\alpha\cos\beta = \frac{1}{2}\left(\cos(\alpha + \beta) + \cos(\alpha - \beta)\right) \tag{5.6}$$

(c) Sum of (5.1) and (5.2) gives

$$\sin(\alpha + \beta) + \sin(\alpha - \beta) = \sin\alpha\cos\beta + \underline{\cos\alpha\sin\beta} + \sin\alpha\cos\beta - \underline{\cos\alpha\sin\beta}$$
$$= 2\sin\alpha\cos\beta$$

$$\therefore$$

$$\sin\alpha\cos\beta = \frac{1}{2}\left(\sin(\alpha + \beta) + \sin(\alpha - \beta)\right) \tag{5.7}$$

Identities (5.5)–(5.7) are extensively used in RF electronics and communication theory.

2. Given the sum identities (5.5)–(5.7)

(a) From (5.7) we write

$$\sin \alpha \cos \beta = \frac{1}{2} \left(\sin(\alpha + \beta) + \sin(\alpha - \beta) \right)$$

$$\therefore$$

$$\sin(\alpha + \beta) + \sin(\alpha - \beta) = 2 \sin \alpha \cos \beta$$

$$\left.\begin{cases} \alpha + \beta = x \\ \alpha - \beta = y \end{cases}\right\} \quad \therefore \quad \left.\begin{cases} \alpha = \dfrac{x + y}{2} \\ \beta = \dfrac{x - y}{2} \end{cases}\right\}$$

$$\therefore$$

$$\sin x + \sin y = 2 \sin \left(\frac{x + y}{2} \right) \cos \left(\frac{x - y}{2} \right)$$

(b) Similarly, from (5.6) we find

$$\cos x + \cos y = 2 \cos \left(\frac{x + y}{2} \right) \cos \left(\frac{x - y}{2} \right)$$

Using the same technique, other similar identities are easily derived.

3. Using the product to sum identities and the unit circle, we find

(a)

$$-4 \sin \frac{5\pi}{12} \cos \frac{7\pi}{12} = -\cancel{4} \; \cancel{2}\frac{1}{\cancel{2}} \left[\sin \left(\frac{5\pi}{12} + \frac{7\pi}{12} \right) + \sin \left(\frac{5\pi}{12} - \frac{7\pi}{12} \right) \right]$$

$$= -2 \left[\sin \left(\frac{\cancel{12}\pi}{\cancel{12}} \right) + \sin \left(-\frac{\cancel{2}\pi}{\cancel{12} \; 6} \right) \right]$$

$$= -2 \left[0 - \sin \left(\frac{\pi}{6} \right) \right] = -2 \left(-\frac{1}{2} \right)$$

$$= 1$$

(b)

$$\frac{1}{2} + 2\sin\left(1 + \frac{\pi}{6}\right)\cos\left(1 + \frac{\pi}{3}\right) = \frac{1}{2} + \left[\sin\left(1 + \frac{\pi}{6} + 1 + \frac{\pi}{3}\right)\right.$$

$$\left. + \sin\left(1 + \frac{\pi}{6} - 1 - \frac{\pi}{3}\right)\right]$$

$$= \frac{1}{2} + \left[\sin\left(2 + \frac{\pi}{2}\right) + \sin\left(-\frac{\pi}{6}\right)\right]$$

$$= \frac{1}{2} + \left[\sin\left(\frac{\pi}{2} + 2\right) - \sin\left(\frac{\pi}{6}\right)\right]$$

$$= \frac{1}{2} + \left[\cos\left(2\right) - \frac{1}{2}\right]$$

$$= \cos(2)$$

(c)

$$8\sin 20° \sin 40° \sin 80° = 4\left[\cos(20° - 40°) - \cos(20° + 40°)\right]\sin 80°$$

$$= 4\left[\sin 80° \cos(-20°) - \sin 80° \cos 60°\right]$$

$$= 4\left[\sin 80° \cos(20°) - \frac{\sin 80°}{2}\right]$$

$$= 4\left[\frac{\sin 100°}{2} + \frac{\sin 60°}{2} - \frac{\sin 80°}{2}\right]$$

$$\left\{\sin 100° = \sin(90° + 10°) = \sin(90° - 10°) = \sin 80°\right\}$$

$$= 4\left[\frac{\sin 80°}{2} + \frac{\sqrt{3}}{4} - \frac{\sin 80°}{2}\right]$$

$$= \sqrt{3}$$

(d)

$$8\cos 10° \cos 50° \cos 70° = 4\left[\cos 60° + \cos 40°\right]\cos 70°$$

$$= 4\left[\cos 60° \cos 70° + \cos 40° \cos 70°\right]$$

$$= 4\left[\frac{\cos 70°}{2} + \frac{\cos 110°}{2} + \frac{\cos 30°}{2}\right]$$

$$\left\{\cos 110° = \cos(90° + 20°) = -\cos(90° - 20°) = -\cos 70°\right\}$$

$$= 4\left[\frac{\cos 70°}{2} - \frac{\cos 70°}{2} + \frac{\sqrt{3}}{4}\right]$$

$$= \sqrt{3}$$

4. With the help of identities and unit circle, we write

$$\tan\alpha = \sqrt{6} + \sqrt{3} - \sqrt{2} - 2 = \sqrt{3}\sqrt{2} + \sqrt{3} - \sqrt{2} - (\sqrt{2})^2$$

$$= \sqrt{3}\left(\sqrt{2}+1\right) - \sqrt{2}\left(\sqrt{2}+1\right) = \frac{\sqrt{2}-1}{\sqrt{2}-1}\ \left(\sqrt{2}+1\right)\left(\sqrt{3}-\sqrt{2}\right)$$

$$\left\{(a - b)(a + b) = a^2 - b^2\right\}$$

$$= \frac{\sqrt{3}-\sqrt{2}}{\sqrt{2}-1}\ \frac{2}{2} = \frac{\frac{\sqrt{3}}{2} - \frac{\sqrt{2}}{2}}{\frac{\sqrt{2}}{2} - \frac{1}{2}} = \frac{\sin 60° - \sin 45°}{\sin 45° - \sin 30°}$$

$$\left\{\sin x - \sin y = 2\cos\left(\frac{x+y}{2}\right)\sin\left(\frac{x-y}{2}\right)\right\}$$

$$= \frac{2\cos\left(\frac{105°}{2}\right)\sin\left(\frac{15°}{2}\right)}{2\cos\left(\frac{75°}{2}\right)\sin\left(\frac{15°}{2}\right)} = \frac{\cos\left(\frac{(180-75)°}{2}\right)}{\cos\left(\frac{75°}{2}\right)} = \frac{\cos\left(90° - \frac{75°}{2}\right)}{\cos\left(\frac{75°}{2}\right)}$$

$$\left\{\cos(90° - \alpha) = \sin\alpha\right\}$$

$$= \frac{\sin\left(\frac{75°}{2}\right)}{\cos\left(\frac{75°}{2}\right)} = \tan\left(\frac{75°}{2}\right)$$

$$\therefore$$

$$\alpha = \frac{75°}{2} = 37.5°$$

Exercise 5.5, page 71

1. Equations that include periodic functions have periodic solutions.

   (a) $\sin x = 0 \quad \therefore \quad x = n\pi, \quad$ where $\quad n = 0, \pm 1, \pm 2, \ldots$

   (b) $\cos x = 0 \quad \therefore \quad x = (2k + 1)\dfrac{\pi}{2}, \quad$ where $\quad k = 0, \pm 1, \pm 2, \ldots$

2. The factorized form of equations helps us to exploit the property if $A \times B \times C = 0$, then either $A = 0$ or $B = 0$ or $C = 0$. Thus, we convert sums into products as

   (a)   We convert the equation into its equivalent factorized form as follows.

   $$\sin x = \sin \alpha \quad \therefore \quad \sin x - \sin \alpha = 0$$

   $$\therefore$$

   $$\left\{ \sin \alpha - \sin \beta = 2 \cos \left(\frac{\alpha + \beta}{2} \right) \sin \left(\frac{\alpha - \beta}{2} \right) \right\}$$

   $$2 \cos \left(\frac{x + \alpha}{2} \right) \sin \left(\frac{x - \alpha}{2} \right) = 0$$

   Therefore,

   $$\cos \left(\frac{x + \alpha}{2} \right) = 0 \quad \therefore \quad \frac{x + \alpha}{2} = (2n + 1)\frac{\pi}{2} \quad \therefore \quad x = -\alpha + (2n + 1)\pi$$

   $$\sin \left(\frac{x - \alpha}{2} \right) = 0 \quad \therefore \quad \frac{x - \alpha}{2} = k\pi \quad \therefore \quad x = \alpha + 2k\pi$$

   where $(k, n) = 0, \pm 1, \pm 2, \ldots$

   (b)   We could also exploit the equivalence relationship between $\sin x$ and $\cos x$ functions (i.e. their argument difference of $\pi/2$) as follows.

   $$\sin x = \cos x \quad \therefore \quad \cos \left(x - \frac{\pi}{2} \right) = \cos x \quad \therefore \quad x - \frac{\pi}{2} = \pm x + 2n\pi$$

   because x takes both positive and negative signs. Therefore, there are two possible solutions that must be reviewed as follows.

   $$\text{case } ' + x' : \quad \therefore \quad \cancel{x} - \frac{\pi}{2} \neq \cancel{x} + 2n\pi \quad \therefore \quad \text{no solution}$$

   $$\text{case } ' - x' : \quad \therefore \quad x - \frac{\pi}{2} = -x + 2n\pi \quad \therefore \quad 2x = 2n\pi + \frac{\pi}{2}$$

   $$\therefore$$

   $$x = n\pi + \frac{\pi}{4}$$

where $n = 0, \pm 1, \pm 2, \ldots$ The solutions are shown in the graph of $\sin x$ and $\cos x$, Fig. 5.9.

Fig. 5.9 Example 5.5-2(b)

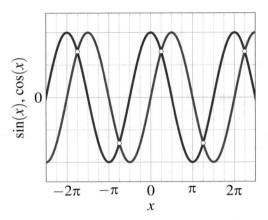

(c)   Factorized form of this equation is solved as follows.

$$\sin x = \sin 2x \quad \therefore \quad \sin 2x - \sin x = 0 \quad \therefore \quad 2\cos\left(\frac{3x}{2}\right)\sin\left(\frac{x}{2}\right) = 0$$

$$\therefore$$

$$\cos\left(\frac{3x}{2}\right) = 0 \quad \therefore \quad \frac{3x}{\not2} = (2n+1)\frac{\pi}{\not2} \quad \therefore \quad \underline{x = \frac{\pi}{3} + 2n\pi}$$

$$\sin\left(\frac{x}{2}\right) = 0 \quad \therefore \quad \frac{x}{2} = k\pi \quad \therefore \quad \underline{x = 2k\pi}$$

where $n, k = 0, \pm 1, \pm 2, \ldots$ The solutions are shown in the graph of $\sin x$ and $\sin 2x$, Fig. 5.10, where the solutions (i.e. the intersect points) are generated by two streams: for $n = 0, \pm 1, \pm 2, \ldots$ and for $k = 0, \pm 1, \pm 2, \ldots$. In the factorized form of the equation (i.e. the equality to zero), roots are found at the intersect with the horizontal axis (i.e $y = 0$).

Fig. 5.10 Example 5.5-2(c)

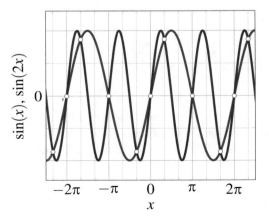

(d) Using the same idea, we factorize this equation and solve as follows.

$$2 \sin^2 x + \sin x = 0 \quad \therefore \quad (2 \sin x + 1) \sin x = 0$$

$$\therefore \quad (2 \sin x + 1) = 0 \quad \text{or} \quad \sin x = 0$$

We solve the two equations separately.

Equation: $\sin x = 0$

$$\sin x = 0 \quad \therefore \quad x = n\pi \quad \text{where} \quad n = 0, \pm 1, \pm 2, \dots$$

Equation: $(2 \sin x + 1) = 0$

$$(2 \sin x + 1) = 0 \quad \therefore \quad \sin x + \frac{1}{2} = 0 \quad \therefore \quad \sin x + \sin\left(\frac{\pi}{6}\right) = 0$$

$$\therefore$$

$$2 \sin\left[\frac{x + \pi/6}{2}\right] \cos\left[\frac{x - \pi/6}{2}\right] = 0$$

Therefore, we search solutions for two equations.

1. Roots of sin function are found at multiples of π, i.e. $0, \pm\pi, \pm2\pi, \dots$

$$\sin\left[\frac{x + \pi/6}{2}\right] = 0 \quad \therefore \quad \frac{x + \pi/6}{2} = k\pi$$

$$\therefore$$

$$x = -\frac{\pi}{6} + 2k\pi \quad \text{where} \quad k = 0, \pm 1, \pm 2, \dots$$

2. Roots of cos function are at odd multiples of $\pi/2$, i.e. $\pm(2m+1)\pi/2$

$$\cos\left[\frac{x - \pi/6}{2}\right] = 0 \quad \therefore \quad \frac{x - \pi/6}{2} = (2m+1)\frac{\pi}{2}$$

$$\therefore$$

$$x = \frac{7\pi}{6} + 2m\pi \quad \text{where} \quad m = 0, \pm 1, \pm 2, \dots$$

These solutions are clearly visible in graph of $f(x) = 2 \sin^2 x + \sin x$ function, Fig. 5.11. One stream of the solutions (i.e. the intersect points) is generated for $n = 0, \pm 1, \pm 2, \dots$, second for $k = 0, \pm 1, \pm 2, \dots$, and third for $m = 0, \pm 1, \pm 2, \dots$

Fig. 5.11 Example 5.5-2(d)

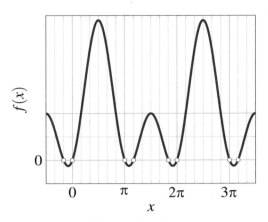

(e)   Knowing periodicity of $\tan x = \sin x / \cos x$ function (it is periodic for $n\pi$), we can solve this equation as follows

$$\cos (x + \pi/6) = \sin (x - \pi/3)$$

$$\therefore$$

$$\cos (x) \cos \left(\frac{\pi}{6}\right) - \sin (x) \sin \left(\frac{\pi}{6}\right) = \sin (x) \cos \left(\frac{\pi}{3}\right) - \cos (x) \sin \left(\frac{\pi}{3}\right)$$

$$\therefore$$

$$\cos (x) \frac{\sqrt{3}}{2} - \sin (x) \frac{1}{2} = \sin (x) \frac{1}{2} - \cos (x) \frac{\sqrt{3}}{2}$$

$$\therefore$$

$$2\sqrt{3} \cos x = 2 \sin x \quad \therefore \quad \tan x = \sqrt{3} = \frac{\sqrt{3}/2}{1/2} = \frac{\sin(\pi/3)}{\cos(\pi/3)}$$

$$\therefore$$

$$x = \frac{\pi}{2} + n\pi \quad \text{where } n = 0, \pm 1, \pm 2, \dots$$

3. Trigonometric equations are specific because each equality periodically repeats both in positive and negative directions. For that reason, given the interval of interest it is necessary to search *all solutions* within the interval. In addition, we exploit factorized form of the equations by using sum to product trigonometric identities,

$$\cos \alpha - \cos \beta = -2 \sin \left(\frac{\alpha + \beta}{2}\right) \sin \left(\frac{\alpha - \beta}{2}\right)$$

$$\sin \alpha - \sin \beta = 2 \cos \left(\frac{\alpha + \beta}{2}\right) \sin \left(\frac{\alpha - \beta}{2}\right)$$

The factorized form of equations helps us to exploit the property if $A \times B \times C = 0$, then either $A = 0$ or $B = 0$ or $C = 0$.

(a)   In the interval $0 \leq x < 2\pi$ we search all solution as follows.

$$\cos\left(\frac{\pi}{3} - x\right) = \cos\left(\frac{\pi}{6}\right)$$

$$\therefore$$

$$\cos\left(\frac{\pi}{3} - x\right) - \cos\left(\frac{\pi}{6}\right) = 0$$

$$\therefore$$

$$\underbrace{-2}_{A} \times \underbrace{\sin\left(\frac{\frac{\pi}{3} - x + \frac{\pi}{6}}{2}\right)}_{B} \times \underbrace{\sin\left(\frac{\frac{\pi}{3} - x - \frac{\pi}{6}}{2}\right)}_{C} = 0$$

Obviously, $-2 \neq 0$ leaving either $B = 0$ or $C = 0$ terms. Both A and B are in the form of $\sin\theta$, a sine function equals to zero for all arguments

$$\sin\theta = 0 \quad \therefore \quad \theta = 0, \pm\pi, \pm2\pi, \ldots \quad \text{in general:} \quad \theta = n\pi, \ n = 0, \pm1, \pm2, \ldots$$

Therefore, we systematically write the cases as

$\underline{n = 0:}$

$$B: \quad \frac{\frac{\pi}{3} - x + \frac{\pi}{6}}{2} = 0 \quad \therefore \quad \frac{\pi}{2} - x = 0 \quad \therefore \quad x = \frac{\pi}{2} \in [0, 2\pi] \ \checkmark$$

$$C: .\frac{\frac{\pi}{3} - x - \frac{\pi}{6}}{2} = 0 \quad \therefore \quad \frac{\pi}{6} - x = 0 \quad \therefore \quad x = \frac{\pi}{6} \in [0, 2\pi] \ \checkmark$$

$\underline{n = 1:}$

$$B: \quad \frac{\frac{\pi}{3} - x + \frac{\pi}{6}}{2} = \pi \quad \therefore \quad \frac{\frac{\pi}{2} - x}{2} = \pi \quad \therefore \quad x = -\frac{3\pi}{2} < 0 \ \times$$

$$C: \quad \frac{\frac{\pi}{3} - x - \frac{\pi}{6}}{2} = \pi \quad \therefore \quad \frac{\frac{\pi}{6} - x}{2} = \pi \quad \therefore \quad x = -\frac{11\pi}{6} < 0 \ \times$$

$\underline{n = -1:}$

$$B: \quad \frac{\frac{\pi}{3} - x + \frac{\pi}{6}}{2} = -\pi \quad \therefore \quad \frac{\frac{\pi}{2} - x}{2} = -\pi \quad \therefore \quad x = \frac{5\pi}{2} > 2\pi \ \times$$

$$C: \quad \frac{\frac{\pi}{3} - x - \frac{\pi}{6}}{2} = -\pi \quad \therefore \quad \frac{\frac{\pi}{6} - x}{2} = -\pi \quad \therefore \quad x = \frac{13\pi}{6} > 2\pi \ \times$$

In conclusion, in interval $0 < x \leq 2\pi$, there are two solutions, see Fig. 5.12 (left).

$$x = \left\{\frac{\pi}{6}, \frac{\pi}{2}\right\}$$

(b)   Similarly, within the interval $-\pi < x \le \pi$, we write,

$$\sin(3x) = \sin\left(x - \frac{\pi}{2}\right)$$

$$\therefore$$

$$\sin(3x) - \sin\left(x - \frac{\pi}{2}\right) = 0$$

$$\therefore$$

$$\underbrace{2}_{A} \times \underbrace{\cos\left(\frac{3x + x - \frac{\pi}{2}}{2}\right)}_{B} \times \underbrace{\sin\left(\frac{3x - x + \frac{\pi}{2}}{2}\right)}_{C} = 0$$

Obviously, $2 \ne 0$ leaving either $B = 0$ or $C = 0$ terms. Term C is a sine function, and B is in the form of $\cos\theta$. A cosine function equals to zero for all arguments

$$\cos\theta = 0 \quad \therefore \quad \theta = \pm\frac{\pi}{2}, \pm\frac{3\pi}{2}, \ldots \quad \text{in general:} \quad \theta = \frac{\pi}{2} \pm n\pi, \quad n = 0, 1, 2, \ldots$$

Therefore, we systematically write the cases as

$\underline{n = 0:}$

$$B: \quad \frac{3x + x - \frac{\pi}{2}}{2} = \frac{\pi}{2} \quad \therefore \quad 4x - \frac{\pi}{2} = \pi \quad \therefore \quad x = \frac{3\pi}{8} \in [-\pi, \pi] \checkmark$$

$$C: \quad \frac{3x - x + \frac{\pi}{2}}{2} = 0 \quad \therefore \quad 2x + \frac{\pi}{2} = 0 \quad \therefore \quad x = -\frac{\pi}{4} \in [-\pi, \pi] \checkmark$$

$\underline{n = 1:}$

$$B: \quad \frac{3x + x - \frac{\pi}{2}}{2} = \frac{\pi}{2} + \pi \quad \therefore \quad 4x - \frac{\pi}{2} = 3\pi \quad \therefore \quad x = \frac{7\pi}{8} \in [-\pi, \pi] \checkmark$$

$$C: \quad \frac{3x - x + \frac{\pi}{2}}{2} = \pi \quad \therefore \quad 2x + \frac{\pi}{2} = 2\pi \quad \therefore \quad x = \frac{3\pi}{4} \in [-\pi, \pi] \checkmark$$

$\underline{n = -1:}$

$$B: \quad \frac{3x + x - \frac{\pi}{2}}{2} = \frac{\pi}{2} - \pi \quad \therefore \quad 4x - \frac{\pi}{2} = -\pi \quad \therefore \quad x = -\frac{\pi}{8} \in [-\pi, \pi] \checkmark$$

$$C: \quad \frac{3x - x + \frac{\pi}{2}}{2} = -\pi \quad \therefore \quad 2x + \frac{\pi}{2} = -2\pi \quad \therefore \quad x = -\frac{5\pi}{4} < -\pi \; ✗$$

$\underline{n = 2:}$

$$B: \quad \frac{3x + x - \frac{\pi}{2}}{2} = \frac{\pi}{2} + 2\pi \quad \therefore \quad 4x - \frac{\pi}{2} = 5\pi \quad \therefore \quad x = \frac{11\pi}{8} > \pi \; ✗$$

$$C: \quad \frac{3x - x + \frac{\pi}{2}}{2} = 2\pi \quad \therefore \quad 2x + \frac{\pi}{2} = 4\pi \quad \therefore \quad x = \frac{7\pi}{4} > \pi \; ✗$$

(b)   *(cont.)*

$n = -2$:

$$B : \quad \frac{3x + x - \frac{\pi}{2}}{2} = \frac{\pi}{2} - 2\pi \quad \therefore \quad 4x - \frac{\pi}{2} = -3\pi \quad \therefore \quad x = -\frac{5\pi}{8} \in [-\pi, \pi] \checkmark$$

$$C : \quad \frac{3x - x + \frac{\pi}{2}}{2} = -2\pi \quad \therefore \quad 2x + \frac{\pi}{2} = -4\pi \quad \therefore \quad x = -\frac{9\pi}{2} < -\pi \; \boldsymbol{X}$$

In conclusion, within the interval $-\pi < x \leq \pi$, there are six solutions (see Fig. 5.12 (right)),

$$x = \left\{ -\frac{5\pi}{8}, -\frac{2\pi}{8}, -\frac{\pi}{8}, \frac{3\pi}{8}, \frac{6\pi}{8}, \frac{7\pi}{8} \right\}$$

Fig. 5.12
Example 5.5-3(a)(b)

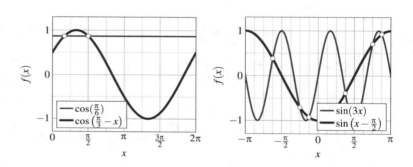

Exercise 5.6, page 71

1. Inequalities are solved by calculating the limits of intervals.

(a)   We convert given equation and use unit circle as follows

$$2 \cos x + \sqrt{3} < 0 \quad \therefore \quad \cos x < -\frac{\sqrt{3}}{2}$$

In other words, there are two angles, see Fig. 5.13 (right), that satisfy equality $x_1 = 5\pi/6$ and $x_2 = -5\pi/6$. Marked region is where $-1 \leq \cos x < -\sqrt{3}/2$ (note that inequality $\cos x < -1$ does not have real solution, i.e. amplitude of $-1 \leq \sin x, \cos x \leq 1$). Therefore, we conclude that periodic solution is the interval, see Fig. 5.13,

$$-\frac{5\pi}{6} + 2n\pi < x < \frac{5\pi}{6} + 2n\pi \quad \text{where} \quad n = 0, \pm 1, \pm 2, \ldots$$

Fig. 5.13 Example 5.6-1(a)

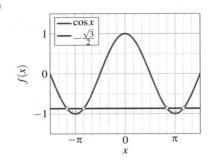 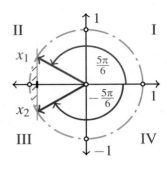

(b) Using the same technique as in (a) we write,

$$2\sin x - 1 > 0 \quad \therefore \quad \sin x > \frac{1}{2} \quad \therefore \quad \frac{\pi}{6} + 2n\pi < x < \frac{5\pi}{6} + 2n\pi$$

$$\therefore$$

$$\text{where } n = 0, \pm 1, \pm 2, \ldots$$

as illustrated in Fig. 5.14.

Fig. 5.14 Example 5.6-1(b)

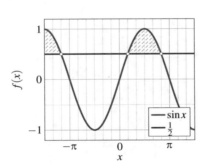 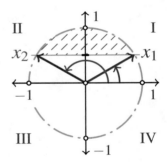

(c) The idea is to convert given inequality that contains multiple trigonometric functions into its equivalent form where all trigonometric functions are combined into a single either $\sin(f(x))$ or $\cos(f(x))$. For example, we can enforce the sum of two arguments form as

$$\cos x - \sin x < 1 \quad \left| \times \frac{\sqrt{2}}{2} \right.$$

$$\frac{\sqrt{2}}{2} \cos x - \frac{\sqrt{2}}{2} \sin x < \frac{\sqrt{2}}{2}$$

$$\sin \frac{\pi}{4} \cos x - \cos \frac{\pi}{4} \sin x < \frac{\sqrt{2}}{2}$$

$$\therefore$$

$$\sin \left(\frac{\pi}{4} - x \right) < \frac{\sqrt{2}}{2}$$

From unit circle we know that $\sin x = \sqrt{2}/2 \Rightarrow x = \pi/4$ or $x = -5\pi/4$, see Fig. 5.15. Therefore we conclude that

$$\frac{\pi}{4} - x < \frac{\pi}{4} \quad \therefore \quad x > 0 + 2n\pi, \quad n = 0, \pm 1, \pm 2, \ldots$$

and

$$\frac{\pi}{4} - x > -\frac{5\pi}{4} \quad \therefore \quad x < \frac{3\pi}{2} + 2n\pi, \quad n = 0, \pm 1, \pm 2, \ldots$$

Fig. 5.15 Example 5.6-1(c)

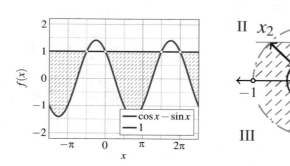

(d)   Knowing that periodicity of $\sin 3x$ is $2\pi/3$, and from unit circle we find that if $\sin x = \sqrt{3}/2$ then $x = \pi/3$ or $x = 2\pi/3$. Thus, we solve given inequality as follows.

$$\sin 3x - \frac{\sqrt{3}}{2} \geq 0 \quad \therefore \quad \sin 3x \geq \frac{\sqrt{3}}{2}$$

$$\therefore$$

$$3x \geq \frac{\pi}{3} \quad \therefore \quad x \geq \frac{\pi}{9} + \frac{2n\pi}{3}, \quad n = 0, \pm 1, \pm 2, \ldots$$

or,

$$3x \le \frac{2\pi}{3} \quad \therefore \quad x \le \frac{2\pi}{9} + \frac{2n\pi}{3}, \quad n = 0, \pm 1, \pm 2, \ldots$$

(e)   Inequalities that include absolute functions are equivalent to two inequalities, thus define and interval.

$$|\sin x| \ge \frac{1}{2} \quad \therefore \quad -\frac{1}{2} \ge \sin x \ge \frac{1}{2}$$

By inspection of Fig. 5.16 we write the solutions as

$$-\frac{5\pi}{6} + 2n\pi \le x \le -\frac{\pi}{6} + 2n\pi$$

and,

$$\frac{\pi}{6} + 2n\pi \le x \le \frac{5\pi}{6} + 2n\pi$$

where $n = 0, \pm 1, \pm 2, \ldots$

Fig. 5.16 Example 5.6-1(e)

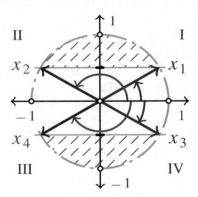

(f)   We work out absolute values by separating them into two cases.

Case 1: $\sin x > 0 \quad \therefore \quad |\sin x| = \sin x$, thus we find the interval's boundary as

$$\cancel{\sin x} = \cancel{\sin x} + 2 \cos x \quad \therefore \quad 2 \cos x = 0 \quad \therefore \quad x = \frac{\pi}{2} + n\pi$$

However, the initial condition is that $\sin x > 0$, which is valid only for

$$x = \frac{\pi}{2} + 2n\pi, \quad \text{where} \quad n = 0, \pm 1, \pm 2, \ldots$$

Case 2: $\sin x < 0$ $\therefore$ $|\sin x| = -\sin x$, thus we find the interval's boundary as

$$-\sin x = \sin x + 2\cos x \quad \therefore \quad 2\sin x + 2\cos x = 0 \quad \therefore \quad \sin x + \cos x = 0$$

We solve this equation using same technique as in (c),

$$\sin x + \cos x = 0 \quad \bigg|\times \frac{\sqrt{2}}{2}$$

$$\frac{\sqrt{2}}{2}\sin x + \frac{\sqrt{2}}{2}\cos x = 0$$
$$\cos \frac{\pi}{4}\sin x + \sin \frac{\pi}{4}\cos x = 0$$

$$\therefore$$

$$\sin \left(x + \frac{\pi}{4}\right) = 0 \quad \therefore \quad x + \frac{\pi}{4} = 0 \quad \therefore \quad x = -\frac{\pi}{4} + n\pi$$

However, the initial condition is that $\sin x < 0$, which is valid only for

$$x = -\frac{\pi}{4} + 2n\pi, \quad \text{where} \quad n = 0, \pm 1, \pm 2, \ldots$$

The two streams of periodic solutions are illustrated in Fig. 5.17. We conclude that x found in the periodic interval

$$-\frac{\pi}{4} + 2n\pi \leq x \leq \frac{\pi}{2} + 2n\pi$$

satisfy given inequality.

Fig. 5.17 Example 5.6-1(f)

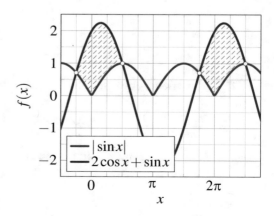

Complex Algebra

6

Important to Know

$$z = a + ib \quad \therefore \quad z^* = a - ib \quad \therefore \quad |z|^2 = z\,z^* = a^2 + b^2$$

$$z = |z|e^{i\theta} = |z|\,(\cos\theta + i\sin\theta) \quad \therefore \quad \theta = \arctan\frac{b}{a}$$

$$z_1 = a + ib \ \text{ and } \ z_2 = c + id \quad \therefore \quad \text{if } z_1 = z_2 \Rightarrow a = c \ \text{ and } \ b = d$$

Geometrical interpretation of the equivalence between Pythagorean triangle, complex numbers, and vector addition, see Fig. 6.1.

Fig. 6.1 Pythagorean triangle, vector addition, and complex plane

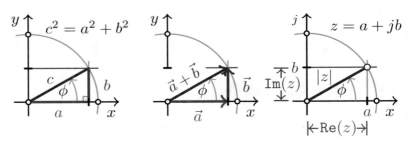

6.1 Exercises

6.1 * Complex Numbers

1. Calculate $|z|$ and $\angle(z)$,

(a) $z = j$

(b) $z = j^2$

(c) $z = j^3$

(d) $z = j^4$

(e) $z = j^5$

(f) $z = j^{2020}$

© Springer Nature Switzerland AG 2021
R. Sobot, *Engineering Mathematics by Example*,
https://doi.org/10.1007/978-3-030-79545-0_6

2. Calculate $|z|$ and $\angle(z)$ in each case and show the following complex numbers z as vectors in the complex plane.

(a) $z = 1$

(b) $z = \dfrac{\sqrt{3}}{2} + j\dfrac{1}{2}$

(c) $z = \dfrac{\sqrt{2}}{2} + j\dfrac{\sqrt{2}}{2}$

(d) $z = \dfrac{1}{2} + j\dfrac{\sqrt{3}}{2}$

(e) $z = j$

(f) $z = -\dfrac{\sqrt{2}}{2} + j\dfrac{\sqrt{2}}{2}$

(g) $z = -\dfrac{\sqrt{3}}{2} + j\dfrac{1}{2}$

(h) $z = -1$

(i) $z = -\dfrac{\sqrt{2}}{2} - j\dfrac{\sqrt{2}}{2}$

(j) $z = -j$

(k) $z = \dfrac{\sqrt{2}}{2} - j\dfrac{\sqrt{2}}{2}$

(l) $z = \dfrac{\sqrt{3}}{2} - j\dfrac{1}{2}$

3. Show the following complex numbers as vectors in the complex plane. Then calculate $|z|$ and $\angle(z)$ in each case.

(a) $z = e^{j\,0}$

(b) $z = e^{j\,\frac{\pi}{6}}$

(c) $z = e^{j\,\frac{\pi}{4}}$

(d) $z = e^{j\,\frac{\pi}{3}}$

(e) $z = e^{j\,\frac{\pi}{2}}$

(f) $z = e^{j\,\frac{3\pi}{4}}$

(g) $z = e^{j\,\frac{5\pi}{6}}$

(h) $z = e^{j\,\pi}$

(i) $z = e^{j\,\frac{5\pi}{4}}$

(j) $z = e^{j\,\frac{3\pi}{2}}$

(k) $z = e^{j\,\frac{7\pi}{4}}$

(l) $z = e^{j\,\frac{11\pi}{6}}$

4. Calculate $\angle(z)$ and $|z|$ of the following complex numbers.

(a) $z = 2\,e^{j\,\frac{\pi}{4}}$

(b) $z = \dfrac{1}{2}\,e^{j\,\frac{\pi}{3}}$

(c) $z = -2\,e^{-j\,\frac{5\pi}{6}}$

(d) $z = -\dfrac{1}{2}\,e^{j\,\frac{7\pi}{4}}$

(e) $z = \left(\sqrt{2}\,e^{j\,\frac{\pi}{4}}\right)^2$

(f) $z = \left(3\,e^{j\,\frac{\pi}{3}}\right)^3$

(g) $z = \left(4\,e^{-j\,\frac{5\pi}{6}}\right)^5$

(h) $z = \left(0.2\,e^{j\,\frac{7\pi}{4}}\right)^4$

(i) $z = \left(-3\,e^{-j\,\frac{\pi}{6}}\right)^3$

6.2 *   Basic Calculations

1. Given that $z = i$, calculate the following complex numbers and show each result in the complex plane:

(a) $z_1 = z^2$

(b) $z_2 = z^3$

(c) $z_3 = z^4$

(d) $z_4 = z^5$

(e) $z_5 = \sqrt{z}$

(f) $z_6 = \sqrt[3]{z}$

(g) $z_7 = \sqrt[4]{z}$

(h) $z_8 = \sqrt[5]{z}$

2. Given that $z = 2i$, calculate the following complex numbers and show each result in the complex plane:

(a) $z_1 = z^2$

(b) $z_2 = z^3$

(c) $z_3 = z^4$

(d) $z_4 = z^5$

3. Calculate the following:

(a) $i^2 + i^3 + i^4$

(b) $i^{-5} + i^{-17} + i^{36}$

(c) $(1+i)^4 + (1-i)^4$

(d) $i^5 + i^{-4} + i^{121}$

4. Calculate the following:

(a) $(3+4i)(3-4i)$

(b) $i^{125} + (-i)^{60} + i^{83}$

(c) $(2i)^2 + (-2i)^{-4}$

(d) $\dfrac{1}{i} + \left(\dfrac{1+i}{\sqrt{2}}\right)^4$

(e) $\left(\dfrac{1+i}{\sqrt{2}}\right)^{1000} + \left(\dfrac{1-i}{\sqrt{2}}\right)^{1000}$

(f) $\left(\dfrac{1+i\sqrt{3}}{2}\right)^{3000} + \left(\dfrac{1-i\sqrt{3}}{2}\right)^{3000}$

5. Calculate the following:

(a) $(1-i)^{100}$

(b) $(1+i)^{50}$

(c) $(2-i)^6$

(d) $\dfrac{1+i\sqrt{3}}{2}\dfrac{1-i\sqrt{3}}{2}$

(e) $\dfrac{(1+i)^{1000}}{(1-i)^{500}}$

(f) $\left(\dfrac{4}{i\sqrt{3}-1}\right)^{12}$

(g) $\dfrac{i^{102} + i^{101}}{i^{100} - i^{99}}$

(h) $\dfrac{(i\sqrt{3}-1)^2}{(2i)^2}$

(i) $\dfrac{-41+63i}{50} - \dfrac{6i+1}{1-7i}$

6. Express z in polar form:

(a) $z = i$

(b) $z = 2 - i$

(c) $z = \dfrac{(1+i)^{200}(6+2i) - (1-i)^{198}(3-i)}{(1+i)^{196}(23-7i) + (1-i)^{194}(10+2i)}$

6.3 *   Complex Equations

1. Solve the following equations in $\mathbb{C}$:

(a) $(2+3i)x + (3+2i)y = 1$

(b) $\dfrac{1}{x+iy} = \dfrac{1}{2+i} + \dfrac{1}{-2+4i}$

2. Solve the following equations for z in $\mathbb{C}$:

(a) $z^2 = i$

(b) $z^2 = -i$

(c) $z^2 = 5 + 12i$

3. Solve the equation $z^4 + 8 + 8\sqrt{3}\,i = 0$ for z in $\mathbb{C}$.

Solutions

Exercise 6.1, page 105

1. We note that powers of a complex number in effect force z to rotate around the circle, therefore there is periodicity in the results. Calculate $|z|$ and $\angle(z)$,

(a) $z = j$ therefore, $|z| = 1$,   and $\angle z = \pi/2$

(b) $z = j^2 = -1$ therefore, $|z| = 1$,   and $\angle z = \pi$

(c) $z = j^3 = j^2 \times j = -j$ therefore, $|z| = 1$,   and $\angle z = 3\pi/2$

(d) $z = j^4 = j^2 \times j^2 = 1$ therefore, $|z| = 1$,   and $\angle z = 2\pi = 0$

(e) $z = j^5 = j^4 \times j = j$ therefore, $|z| = 1$,   and $\angle z = \pi/2$

(f)
$$z = j^{2020} = j^{4 \times 505} = \left(j^4\right)^{505} = 1^{505} = 1$$

$$\text{therefore,} \quad |z| = 1, \quad \text{and} \quad \angle z = 2\pi = 0$$

2. Here, complex number modules are found simply by Pythagoras's theorem, i.e. $|z| = \sqrt{\Re(z)^2 + \Im(z)^2}$.

(a)   Given: $z = 1$, we write

$$\left.\begin{array}{l} \Re(z) = 1 > 0 \\[6pt] \Im(z) = 0 \end{array}\right\} \text{(I)}$$

then (Fig. 6.2), module is

$$|z| = \sqrt{1^2 + 0^2} = 1$$

and phase is

$$\angle z = \arctan \frac{\Im(z)}{\Re(z)} = \arctan \frac{0}{1}$$

$$= \arctan(0) = 0°$$

(b)   Given: $z = \sqrt{3}/2 + j^{1}/2$, we write

$$\left.\begin{array}{l} \Re(z) = \sqrt{3}/2 > 0 > 0 \\[6pt] \Im(z) = 1/2 > 0 \end{array}\right\} \text{(I)}$$

then (Fig. 6.3), module is

$$|z| = \sqrt{\left(\sqrt{3}/2\right)^2 + (1/2)^2} = 1$$

$$\angle z = \arctan \frac{\Im(z)}{\Re(z)} = \arctan \frac{1/2}{\sqrt{3}/2}$$

$$= 30° = \frac{\pi}{6}$$

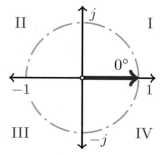

Fig. 6.2 Example 6.1-2(a)

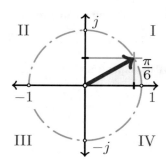

Fig. 6.3 Example 6.1-2(b)

(c)   Given: $z = \sqrt{2}/2 + j\sqrt{2}/2$, we write

$$\left.\begin{array}{l} \Re(z) = \sqrt{2}/2 > 0 \\[6pt] \Im(z) = \sqrt{2}/2 > 0 \end{array}\right\} \text{ (I)}$$

then (Fig. 6.4), module is

$$|z| = \sqrt{\left(\sqrt{2}/2\right)^2 + \left(\sqrt{2}/2\right)^2} = 1$$

$$\angle z = \arctan\frac{\Im(z)}{\Re(z)} = \arctan\frac{\frac{\sqrt{2}}{2}}{\frac{\sqrt{2}}{2}}$$

$$= \arctan(1) = 45° = \frac{\pi}{4}$$

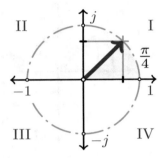

(d)   Given: $z = 1/2 + j\sqrt{3}/2$, we write

$$\left.\begin{array}{l} \Re(z) = 1/2 > 0 \\[6pt] \Im(z) = \sqrt{3}/2 > 0 \end{array}\right\} \text{ (I)}$$

then (Fig. 6.5), module is

$$|z| = \sqrt{(1/2)^2 + \left(\sqrt{3}/2\right)^2} = 1$$

$$\angle z = \arctan\frac{\Im(z)}{\Re(z)} = \arctan\frac{\frac{\sqrt{3}}{2}}{\frac{1}{2}}$$

$$= \arctan(\sqrt{3}) = 60° = \frac{\pi}{3}$$

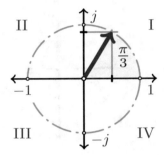

Fig. 6.4   Example 6.1-2(c)

Fig. 6.5   Example 6.1-2(d)

(e)   Given: $z = j$, we write

$$\left.\begin{array}{l} \Re(z) = 0 \\ \Im(z) = 1 > 0 \end{array}\right\} \text{(I)}$$

then (Fig. 6.6), module is

$$|z| = \sqrt{(0)^2 + (1)^2} = 1$$

$$\angle z = \arctan\frac{\Im(z)}{\Re(z)} = \arctan\frac{1}{0}$$

$$= \arctan(\infty) = 90° = \frac{\pi}{2}$$

(f)   Given: $z = -\sqrt{2}/2 + j\sqrt{2}/2$, we write

$$\left.\begin{array}{l} \Re(z) = -\sqrt{2}/2 < 0 \\ \Im(z) = \sqrt{2}/2 > 0 \end{array}\right\} \text{(II)}$$

then (Fig. 6.7), module is

$$|x| = \sqrt{\left(-\sqrt{2}/2\right)^2 + \left(\sqrt{2}/2\right)^2} = 1$$

$$\angle z = \arctan\frac{\Im(z)}{\Re(z)} = \arctan\frac{\frac{\sqrt{2}}{2}}{-\frac{\sqrt{2}}{2}}$$

$$= \arctan\left(\frac{1}{-1}\right) = 135° = \frac{3\pi}{4}$$

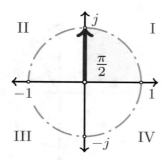

Fig. 6.6 Example 6.1-2(e)

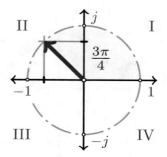

Fig. 6.7 Example 6.1-2(f)

(g) Given: $z = -\sqrt{3}/2 + j\,1/2$, we write

$$\left.\begin{array}{l}\Re(z) = -\sqrt{3}/2 < 0 \\[4pt] \Im(z) = 1/2 > 0\end{array}\right\} \text{(II)}$$

then (Fig. 6.8), module is

$$|x| = \sqrt{(1/2)^2 + \left(-\sqrt{3}/2\right)^2} = 1$$

$$\angle z = \arctan\frac{\Im(z)}{\Re(z)} = \arctan\frac{\frac{1}{2}}{-\frac{\sqrt{3}}{2}}$$

$$= \arctan\left(1/-\sqrt{3}\right) = 150° = \frac{5\pi}{6}$$

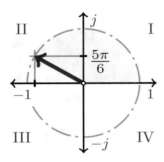

Fig. 6.8 Example 6.1-2(g)

(h) Given: $z = -1$, we write

$$\left.\begin{array}{l}\Re(z) = -1 < 0 \\[4pt] \Im(z) = 0\end{array}\right\} \text{(II)}$$

then (Fig. 6.9), module is

$$|x| = \sqrt{(-1)^2 + (0)^2} = 1$$

$$\angle z = \arctan\frac{\Im(z)}{\Re(z)} = \arctan\frac{0}{-1}$$

$$= 180° = \pi$$

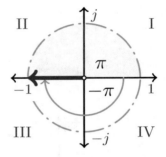

Fig. 6.9 Example 6.1-2(h)

(i)   Given: $z = -\sqrt{2}/2 - j\sqrt{2}/2$, we write
$$\left.\begin{array}{l} \Re(z) = -\sqrt{2}/2 < 0 \\[2mm] \Im(z) = -\sqrt{2}/2 < 0 \end{array}\right\} \text{(III)}$$
then (Fig. 6.10), module is

$$|x| = \sqrt{\left(-\sqrt{2}/2\right)^2 + \left(-\sqrt{2}/2\right)^2} = 1$$

$$\angle z = \arctan \frac{\Im(z)}{\Re(z)} = \arctan \frac{-\sqrt{2}/2}{-\sqrt{2}/2}$$

$$= \arctan\left(\frac{-1}{-1}\right) = 225° = -135°$$

$$= \frac{5\pi}{4} = -\frac{3\pi}{4}$$

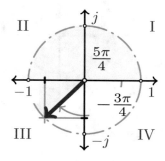

Fig. 6.10 Example 6.1-2(i)

(j)   Given: $z = -j$, we write
$$\left.\begin{array}{l} \Re(z) = 0 \\[2mm] \Im(z) = -1 < 0 \end{array}\right\} \text{(III)}$$
then (Fig. 6.11), module is

$$|x| = \sqrt{(0)^2 + (-1)^2} = 1$$

$$\angle z = \arctan \frac{\Im(z)}{\Re(z)} = \arctan \frac{-1}{0}$$

$$= 270° = -90° = \frac{3\pi}{2} = -\frac{\pi}{2}$$

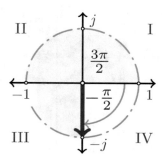

Fig. 6.11 Example 6.1-2(j)

(k) Given: $z = \sqrt{2}/2 - j\sqrt{2}/2$, we write
$$\left.\begin{array}{l} \Re(z) = \sqrt{2}/2 > 0 \\[4pt] \Im(z) = -\sqrt{2}/2 < 0 \end{array}\right\} \text{(IV)}$$
then (Fig. 6.12), module is

$$|x| = \sqrt{\left(\sqrt{2}/2\right)^2 + \left(-\sqrt{2}/2\right)^2} = 1$$

$$\angle z = \arctan \frac{\Im(z)}{\Re(z)} = \arctan \frac{-\sqrt{2}/2}{\sqrt{2}/2}$$

$$= \arctan\left(\frac{-1}{1}\right) = 315° = -45°$$

$$= \frac{7\pi}{4} = -\frac{\pi}{4}$$

(l) Given: $z = \sqrt{3}/2 - j1/2$, we write
$$\left.\begin{array}{l} \Re(z) = \sqrt{3}/2 > 0 \\[4pt] \Im(z) = -1/2 < 0 \end{array}\right\} \text{(IV)}$$
then (Fig. 6.13), module is

$$|x| = \sqrt{\left(\sqrt{3}/2\right)^2 + \left(-1/2\right)^2} = 1$$

$$\angle z = \arctan \frac{\Im(z)}{\Re(z)} = \arctan \frac{-1/2}{\sqrt{3}/2}$$

$$= \arctan \frac{-1}{\sqrt{3}} = 330° = -30°$$

$$= \frac{11\pi}{6} = -\frac{\pi}{6}$$

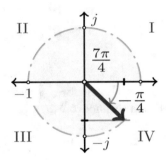

Fig. 6.12 Example 6.1-2(k)

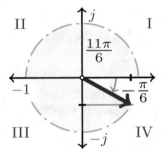

Fig. 6.13 Example 6.1-2(l)

3. Complex numbers in this example given in exponential form are identical to those in Example 6.2-2. Thus, by exploiting Euler's equation, phase and module of a complex number are found by inspection.

Aside from Pythagorean theorem used to calculate $|z|$ it is also important to know how to calculate $|z|$ in the exponential form,

$$|z|^2 = z\,z^* = e^{j\frac{\pi}{6}}\,e^{-j\frac{\pi}{6}} = e^{j\frac{\pi}{6} - j\frac{\pi}{6}}$$

$$= e^0 = 1 \quad \therefore \quad |z| = 1$$

which illustrates that module of any complex number found on the unity circle consequently equals one. That is to say, complex exponent by itself keeps only the phase, while numbers whose module $|z| = A \neq 1$ are written in the form

$$z = A\,e^{j\phi} \quad \therefore \quad |z| = \left| A\,e^{j\phi} \right|$$

$$= |A| \left| e^{j\phi} \right|^1$$

where A is any real number.

(a) $z = e^{j\,0}$, therefore, $\phi = 0\,\text{rad}$, its vector graph is in Example 6.2-2(a):

$$\Re(z) = \cos 0 = 1$$

$$\Im(z) = \sin 0 = 0$$

$$|z| = \sqrt{\Re(z)^2 + \Im(z)^2} = 1$$

(b) $z = e^{j\frac{\pi}{6}}$ therefore, $\phi = \pi/6$, its vector graph is in Example 6.2-2(b):

$$\Re(z) = \cos \pi/6 = \sqrt{3}/2$$

$$\Im(z) = \sin \pi/6 = 1/2$$

$$|z| = \sqrt{\Re(z)^2 + \Im(z)^2} = 1$$

(c) $z = e^{j\frac{\pi}{4}}$ therefore, $\phi = \pi/4$, its vector graph is in Example 6.2-2(a):

$$\Re(z) = \cos \pi/4 = \sqrt{2}/2$$

$$\Im(z) = \sin \pi/4 = \sqrt{2}/2$$

$$|z|^2 = z\,z^* = e^{j\frac{\pi}{4}}\,e^{-j\frac{\pi}{4}} = e^0 = 1$$

(d) $z = e^{j\frac{\pi}{3}}$ therefore, $\phi = \pi/3$, its vector graph is in Example 6.2-2(d):

$$\Re(z) = \cos \pi/3 = 1/2$$

$$\Im(z) = \sin \pi/3 = \sqrt{3}/2$$

$$|z|^2 = z\,z^* = e^{j\left(\frac{\pi}{3} - \frac{\pi}{3}\right)} = e^0 = 1$$

(e) $z = e^{j\frac{\pi}{2}}$ therefore, $\phi = \pi/2$, its vector graph is in Example 6.2-2(e):

$$\Re(z) = \cos \pi/2 = 0$$

$$\Im(z) = \sin \pi/3 = 1$$

$$|z|^2 = z\,z^* = e^{j\left(\frac{\pi}{2} - \frac{\pi}{2}\right)} = e^0 = 1$$

(f) $z = e^{j\frac{3\pi}{4}}$ therefore, $\phi = 3\pi/4$, its vector graph is in Example 6.2-2(f):

$$\Re(z) = \cos 3\pi/4 = -\sqrt{2}/2$$

$$\Im(z) = \sin 3\pi/4 = \sqrt{2}/2$$

$$|z|^2 = z\,z^* = e^{j\left(\frac{3\pi}{4} - \frac{3\pi}{4}\right)} = e^0 = 1$$

(g) $z = e^{j\frac{5\pi}{6}}$ therefore, $\phi = 5\pi/6$, its vector graph is in Example 6.2-2(g):

$$\Re(z) = \cos 5\pi/6 = -\sqrt{3}/2$$

$$\Im(z) = \sin 5\pi/6 = 1/2$$

$$|z|^2 = z\,z^* = e^{j\left(\frac{5\pi}{6} - \frac{5\pi}{6}\right)} = e^0 = 1$$

(h) $z = e^{j\,\pi}$ therefore, $\phi = \pi$, its vector graph is in Example 6.2-2(h):

$$\Re(z) = \cos \pi = -1$$

$$\Im(z) = \sin \pi = 0$$

$$|z|^2 = z\,z^* = e^{j\,(\pi - \pi)} = e^0 = 1$$

(i) $z = e^{j\frac{5\pi}{4}}$ therefore, $\phi = 5\pi/4 = -3\pi/4$, its vector graph is in Example 6.2-2(i):

$$\Re(z) = \cos 5\pi/4 = -\sqrt{2}/2$$

$$\Im(z) = \sin 5\pi/4 = -\sqrt{2}/2$$

$$|z|^2 = z\,z^* = e^{j\left(\frac{5\pi}{4} - \frac{5\pi}{4}\right)} = e^0 = 1$$

(j) $z = e^{j\frac{3\pi}{2}}$ therefore, $\phi = 3\pi/2 = -\pi/2$, its vector graph is in Example 6.2-2(j):

$$\Re(z) = \cos 3\pi/2 = 0$$

$$\Im(z) = \sin 3\pi/2 = -1$$

$$|z|^2 = z\,z^* = e^{j\left(\frac{3\pi}{2} - \frac{3\pi}{2}\right)} = e^0 = 1$$

(k) $z = e^{j\frac{7\pi}{4}}$ therefore, $\phi = 7\pi/4 = -\pi/4$, its vector graph is in Example 6.2-2(k):

$$\Re(z) = \cos 7\pi/4 = \sqrt{2}/2$$

$$\Im(z) = \sin 7\pi/4 = -\sqrt{2}/2$$

$$|z|^2 = z\,z^* = e^{j\left(\frac{7\pi}{4} - \frac{7\pi}{4}\right)} = e^0 = 1$$

(l) $z = e^{j\frac{11\pi}{6}}$ therefore, $\phi = 11\pi/6 = -\pi/6$, its vector graph is in Example 6.2-2(l):

$$\Re(z) = \cos 11\pi/6 = \sqrt{3}/2$$

$$\Im(z) = \sin 11\pi/6 = -1/2$$

$$|z|^2 = z\,z^* = e^{j\left(\frac{11\pi}{6} - \frac{11\pi}{6}\right)} = e^0 = 1$$

4. By exploiting Euler's exponential form of a complex number, phase and module are found as:

(a) $z = 2\,e^{j\frac{\pi}{4}}$ therefore, $\phi = \pi/4$ and:

$$|z|^2 = z\,z^* = 2e^{j\frac{\pi}{4}}\,2e^{-j\frac{\pi}{4}} = 4e^0$$

$$|z| = 2$$

(b) $z = 1/2\,e^{j\frac{\pi}{3}}$ therefore, $\phi = \pi/3$ and $|z| = 1/2$

(c) $z = -2\,e^{-j\frac{5\pi}{6}}$ therefore, $\phi = -5\pi/6$ and $|z| = 2$

(d) $z = -1/2\,e^{j\frac{7\pi}{4}}$ therefore, $\phi = 7\pi/4 = -\pi/4$ and $|z| = 1/2$

In the following examples, we note that powers of a complex number, in effect, aside from increasing its module also *rotate* the associated vector by multiples of the original phase.

(e) First, we must calculate the exponential form of z as,

$$z = \left(\sqrt{2}\,e^{j\frac{\pi}{4}}\right)^2$$

$$= \left(\sqrt{2}\right)^2\left(e^{j\frac{\pi}{4}}\right)^2$$

$$= 2\,e^{j\frac{\pi}{4}\times 2} = 2\,e^{j\frac{\pi}{2}}$$

therefore, $\phi = \pi/2$ and $|z| = 2$.

(f) First, we must calculate the exponential form of z as,

$$z = \left(3\,e^{j\frac{\pi}{3}}\right)^3$$

$$= 9\,e^{j\pi} \overset{-1}{} = 9\,(-1) = -9$$

therefore, $\phi = \pi$ and $|z| = 9$. We note that this complex number rotated horizontal position, thus into a real number.

(g) First, we must calculate the exponential form of z as,

$$z = \left(4\,e^{j\frac{5\pi}{6}}\right)^5 = 4^5\,e^{j\frac{25\pi}{6}}$$

$$= 2^{10}\,e^{j\frac{\pi}{6}} = 1024\,e^{j\frac{\pi}{6}}$$

therefore, after making four turns around the unity circle $\phi = \pi/6$, and $|z| = 1024$.

(h) First, we must calculate the exponential form of z as,

$$z = \left(0.2\,e^{j\frac{7\pi}{4}}\right)^4 = 0.2^4\,e^{j7\pi}$$

$$= 5^{-4}\,e^{j\pi} = 1.6 \times 10^{-3}\,(-1)$$

$$= -1.6 \times 10^{-3}$$

therefore, after making two turns around the unity circle $\phi = \pi$ the complex number rotated into its position, thus into a real number, and $|z| = 1.6 \times 10^{-3}$ (in engineering units).

(i) First, we must calculate the exponential form of z as,

$$z = \left(-3\,e^{-j\frac{\pi}{6}}\right)^3 = (-3)^3\,e^{-j\frac{\pi}{2}}$$

$$= -27\,(-j) = 27\,j$$

therefore, $\phi = -\pi/2$ and $|z| = 27$.

Exercise 6.2, page 106

1. We use the equivalent forms or roots and fractional powers in the exponential form of a complex number. Therefore, $z = i = e^{i\pi/2} = \cos\pi/2 + i\,\sin\pi/2$.

(a) $z^2 = i^2 = \left(e^{i\frac{\pi}{2}}\right)^2 = e^{i\pi} = \left(\cos\pi + i\,\underset{0}{\underbrace{\sin\pi}}\right) = -1$

(b) $z^3 = i^3 = \left(e^{i\frac{\pi}{2}}\right)^3 = e^{i\frac{3\pi}{2}} = \left(\underset{0}{\underbrace{\cos\frac{3\pi}{2}}} + i\,\sin\frac{3\pi}{2}\right) = -i$

(c) $z^4 = i^4 = \left(e^{i\frac{\pi}{2}}\right)^4 = e^{i\,2\pi} = \left(i^2\right)^2 = (-1)^2 = 1$

(d) $z^5 = i^5 = i \times i^4 = i$ $\therefore$ $i^6 = \underset{-1}{\underbrace{i^2}} \times \underset{1}{\underbrace{i^4}} = -1$, etc.

(e) $\sqrt{z} = \sqrt{i} = \left(e^{i\frac{\pi}{2}}\right)^{\frac{1}{2}} = e^{i\frac{\pi}{4}} = \left(\cos\frac{\pi}{4} + i\,\sin\frac{\pi}{4}\right) = \frac{\sqrt{2}}{2} + i\,\frac{\sqrt{2}}{2}$

(f) $\sqrt[3]{z} = \sqrt[3]{i} = \left(e^{i\frac{\pi}{2}}\right)^{\frac{1}{3}} = e^{i\frac{\pi}{6}} = \left(\cos\frac{\pi}{6} + i\,\sin\frac{\pi}{6}\right) = \frac{\sqrt{3}}{2} + i\,\frac{1}{2}$

(g) $\sqrt[4]{z} = \sqrt[4]{i} = \left(e^{i\frac{\pi}{2}}\right)^{\frac{1}{4}} = e^{i\frac{\pi}{8}} = \left(\cos\frac{\pi}{8} + i\,\sin\frac{\pi}{8}\right)$

(h) $\sqrt[5]{z} = \sqrt[5]{i} = \left(e^{i\frac{\pi}{2}}\right)^{\frac{1}{5}} = e^{i\frac{\pi}{10}} = \left(\cos\frac{\pi}{10} + i\ \sin\frac{\pi}{10}\right)$

Fig. 6.14 Example 6.2-1

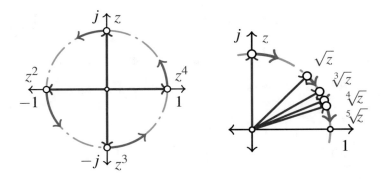

We note that powers of a complex number whose $|z| = 1$ effectively rotate its vector, Fig. 6.14.

2. Radical and fractional power forms are equivalent, thus the exponential form of a complex number results in $z = 2i = 2e^{i\pi/2} = 2\left(\cos{\pi/2} + i\ \sin{\pi/2}\right)$.

(a) $z^2 = (2i)^2 = \left(2e^{i\frac{\pi}{2}}\right)^2 = 4e^{i\pi} = 4\left(\cos\pi + i\ \underset{0}{\sin\pi}\right) = -4$

(b) $z^3 = (2i)^3 = \left(2e^{i\frac{\pi}{2}}\right)^3 = 8e^{i\frac{3\pi}{2}} = 8\left(\underset{0}{\cos\frac{3\pi}{2}} + i\ \sin\frac{3\pi}{2}\right) = -8i$

(c) $z^4 = (2i)^4 = \left(2e^{i\frac{\pi}{2}}\right)^4 = 16e^{i\,2\pi} = 16\left(i^2\right)^2 = 16(-1)^2 = 16$

(d) $z^5 = (2i)^5 = 2^5 \times i \times i^4 = 32\,i$

We note that powers of a complex number whose $|z| \neq 1$ effectively rotate its vector and increase its module, thus create spiral function, Fig. 6.15.

Fig. 6.15 Example 6.2-2(d)

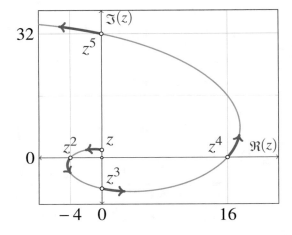

3. By definition of the imaginary number $i^2 = -1$ we find

(a) $i^2 + i^3 + i^4 = i^2 + i \times i^2 + \left(i^2\right)^2 = i^2 + i \times (-1) + (-1)^2$

$$= -1 - i + 1 = -i$$

(b) $i^{-5} + i^{-17} + i^{36} = \dfrac{1}{i^5} + \dfrac{1}{i^{17}} + i^{36} = \dfrac{1}{i \times i\!\!\!/^4} + \dfrac{1}{i \times \left(i\!\!\!/^4\right)^4} + \left(i\!\!\!/^4\right)^9$

$$= \dfrac{2}{i} + 1 = \dfrac{i}{i}\dfrac{2}{i} + 1 = 1 - 2i$$

(c) $(1+i)^4 + (1-i)^4 = \left[(1+i)^2\right]^2 + \left[(1-i)^2\right]^2$

$$= (\cancel{1} + 2i - \cancel{1})^2 + (\cancel{1} - 2i - \cancel{1})^2 = 4i^2 + 4i^2$$

$$= -4 - 4 = -8$$

(d) $i^5 + i^{-4} + i^{121} = i \times i\!\!\!/^4 + \dfrac{1}{i\!\!\!/^4} + i \times \left(i\!\!\!/^4\right)^{30} = i + 1 + i = 1 + 2i$

4. By definition we find

(a) $(3 + 4i)(3 - 4i) = 3^2 - (4i)^2 = 9 + 16 = 25$

(b) $i^{125} + (-i)^{60} + i^{83} = i \times \left(i\!\!\!/^4\right)^{31} + (-1)^{60}\left(i\!\!\!/^4\right)^{15} + i \times i^2 \times \left(i\!\!\!/^4\right)^{20}$

$$= i\!\!\!/ + 1 - i\!\!\!/ = 1$$

(c) $(2i)^2 + (-2i)^{-4} = -4 + \dfrac{1}{(2i)^4} = -4 + \dfrac{1}{16\,i\!\!\!/^4} = -\dfrac{63}{16}$

(d) $\dfrac{1}{i} + \left(\dfrac{1+i}{\sqrt{2}}\right)^4 = \dfrac{i}{i}\dfrac{1}{i} + \dfrac{\left[(1+i)^2\right]^2}{2^2} = -i + \dfrac{(2i)^2}{4} = -i - \dfrac{4}{4} = -1 - i$

(e) Large power in complex number indicate large number of its rotations

$$\left(\dfrac{1+i}{\sqrt{2}}\right)^{1000} + \left(\dfrac{1-i}{\sqrt{2}}\right)^{1000} = \left(\dfrac{1+i}{\sqrt{2}}\right)^{2\times500} + \left(\dfrac{1-i}{\sqrt{2}}\right)^{2\times500}$$

$$= \left(\dfrac{2i}{2}\right)^{500} + \left(-\dfrac{2i}{2}\right)^{500} = 2\left(i\!\!\!/^4\right)^{125} = 2$$

(f) Often, symmetrical forms of complex numbers may be simplified

$$\left(\dfrac{1 + i\sqrt{3}}{2}\right)^{3000} + \left(\dfrac{1 - i\sqrt{3}}{2}\right)^{3000}$$

$$= \left(\dfrac{(1 + i\sqrt{3})^2(1 + i\sqrt{3})}{2^3}\right)^{1000} + \left(\dfrac{(1 - i\sqrt{3})^2(1 - i\sqrt{3})}{8}\right)^{1000}$$

$$= \left(\dfrac{(-2 + 2i\sqrt{3})(1 + i\sqrt{3})}{8}\right)^{1000} + \left(\dfrac{(-2 - 2i\sqrt{3})(1 - i\sqrt{3})}{8}\right)^{1000}$$

$$= \left(\dfrac{-\cancel{2}(1 - i\sqrt{3})(1 + i\sqrt{3})}{\cancel{8}}\right)^{1000} + \left(\dfrac{-\cancel{2}(1 + i\sqrt{3})(1 - i\sqrt{3})}{\cancel{8}}\right)^{1000}$$

$$= \left(-\dfrac{1 - (-3)}{4}\right)^{1000} + \left(-\dfrac{1 - (-3)}{4}\right)^{1000}$$

$$= (-1)^{1000} + (-1)^{1000} = 1 + 1 = 2$$

5. By exploiting square and fourth power of i we find

(a) $(1-i)^{100} = (1-i)^{2 \times 50} = (-2i)^{50} = (2i)^{50} = 2^{50} \times i^2 \times i^{4 \times 12} = -2^{50}$

(b) $(1+i)^{50} = (1+i)^{2 \times 25} = (2i)^{25} = 2^{25} \times i \times i^{4 \times 6} = 2^{25} \times i$

(c) $(2-i)^6 = (2-i)^{2 \times 3} = (4-4i-1)^3 = (3-4i)^2 \times (3-4i)$
$$= (9-24i-16)(3-4i) = (-7-24i)(3-4i)$$
$$= -21 + 28i - 72i - 96 = -117 - 44i$$

(d) $\dfrac{1+i\sqrt{3}}{2} \dfrac{1-i\sqrt{3}}{2} = \dfrac{1-(i\sqrt{3})^2}{4} = \dfrac{1+3}{4} = 1$

(e) $\dfrac{(1+i)^{1000}}{(1-i)^{500}} = \dfrac{(1+i)^{2 \times 500}}{(1-i)^{2 \times 250}} = \dfrac{(2i)^{500}}{(-2i)^{250}} = \dfrac{2^{500} \times i^{4 \times 125}}{2^{250} \times i^2 \times i^{4 \times 62}}$
$$= \dfrac{2^{500-250}}{i^2} = -2^{250}$$

(f) $\left(\dfrac{4}{i\sqrt{3}-1}\right)^{12} = \left(-\dfrac{4}{1-i\sqrt{3}}\right)^{12} = \dfrac{4^{12}}{(1-i\sqrt{3})^{12} \left(\frac{2}{2}\right)^{12}} = \dfrac{\frac{4^{12}}{2^{12}}}{\frac{(1-i\sqrt{3})^{12}}{2^{12}}}$
$$= \dfrac{\frac{4^{12}}{2^{2 \times 6}}}{\left(\frac{1-i\sqrt{3}}{2}\right)^{3 \times 4}} = \dfrac{4^{12-6}}{(-1)^4} = 4^6 = 4096$$

(g) $\dfrac{i^{102} + i^{101}}{i^{100} - i^{99}} = \dfrac{i^2 \times i^{4 \times 25} + i \times i^{4 \times 25}}{i^{4 \times 25} - i^3 \times i^{4 \times 24}} = \dfrac{i^2 + i}{1 - i^3}$
$$= \{z z^* = (a+ib)(a-ib) = a^2 + b^2 = |z|^2\}$$
$$= \dfrac{-1+i}{1+i} \dfrac{1-i}{1-i} = \dfrac{-1+i+i+1}{1+1} = \dfrac{2i}{2} = i$$

Alternative solution:

$\dfrac{i^{102} + i^{101}}{i^{100} - i^{99}} = \dfrac{i^{99}\left[i^3{}^{-i} + i^2{}^{-1}\right]}{i^{99}[i-1]} = \dfrac{-1-i}{-1+i} \dfrac{-1-i}{-1-i} = \dfrac{2i}{2} = i$

(h) $\dfrac{(i\sqrt{3}-1)^2}{(2i)^2} = \dfrac{1 - 2i\sqrt{3} - 3}{-4} = \dfrac{-2 - 2i\sqrt{3}}{-4} = \dfrac{1}{2} + i\dfrac{\sqrt{3}}{2}$

(i) $\dfrac{-41 + 63i}{50} - \dfrac{6i+1}{1-7i} = -\dfrac{41}{50} + \dfrac{63}{50}i + \dfrac{6i+1}{1-7i} \dfrac{1+7i}{1+7i}$
$$= -\dfrac{41}{50} + \dfrac{63}{50}i - \dfrac{6i - 42 + 1 + 7i}{1 + 49}$$
$$= \dfrac{-41 + 63i + 41 - 13i}{50} = \dfrac{63 - 13}{50}i = i$$

6. Express z in polar form:

(a) $z = i$ $\therefore$ $|z| = 1$, $\arg(z) = \dfrac{\pi}{2} \Rightarrow z = e^{i\frac{\pi}{2}}$

(b)   By definition,

$z = 2 - i$ $\therefore$ $|z| = \sqrt{(2^2 + 1)} = \sqrt{5}$

$$\tan(\arg(z)) = \frac{\Im(z)}{\Re(z)} = -\frac{1}{2}$$

$$\therefore \quad \arg(z) = \arctan\left(-\frac{1}{2}\right) \approx -26.56°$$

$$\therefore$$

$$z = \sqrt{5}\, e^{-j \times 26.56°}$$

(c) $z = \dfrac{(1 + i)^{200}(6 + 2i) - (1 - i)^{198}(3 - i)}{(1 + i)^{196}(23 - 7i) + (1 - i)^{194}(10 + 2i)}$

$= \dfrac{(1 + i)^{2 \times 100}(6 + 2i) - (1 - i)^{2 \times 99}(3 - i)}{(1 + i)^{2 \times 98}(23 - 7i) + (1 - i)^{2 \times 97}(10 + 2i)}$

$= \dfrac{(2i)^{100}(6 + 2i) - (-2i)^{99}(3 - i)}{(2i)^{98}(23 - 7i) + (-2i)^{97}(10 + 2i)} = \dfrac{2^{99} \times 4(3 + i) - 2^{99}i(3 - i)}{-2^{98}(23 - 7i) - 2^{98}i(5 + i)}$

$= \dfrac{2^{99}(12 + 4i - 3i - 1)}{-2^{98}(23 - 7i + 5i - 1)} = \dfrac{\cancel{2}(11 + i)}{\cancel{2}(-11 + i)}\, \dfrac{11 + i}{11 + i} = \dfrac{(11 + i)^2}{(i - 11)(11 + i)}$

$= \dfrac{121 + 22i - 1}{-1 - 121} = \dfrac{120 + 22i}{-122} = -\dfrac{60 + 11i}{61} = -\dfrac{60}{61} - \dfrac{11}{61}i$

(c)   *(cont.)*

$|z| = \dfrac{|11 + i|}{|-1||11 - i|} = \dfrac{\sqrt{122}}{\sqrt{122}} = 1$,

$\tan(\arg(z)) = \dfrac{\Im(z)}{\Re(z)} = \dfrac{-\frac{11}{\cancel{61}}}{-\frac{60}{\cancel{61}}}$

$\arg(z) = \arctan \dfrac{-11}{-60} \approx -170°$ (in III quadrant)

$$\therefore$$

$z = e^{-170°}$

Exercise 6.3, page 107

1. The necessary condition that two numbers are equal is that, respectively, their real parts are equal and their imaginary parts are equal.

(a)   We explicitly separate real and complex parts of complex equations, then recall that imaginary part of a real number, by definition, equals zero:

$$(2 + 3i)x + (3 + 2i)y = 1$$

$$2x + 3ix + 3y + 2iy = 1$$

$$(2x + 3y) + (3x + 2y)i = 1 + 0i$$

$$\therefore$$

$$2x + 3y = 1$$

$$3x + 2y = 0$$

This system of equation is solved, for example, by Cramer's rule as

$$\left. \begin{array}{l} \Delta = \begin{vmatrix} 2 & 3 \\ 3 & 2 \end{vmatrix} = 4 - 9 = -5 \neq 0 \\[2mm] \Delta x = \begin{vmatrix} 1 & 3 \\ 0 & 2 \end{vmatrix} = 2 - 0 = 2 \\[2mm] \Delta y = \begin{vmatrix} 2 & 1 \\ 3 & 0 \end{vmatrix} = 0 - 3 = -3 \end{array} \right\} \quad \therefore \ \ x = \frac{\Delta x}{\Delta} = -\frac{2}{5} \ \text{ and } \ y = \frac{\Delta y}{\Delta} = \frac{3}{5}$$

(b)   This equation may be solved by enforcing same form on both sides of the equation, as

$$\frac{1}{x + iy} = \frac{1}{2 + i} + \frac{1}{-2 + 4i} = \frac{-2 + 4i + 2 + i}{(2 + i)(-2 + 4i)} = \frac{5i}{-4 + 8i - 2i - 4}$$

$$= \frac{5i}{-8 + 6i} = \frac{1}{\frac{-8}{5i} + \frac{6i}{5i}} = \frac{1}{\frac{6}{5} - \frac{8}{5i}\frac{i}{i}}$$

$$\therefore$$

$$\frac{1}{x + i\, y} = \frac{1}{\dfrac{6}{5} + i\, \dfrac{8}{5}} \tag{6.1}$$

By inspection of (6.1) we write

$$x = \frac{6}{5} \ \text{ and } \ y = \frac{8}{5}$$

2. Solutions of complex equations are symmetrically distributed in the complex plane.

(a)   We exploit the equivalence between radicals and fractional powers in the exponential form of a complex number. We note that the two solutions of quadratic equation, z_1 and z_2, are located symmetrically in the complex plane, Fig. 6.16.

$$z^2 = i \quad \therefore \quad \sqrt{z^2} = \sqrt{i}$$

$$|z| = i^{\frac{1}{2}} = \left(e^{i\frac{\pi}{2}}\right)^{\frac{1}{2}} = e^{i\frac{\pi}{4}}$$

$$\therefore$$

$$z_{1,2} = \pm e^{i\frac{\pi}{4}}$$

$$= \pm \left(\cos\frac{\pi}{4} + i\ \sin\frac{\pi}{4}\right)$$

$$= \pm \left(\frac{\sqrt{2}}{2} + i\frac{\sqrt{2}}{2}\right)$$

Fig. 6.16 Example 6.2-2(a)

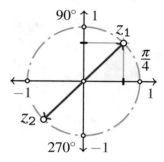

(b)   Similarly, using the same technique, we find two solutions of quadratic equation z_1 and z_2 that are located symmetrically in the complex plane, Fig. 6.17.

$$z^2 = -i \quad \therefore \quad \sqrt{z^2} = \sqrt{-i}$$

$$\therefore$$

$$|z| = (-i)^{\frac{1}{2}} = \left(e^{-i\frac{\pi}{2}}\right)^{\frac{1}{2}} = e^{-i\frac{\pi}{4}}$$

$$\therefore$$

$$z_{1,2} = \pm e^{-i\frac{\pi}{4}}$$

$$= \pm \left(\cos\frac{\pi}{4} - i\ \sin\frac{\pi}{4}\right)$$

$$= \pm \left(\frac{\sqrt{2}}{2} - i\frac{\sqrt{2}}{2}\right)$$

Fig. 6.17 Example 6.2-2(b)

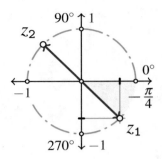

(c)   Two solutions to

$$z^2 = 5 + 12i \tag{6.2}$$

are complex numbers, thus their general algebraic form must be as

$$z = a + bi, \quad (a, b) \in \mathbb{R}$$

$$\therefore$$

$$z^2 = a^2 + 2abi - b^2 = (a^2 - b^2) + (2ab)\,i \tag{6.3}$$

By the equivalence of real and complex parts in (6.2) and (6.3), we write two identities

$$2ab = 12 \quad \therefore \quad ab = 6 \Rightarrow a = \frac{6}{b}$$

$$a^2 - b^2 = 5 \quad \therefore \quad \left(\frac{6}{b}\right)^2 - b^2 = 5$$

Where the last bi-quadratic equation per b is solved as follows, Fig. 6.18,

$$\frac{36}{b^2} - b^2 = 5 \quad \therefore \quad 36 - b^4 = 5b^2$$

$$b^4 + 5b^2 - 36 = 0$$

$$b^4 + 9b^2 - 4b^2 - 36 = 0$$

$$b^2(b^2 + 9) - 4(b^2 + 9) = 0$$

$$(b^2 - 4)(b^2 + 9) = 0$$

$$(b - 2)(b + 2)(b^2 + 9) = 0$$

$$\tag{6.4}$$

$$\therefore$$

$$b_{1,2} = \pm 2, \in \mathbb{R}, \quad \underline{b_{3,4} \in \mathbb{C}} \tag{6.5}$$

$$b_{1,2} = \pm 2 \quad \therefore \quad a = \frac{6}{\pm 2} = \pm 3$$

$$\therefore$$

$$z_{1,2} = \pm(3 + 2i) \tag{6.6}$$

Fig. 6.18 Example 6.2-2(c)

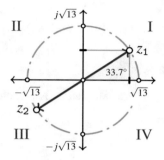

We find the module and arguments of $z_{1,2}$ in (6.6) as

$$|z| = \sqrt{3^2 + 2^2} = \sqrt{13} \quad \text{and} \quad \arg(z) = \arctan\frac{2}{3} \approx 33.69° \tag{6.7}$$

3. One possible way of solving $z^4 + 8 + 8\sqrt{3}\,i = 0$ may be as follows.

$$z^4 + 8 + 8\sqrt{3}\,i = 0 \tag{6.8}$$

$$\therefore$$

$$z = \sqrt[4]{-8 - 8\sqrt{3}\,i}$$

Main idea in this kind of problems is that powers and roots of a complex number are much easier to find if we convert its algebraic into exponential form.

Thus, first we calculate module of a temporary variable $|t|$ and argument θ of the 4th root's argument t, then we write its exponential form as follows.

$$t = -8 - 8\sqrt{3}\,i$$

$$\therefore$$

$$|t| = \sqrt{\Re^2(x) + \Im^2(x)} = \sqrt{(-8)^2 + (-8\sqrt{3})^2} = \sqrt{64 + 64 \times 3} = 16$$

$$\tan\theta = \frac{\Im(x)}{\Re(x)} = \frac{-8\sqrt{3}}{-8} = \sqrt{3} \quad\therefore\quad \theta = \arctan\sqrt{3} = \frac{\pi}{3}$$

We note that $\theta = \pi/3$ is found in the first quadrant; however, both real and imaginary parts of t are negative, thus t is found in the third quadrant, see Fig. 6.19. Consequently, the correct argument of t is actually

$$\theta = \pi + \frac{\pi}{3} = \frac{4\pi}{3}$$

Therefore, we write

$$t = -8 - 8\sqrt{3}\,i = 16\,\exp\left(i\,\frac{4\pi}{3}\right)$$

Fig. 6.19 Example 6.2-3

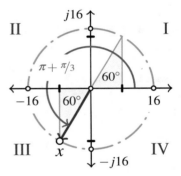

With this conversion, the original problem takes the following form.

$$z = \sqrt[4]{t} = \sqrt[4]{16\,\exp\left(i\,\frac{4\pi}{3}\right)} = 2\left[\exp\left(i\,\frac{4\pi}{3}\right)\right]^{\frac{1}{4}} = 2\,\exp\left(i\,\frac{\pi}{3}\right) \tag{6.9}$$

The original equation (6.8) is 4th order polynomial, therefore there must be, in total, four values of z that are valid solutions. Obviously, (6.9) is only the first of the four values. All solutions of a complex equations are found on circle whose radius equals to z.

Arguments of the solutions are symmetrically distributed within 2π. That is to say, for example, in the case of second order equation, the two solutions are separated by $2\pi/2 = \pi$ angle. In the case of third order equation, the three solutions are separated by $2\pi/3$ angle. Therefore, the four solutions of fourth order equation are separated by $2\pi/4 = \pi/2$ and are found as

$$z_k = 2\,e^{\,i\left(\frac{\pi}{3} + k\frac{\pi}{2}\right)}$$
$$= 2\left[\cos\left(\frac{\pi}{3} + k\frac{\pi}{2}\right) + i\,\sin\left(\frac{\pi}{3} + k\frac{\pi}{2}\right)\right]; \quad k = 0, 1, 2, 3$$

In summary,

$$k = 0: \ z_0 = 2\left[\cos\left(\frac{\pi}{3}\right) + i\,\sin\left(\frac{\pi}{3}\right)\right] = 1 + i\sqrt{3}$$

$$k = 1: \ z_1 = 2\left[\cos\left(\frac{\pi}{3} + \frac{\pi}{2}\right) + i\,\sin\left(\frac{\pi}{3} + \frac{\pi}{2}\right)\right] = -\sqrt{3} + i$$

$$k = 2: \ z_2 = 2\left[\cos\left(\frac{\pi}{3} + \pi\right) + i\,\sin\left(\frac{\pi}{3} + \pi\right)\right] = -1 - i\sqrt{3}$$

$$k = 3: \ z_3 = 2\left[\cos\left(\frac{\pi}{3} + \frac{3\pi}{2}\right) + i\,\sin\left(\frac{\pi}{3} + \frac{3\pi}{2}\right)\right] = \sqrt{3} - i$$

We note that, indeed, all four modules are equal

$$|z_k| = \sqrt{1^2 + \left(\sqrt{3}\right)^2} = 2$$

and all arguments are uniformly distributed on the circle,[1] see Fig. 6.20.

Fig. 6.20 Example 6.2-3

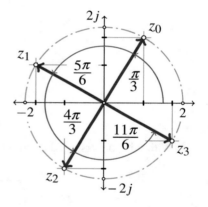

Proof Each of the four solutions $z_{0,1,2,3}$ should be verified against (6.8), for example,

$$z_1^4 + 8 + 8\sqrt{3}\,i = 0$$

$$\left(2\,e^{\,i\left(\frac{5\pi}{6}\right)}\right)^4 + 8 + 8\sqrt{3}\,i = 0$$

$$16\,e^{\,i\left(\frac{10\pi}{3}\right)} + 8 + 8\sqrt{3}\,i = 0$$

$$16\,e^{\,i\left(\frac{6\pi}{3}\,0 + \frac{4\pi}{3}\right)} + 8 + 8\sqrt{3}\,i = 0$$

$$16\left[\cos\left(\frac{4\pi}{3}\right) + i\,\sin\left(\frac{4\pi}{3}\right)\right] + 8 + 8\sqrt{3}\,i = 0$$

$$16\left(-\frac{1}{2} - \frac{\sqrt{3}}{2}\,i\right) + 8 + 8\sqrt{3}\,i = 0$$

$$-8 - 8\sqrt{3}\,i + 8 + 8\sqrt{3}\,i = 0$$

$$0 = 0\ \checkmark$$

Linear Algebra

Important to Know

Column matrix:

$$\begin{bmatrix} a_1 \\ \vdots \\ a_m \end{bmatrix}$$

Raw matrix:

$$\begin{bmatrix} a_1 & \cdots & a_m \end{bmatrix}$$

The transpose of a row vector is a column vector

$$\begin{bmatrix} x_1 & x_2 & \ldots & x_m \end{bmatrix}^{\text{T}} = \begin{bmatrix} x_1 \\ x_2 \\ \vdots \\ x_m \end{bmatrix}$$

The transpose of a column vector is a row vector

$$\begin{bmatrix} x_1 \\ x_2 \\ \vdots \\ x_m \end{bmatrix}^{\text{T}} = \begin{bmatrix} x_1 & x_2 & \ldots & x_m \end{bmatrix}$$

Matrix A_{mn} is written as

$$A_{mn} = \begin{bmatrix} a_{11} & \ldots & a_{1,n} \\ \vdots & \ldots & \vdots \\ a_{m1} & \ldots & a_{mn} \end{bmatrix}$$

© Springer Nature Switzerland AG 2021
R. Sobot, *Engineering Mathematics by Example*,
https://doi.org/10.1007/978-3-030-79545-0_7

The identity matrix I_n is written as

$$I_n = \begin{bmatrix} 1 & 0 & 0 & \cdots & 0 \\ 0 & 1 & 0 & \cdots & 0 \\ 0 & 0 & 1 & \cdots & 0 \\ \vdots & \vdots & \vdots & \ddots & \vdots \\ 0 & 0 & 0 & \cdots & 1 \end{bmatrix}$$

Determinant of matrix A_2 is calculated as

$$|A_2| = \begin{vmatrix} a & b \\ c & d \end{vmatrix} = ad - bc$$

Cramer's rule: Given a system of m linear equations

$$a_{11}x_1 + a_{12}x_2 + \cdots + a_{1n}x_n = b_1$$
$$a_{21}x_1 + a_{22}x_2 + \cdots + a_{2n}x_n = b_2$$
$$\vdots$$
$$a_{m1}x_1 + a_{m2}x_2 + \cdots + a_{mn}x_n = b_m$$

where $x_1, x_2, \ldots, x_n$ are the n unknown variables, $a_{11}, a_{12}, \ldots, a_{mn}$ are the coefficients of the system, and $b_1, b_2, \ldots, b_m$ are the constant terms of the equations. Then,

$$\Delta = |A_{mn}| = \begin{vmatrix} a_{11} & a_{12} & \ldots & a_{1n} \\ \vdots & \vdots & \ldots & \vdots \\ a_{m1} & a_{m2} & \ldots & a_{mn} \end{vmatrix} \quad (\Delta \neq 0)$$

After replacing the i-th column of $|A_{mn}|$ by the column vector b_m, we write

$$\Delta_{x_1} = \begin{vmatrix} b_1 & a_{12} & \ldots & a_{1n} \\ \vdots & \vdots & \ldots & \vdots \\ b_m & a_{m2} & \ldots & a_{mn} \end{vmatrix} \quad \Delta_{x_2} = \begin{vmatrix} a_{11} & b_1 & \ldots & a_{1n} \\ \vdots & \vdots & \ldots & \vdots \\ a_{m1} & b_m & \ldots & a_{mn} \end{vmatrix} \quad \cdots \quad \Delta_{x_n} = \begin{vmatrix} a_{11} & a_{12} & \ldots & b_1 \\ \vdots & \vdots & \ldots & \vdots \\ a_{m1} & a_{m2} & \ldots & b_m \end{vmatrix}$$

and,

$$x_1 = \frac{\Delta_{x_1}}{\Delta}; \quad x_2 = \frac{\Delta_{x_2}}{\Delta}; \quad \cdots; \quad x_n = \frac{\Delta_{x_n}}{\Delta}$$

7.1   Exercises

7.1   Vector Definitions

1. In a simple graph, illustrate the geometrical interpretation of dimensions in 3D, 2D, 1D, and 0D spaces, where "xD" refers to number of dimensions.

2. Show the "preferred" vector representations in physics, mathematics, and informatics.

3. Given three basis vectors $\vec{i}$, $\vec{j}$, $\vec{k}$, sketch these three vectors in a single graph. Comment on the space defined by these basis vectors?

$$\vec{i} = \begin{bmatrix} i_x \\ i_y \\ i_z \end{bmatrix} = \begin{bmatrix} 1 \\ 0 \\ 0 \end{bmatrix} \quad \vec{j} = \begin{bmatrix} j_x \\ j_y \\ j_z \end{bmatrix} = \begin{bmatrix} 0 \\ 1 \\ 0 \end{bmatrix} \quad \vec{k} = \begin{bmatrix} k_x \\ k_y \\ k_z \end{bmatrix} = \begin{bmatrix} 0 \\ 0 \\ 1 \end{bmatrix}$$

4. Given four points in 2D space: $A = (-2, 2)$, $B = (1, 4)$, $C = (5, 6)$, $D = (3, 1)$,

   (a)   Sketch a graph with $\overrightarrow{AB}$ and $\overrightarrow{CD}$ vectors.

   (b)   Write the matrix form of $\overrightarrow{AB}$ and $\overrightarrow{CD}$ vectors

5. Sketch a graph that shows the following vectors

$$\vec{a} = \begin{bmatrix} 3 \\ 3 \end{bmatrix} \quad \vec{b} = \begin{bmatrix} -4 \\ 3 \end{bmatrix} \quad \vec{c} = \begin{bmatrix} -3 \\ -4 \end{bmatrix}$$

6. Calculate magnitude of $\vec{p}$, that is to say $|\vec{p}|$ if

   (a) $\vec{p} = (0, 1)$ (b) $\vec{p} = (3, 4)$

   (c) $\vec{p} = (-3, 1)$ (d) $\vec{p} = (2, 3, 6)$

7. Given coordinates of points in space, calculate coordinates of $\overrightarrow{AB}$ and $\overrightarrow{BA}$ in matrix form, then calculate its magnitude (i.e. module).

   (a) $A(4, 1)$, $B(1, -3)$ (b) $A(2, 3)$, $B(-1, 4)$

   (c) $A(-1, -3)$, $B(4, 2)$ (d) $A(1, -2)$, $B(3, 2)$

8. Given coordinates of points in space, calculate coordinates of $\overrightarrow{AB}$ and $\overrightarrow{BA}$ in matrix form, then calculate its magnitude (i.e. module).

   (a) $A(4, 1, 6)$, $B(2, 4, -2)$

   (b) $\overrightarrow{OA} = (2, 3, 4)$, $\overrightarrow{OB} = (3, 0, -1)$ (Note: $O = (0, 0, 0)$).

9. Given data, calculate coordinates of the following vectors in matrix form.

   (a) $A(2, 3, 1)$, $B(3, -1, 0)$, $C = (-1, -2, 1)$, $D = (-3, 3, -2)$, calculate $\vec{m} = \overrightarrow{AB} + \overrightarrow{CD}$, $\vec{n} = \overrightarrow{AB} - \overrightarrow{CD}$.

   (b) $\vec{a} = (2, -3)$, $\vec{b} = (3, -1)$, calculate $\vec{c} = \vec{a} + \vec{b}$, $\vec{d} = \vec{a} - \vec{b}$.

(c) $\vec{a} = (-1, 2)$, $\vec{b} = (3, 2)$, $\vec{c} = (-2, -3)$, calculate $\vec{p} = \vec{a} + \vec{b}$, $\vec{m} = 2\vec{a} - \vec{b}$, $\vec{n} = 3\vec{a} - \vec{b} - \vec{c}$, $\vec{r} = 2\vec{b} - 1/2\,\vec{c}$

10. Vectors $\vec{a}$ and $\vec{b}$ create angle $\pi/6$, given that $|\vec{a}| = \sqrt{3}$ and $|\vec{b}| = 1$, calculate angle θ between $\vec{p} = \vec{a} + \vec{b}$ et $\vec{q} = \vec{a} - \vec{b}$.

7.2 **   Analytical Geometry

1. Solve the following vector problems.

(a)   Given three points in space $\mathscr{A} = (3, 5)$, $\mathscr{B} = (-1, 3)$, $\mathscr{C} = (1, -7)$, calculate the coordinates of point $\mathscr{D}$, given that $\overrightarrow{AD} = 3\overrightarrow{AB} - 3/2\overrightarrow{BC}$.

(b)   Derive equation of a line that intersects point $\mathscr{A} = (2, 5)$ and is normal to vector $\vec{n} = (1, 3)$.

(c)   Given two points in space $\mathscr{A} = (5, 4)$, $\mathscr{B} = (8, 6)$ derive equation of a line that intersects the origin and is normal to $\overrightarrow{AB}$.

(d)   Given line equation $2x - 3y + 5 = 0$ and point in space $\mathscr{A} = (7, 6)$, derive equation of a line that intersects point $\mathscr{A}$ and is parallel to the given line.

2. Solve the following problems related to circle.

(a)   Derive the equation of a circle whose centre is found at $\mathscr{C} = (4, 0)$ and its radius equals $r = 3$.

(b)   Given equation of a circle $x^2 + y^2 - 10x + 4y + 7 = 0$, calculate its radius.

(c)   Given two points in space $\mathscr{A} = (1, 2)$, $\mathscr{B} = (-3, 4)$ derive equation of circle whose diameter equals $\overrightarrow{AB}$.

(d)   Given coordinates of circle centre $\mathscr{C} = (2, 0)$ and its radius $r = 5$, derive equation of this circle. Then, determine coordinates of the intersect points between this circle and line $x - y - 3 = 0$.

3. Solve the following problems related to triangle.

(a)   Given triangle whose sides are $AB = 4$, $AC = 6$, $BC = 8$ calculate $\cos \alpha$, where angle α is associated with the point A.

(b)   Given triangle as $AB = 3$, $AC = 4$, $\alpha = \pi/3$ calculate side BC.

7.3   Vector Operations

1. Sketch a graph to illustrate the operation of addition in 1D and 2D spaces. As an example, show the following vector additions:

(a) $2 + 3 = 5$ and $2 - 3 = -1$

(b) $\vec{c} = \vec{a} + \vec{b}$ if $\vec{a} = \begin{bmatrix} 1 \\ 2 \end{bmatrix}$ and $\vec{b} = \begin{bmatrix} 3 \\ -1 \end{bmatrix}$

2. Given vector

$$\vec{x} = \begin{bmatrix} x_x \\ x_y \end{bmatrix} = \begin{bmatrix} 3 \\ 2 \end{bmatrix} \tag{7.1}$$

sketch diagram of $\vec{x}$ multiplied by a constant.

(a) $\vec{a} = 2\,\vec{x}$ (b) $\vec{b} = \dfrac{2}{3}\,\vec{x}$ (c) $\vec{c} = -\dfrac{3}{4}\,\vec{x}$

7.4 Linear Transformation

1. By a simple sketch show similarities between a *function* $f(x)$ and a *linear transformation* $L(\vec{v})$ operation, then give a brief explanation of the relation between linear transformations and space.

2. Given vector $\vec{v} = -2\vec{i} + \vec{j}$,

   (a) By sketching a diagram, show that the numerical form of $\vec{v}$ stays the same after an arbitrary linear transformation $L(\vec{v})$.
   (b) With the help of the diagram, calculate the matrix form of $L(\vec{v})$ used in your example.

3. Show the geometrical interpretation of linear transformation $L(\vec{x})$ given that

$$L(\vec{x}) = \begin{bmatrix} 3 & 1 \\ 1 & 2 \end{bmatrix} \quad \text{and} \quad \vec{x} = \begin{bmatrix} -1 \\ 2 \end{bmatrix}$$

4. Find linear transformations to:

   (a) rotate 2D space $90°$ counterclockwise,
   (b) shear 2D space $45°$ clockwise,
   (c) first rotate $90°$ counterclockwise then shear $45°$ clockwise 2D space.

5. Give geometrical interpretation of determinant for the following linear transformations:

   (a) $L(\vec{x}) = \begin{bmatrix} 3 & 0 \\ 0 & 2 \end{bmatrix}$ (b) $L(\vec{x}) = \begin{bmatrix} 1 & 2 \\ 1 & -1 \end{bmatrix}$ (c) $L(\vec{x}) = \begin{bmatrix} 4 & 2 \\ 2 & 1 \end{bmatrix}$

6. Given vectors,

   (a) $\vec{a} = (2, 1)$, $\vec{b} = (1, 0)$, write $\vec{m} = (9, 1)$ as linear combination of $\vec{a}$ and $\vec{b}$.
   (b) $\vec{u} = (3, -1)$, $\vec{v} = (1, -2)$, $\vec{w} = (-1, 7)$, write $\vec{m} = \vec{u} + \vec{v} + \vec{w}$ as linear combination of $\vec{u}$ and $\vec{v}$.

7. First show that $\vec{a}$, $\vec{b}$, and $\vec{c}$ linearly independent, then write $\vec{c}$ as linear combination of $\vec{a}$ and $\vec{b}$ if

   (a) $\vec{a} = (3, -2)$, $\vec{b} = (-2, 1)$, $\vec{c} = (7, -4)$
   (b) $\vec{a} = (1, 2)$, $\vec{b} = (3, 4)$, $\vec{c} = (0, -2)$

8. (a) Calculate parameter λ so that given vectors are linearly independent:
   (b) $\vec{a} = (3, \lambda)$, $\vec{b} = (2, 6)$
   (c) $\vec{a} = (6, 8, 4)$, $\vec{b} = (3, 4, 2)$, $\vec{c} = (\lambda, 0, 1)$

7.5   Determinants

1. Calculate determinants of the following second and third order matrix.

   (a)   (b)

$$A_2 = \begin{bmatrix} 3 & 0 \\ 0 & 2 \end{bmatrix}$$ $$A_2 = \begin{bmatrix} 1 & 2 \\ 1 & -1 \end{bmatrix}$$

   (c)   (d)

$$A_2 = \begin{bmatrix} 4 & 2 \\ 2 & 1 \end{bmatrix}$$ $$A_3 = \begin{bmatrix} 2 & 3 & 4 \\ 5 & -2 & 1 \\ 1 & 2 & 3 \end{bmatrix}$$

2. Calculate the volume of cuboid created by vectors

   (a) $\vec{a} = (1, -3, 1)$, $\vec{b} = (2, 1, -3)$, $\vec{c} = (1, 2, 1)$
   (b) $\vec{a} = (1, 0, 3)$, $\vec{b} = (0, 1, 2)$, $\vec{c} = (3, 4, 0)$

3. Calculate determinants

   (a) $\begin{vmatrix} 2 & -1 \\ 1 & 0 \end{vmatrix}$ (b) $\begin{vmatrix} \sqrt{3} & -3\sqrt{2} \\ \sqrt{3} & 2\sqrt{2} \end{vmatrix}$ (c) $\begin{vmatrix} 1 & i \\ -i & 1 \end{vmatrix}$ (d) $\begin{vmatrix} \sin\alpha & \cos\alpha \\ -\cos\alpha & \sin\alpha \end{vmatrix}$

4. Calculate determinants

   (a) $\begin{vmatrix} 3 & 2 & 1 \\ 4 & 5 & 6 \\ 8 & 9 & 7 \end{vmatrix}$ (b) $\begin{vmatrix} 2 & 3 & 1 \\ 3 & -1 & 2 \\ 1 & 1 & -3 \end{vmatrix}$ (c) $\begin{vmatrix} 5 & 0 & 4 \\ 8 & 0 & -7 \\ 3 & 2 & 1 \end{vmatrix}$ (d) $\begin{vmatrix} 1 & 3 & 5 \\ 7 & 9 & 11 \\ 13 & 15 & 17 \end{vmatrix}$

7.6   Systems of Linear Equations

1. Solve the following system of linear equations for $\{x, y, z\}$:

$$5x - 5y - 15z = 40$$
$$4x - 2y - 6z = 19$$
$$3x - 6y - 17z = 41$$

2. Given coordinates of three points

$$A = (-2, 20); B = (1, 5); C = (3, 25)$$

   determine quadratic function

$$f(x) = ax^2 + bx + c$$

so that it can be used for curve fitting of these three data points, Fig. 7.1.

Fig. 7.1 Example 7.6-2

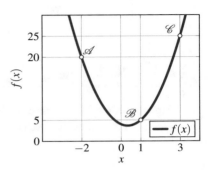

3. Solve the following systems using the Cramer's rule.

(a) $4x_1 - 3x_2 = 0$
$2x_1 + 3x_2 = 18$

(b) $5x - 5y - 15z = 40$
$4x - 2y - 6z = 19$
$3x - 6y - 17z = 41$

4. Solve the following system by using Cramer's rule.

$$x + y = 7$$
$$y + z = 8$$
$$-x + 2z = 7$$

5. Solve the following systems of linear equations

(a) $x + 2y = -5$
$3x - y = 13$

(b) $x + 2y = -8$
$2x - y = -1$

(c) $x + y + z = 36$
$2x - z = -17$
$6x - 5z = 7$

(d) $2x + 3y - z = 5$
$x + y + 2z = 7$
$2x - y + z = 1$

7.7   Matrix-Vector Product

1. Calculate the product and show its geometric interpretation

(a) $\begin{bmatrix} 1 & 1 \\ 2 & -1 \end{bmatrix} \begin{bmatrix} 3 \\ 0 \end{bmatrix}$

(b) $\begin{bmatrix} 2 & -1 \\ 3 & 2 \end{bmatrix} \begin{bmatrix} -5 \\ 3 \end{bmatrix}$

(c) $\begin{bmatrix} 2 & -3 \\ 1 & 2 \end{bmatrix} \begin{bmatrix} 3 \\ -2 \end{bmatrix}$

(d) $\begin{bmatrix} 1 & 2 \\ 2 & -1 \end{bmatrix} \begin{bmatrix} -8 \\ -1 \end{bmatrix}$

2. Calculate the product and show its geometric interpretation

(a) $\begin{bmatrix} 1 & 1 \\ 2 & -1 \end{bmatrix} \begin{bmatrix} 2 & -1 \\ 3 & 2 \end{bmatrix}$

(b) $\begin{bmatrix} 2 & -1 \\ 3 & 2 \end{bmatrix} \begin{bmatrix} 2 & -3 \\ 1 & 2 \end{bmatrix}$

(c) $\begin{bmatrix} 2 & -3 \\ 1 & 2 \end{bmatrix} \begin{bmatrix} 1 & 1 \\ 2 & -1 \end{bmatrix}$

7.8 *   Matrix Product

1. Calculate the product of AB if

$$A = \begin{bmatrix} 3 & 0 & 5 \\ -2 & -1 & 4 \end{bmatrix} \qquad B = \begin{bmatrix} 1 & 2 \\ 5 & -1 \\ 0 & -6 \end{bmatrix}$$

7.9 *   Eigenvalue and Eigenvector

1. Very important special case in linear algebra is when matrix-vector product equals zero, even when neither matrix nor vectors equal zero, i.e.

$$A \cdot \vec{x} = \vec{0} \qquad\qquad (7.2)$$

Verify if (7.2) is satisfied given

(a) Matrix A and $\vec{x_1}$,   (b) The same matrix A and $\vec{x_2}$,

$$A = \begin{bmatrix} -2 & -9 & -1 \\ 2 & -6 & -5 \\ -4 & -3 & 4 \end{bmatrix} \quad \vec{x_1} = \begin{bmatrix} 1 \\ 1 \\ 2 \end{bmatrix} \qquad\qquad A = \begin{bmatrix} -2 & -9 & -1 \\ 2 & -6 & -5 \\ -4 & -3 & 4 \end{bmatrix} \quad \vec{x_2} = \begin{bmatrix} 13 \\ -4 \\ 10 \end{bmatrix}$$

2. Verify if $\vec{v}$ is eigenvector of matrix A, and determine eigenvalue λ.

(a) Matrix A and $\vec{v_1}$,   (b) The same matrix A and $\vec{v_2}$,

$$A = \begin{bmatrix} 3 & -3 \\ 2 & -4 \end{bmatrix} \quad \vec{v_1} = \begin{bmatrix} 3 \\ 1 \end{bmatrix} \qquad\qquad A = \begin{bmatrix} 3 & -3 \\ 2 & -4 \end{bmatrix} \quad \vec{v_2} = \begin{bmatrix} 2 \\ 1 \end{bmatrix}$$

3. Calculate eigenvector(s) $\vec{v}$ and eigenvalue(s) λ of matrix A.

$$A = \begin{bmatrix} 4 & 6 & 10 \\ 3 & 10 & 13 \\ -2 & -6 & -8 \end{bmatrix}$$

7.10 *   Inverse Matrix

1. Calculate inverse A^{-1} of matrix A.

$$A = \begin{bmatrix} 4 & -2 \\ 2 & 3 \end{bmatrix}$$

7.11 *   **Matrix Powers**

1. Calculate the following powers

(a) D^2 if $D = \begin{bmatrix} 2 & 0 \\ 0 & 2 \end{bmatrix}$ (b) D^5 if $D = \begin{bmatrix} 2 & 0 \\ 0 & 5 \end{bmatrix}$ (c) A^5 if $A = \begin{bmatrix} 2 & 0 \\ -1 & 3 \end{bmatrix}$

Solutions

Exercise 7.1, page 131

1. Geometrical interpretation of space is based on definition of *point*, which is assumed to have size of *zero*. A loose visual interoperation would be a sphere whose radius equals zero, Fig. 7.2. Appropriately, assuming nothing else exists, a point would represent "0-order space" annotated as R^0.

1. *Line:* Multiple points aligned next to each other (albeit, the distance between any two points being zero) form *line*. Therefore, it takes infinitely points to create a line of any length L. If nothing else exists, line represents *one-dimensional* space, annotated as R^1. As an analogy, a little ant walking along very long wire can move only along forward–reverse direction of the wire, thus it has only one freedom of movement. Assuming one arbitrary point as the reference for measuring relative distances, it takes only one number to describe distance from any other point to the reference. In addition, sign "$\pm$" shows which side of the reference is measured, forward (i.e. "+") or reverse (i.e. "−")).

2. *Surface:* Multiple lines put next to each other create *surface*, annotated as R^2. Analogy would be a sheet of paper whose thickness equals zero. In this case the little ant walking on the surface has two freedoms of movement: left–right and forward–reverse. Assuming one arbitrary point as the reference for measuring relative distances, it takes two numbers to describe position of any other point on the surface. This pair of numbers is referred to as "coordinates", by convention coordinates are written as (x, y) pair, where each number represents distance from the "origin" in one of two directions. In analogy, assuming lower left corner of a page to be the reference point, we can describe position of each letter on the page by measuring horizontal and vertical distances.

3. *Volume:* Multiple surfaces put on top of each other create *volume*, annotated as R^3. Analogy would be a book, where the stacked pages form the third dimension. In this R^3 space position of any point is described by *three* numbers, (x, y, z), each measuring distance from the origin in the three possible directions. In this space a bee has freedom to fly, that is to say to move in up–down direction. Our physical universe is R^3 space, and our perceptions are limited to 3D space dimensions. In mathematics, however, there is no limit on order of space.

An important observation is that each lower order space can be seen as one of possible projections of the higher order space. That is to say, looked from "straight ahead" a line looks like a point. Looked from "a side" a surface looks like a line. Looked from one direction, a cube (3D) looks like a square (2D), or a sphere (3D) looks like a circle (2D). Mathematically, this reduction is space is found when one or more equations in a system is not independent, i.e. they may be written as the product of a simple factor and another equation. In linear algebra, a matrix may be seen a compact form of writing a system of equations, where the zero value of its determinant indicates that one or more equations is dependent, thus the real order of space is lower than the number of equations in system.

Fig. 7.2 Example 7.1-1

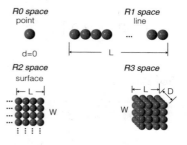

2. A vector signifies a variable that has *two* properties: magnitude and direction. That is to say, a simple number is not vector because it has only magnitude but not direction. For example, "speed" is not vector because it shows only the time rate at which an object is moving along the path. A car can be speeding $100\,$km/h down the highway in both directions. On the other hand, "velocity" shows both the rate of change and direction of an object's movement.

Three commonly used vector representations are as follows. In physics, "vector" takes geometrical interpretation in form of a "directed arrow" that indicates both magnitude (i.e. the arrow length) and direction (i.e. from its tail to the head). In mathematics, assuming 3D space a vector variable is annotated in algebraic form as

$$\vec{a} = a_x \vec{i} + a_y \vec{j} + a_z \vec{k}$$

where (a_x, a_y, a_z) are vector magnitudes as measured along the three directions $(\vec{i}, \vec{j}, \vec{k})$, respectively (also referred to as the "unity vectors").

In informatics, the preferred way of representing a vector is in one-column matrix form, as for example

$$\vec{a} = \begin{bmatrix} 3.14 \\ -5 \\ 2 \end{bmatrix}$$

where the column list contains coordinates of each direction relative to origin $(0, 0, 0)$, in this case 3D space. Of course, this matrix version is abstract and the list can be associated with any physical or nonphysical variable.

3. As defined, the three unity vectors are used to define 3D space and are written in a form of 3×3 matrix as

$$(\vec{i}, \vec{j}, \vec{k}) = \begin{bmatrix} 1 & 0 & 0 \\ 0 & 1 & 0 \\ 0 & 0 & 1 \end{bmatrix}$$

where the first column is for $\vec{i}$, the second column is for $\vec{j}$, and the third column is for $\vec{k}$ unity vector, Fig. 7.3. This resulting matrix where all elements along the main diagonal equal one and all other matrix elements equal zero is known as "unity matrix" with properties similar to number one.

Fig. 7.3 Example 7.1-3

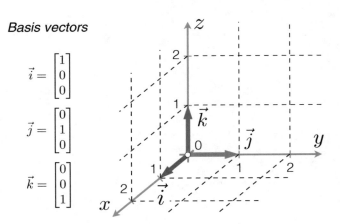

Basis vectors

4. By convention, vectors are oriented in direction from first to second coordinate.

(a) Given four points in 2D space: $A = (-2, 2)$, $B = (1, 4)$, $C = (5, 6)$, $D = (3, 1)$, we draw two vectors as in Fig. 7.4.

Fig. 7.4 Example 7.1-4

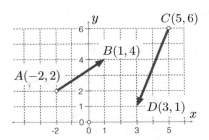

(b) The matrix form assumes that all vectors originate at $(0, 0)$ point. Geometrically, that means vectors are translated so that their origin is at $(0, 0)$ point, when their final position is as for $\overrightarrow{RQ}$ in Fig. 7.5 (top), we write

$$\overrightarrow{RQ} = \begin{bmatrix} 3 \\ 2 \end{bmatrix}$$

Fig. 7.5 Example 7.1-4

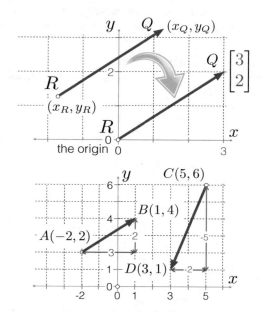

Similarly, after being translated to the origin (see Fig. 7.5 (bottom)), for vectors $\overrightarrow{AB}$ and $\overrightarrow{CD}$ we write

$$\overrightarrow{AB} = \begin{bmatrix} 3 \\ 2 \end{bmatrix} \quad \text{and} \quad \overrightarrow{CD} = \begin{bmatrix} -2 \\ -5 \end{bmatrix}$$

It is important to note that, by definition, translated vector are still identical: they have same magnitude and direction.

5. By convention, the vector matrix form lists coordinates in order, therefore the three vectors are as in Fig. 7.6.

Fig. 7.6 Example 7.1-5

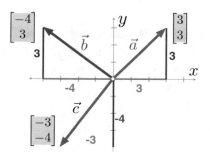

6. Given p_x, p_y, and p_z vector projections, we write,

   (a) $\vec{p} = (0, 1) \quad \therefore \quad |\vec{p}| = \sqrt{0^2 + 1^2} = 1$

(b) $\vec{p} = (3, 4)$ $\therefore$ $|\vec{p}| = \sqrt{3^2 + 4^2} = 5$

(c) $\vec{p} = (-3, 1)$ $\therefore$ $|\vec{p}| = \sqrt{3^2 + 1^2} = \sqrt{10}$

(d) $\vec{p} = (2, 3, 6)$ $\therefore$ $|\vec{p}| = \sqrt{2^2 + 3^2 + 6^2} = 7$

7. Geometrical and matrix interoperation of $\overrightarrow{AB}$ and $\overrightarrow{BA}$ vectors are as follows.

(a) Coordinates of $\overrightarrow{AB}$ we calculate as the end point minus the starting point,

$$\overrightarrow{AB} = \begin{bmatrix} 1 \\ -3 \end{bmatrix} - \begin{bmatrix} 4 \\ 1 \end{bmatrix} = \begin{bmatrix} (1-4) \\ (-3-1) \end{bmatrix} = \begin{bmatrix} -3 \\ -4 \end{bmatrix}$$

while coordinates of $\overrightarrow{BA}$ are

$$\overrightarrow{BA} = \begin{bmatrix} 4 \\ 1 \end{bmatrix} - \begin{bmatrix} 1 \\ -3 \end{bmatrix} = \begin{bmatrix} (4-1) \\ (1-(-3)) \end{bmatrix} = \begin{bmatrix} 3 \\ 4 \end{bmatrix}$$

in both cases, by inspection of graph, as for $\overrightarrow{AB}$ in Fig. 7.7, we calculate $|\overrightarrow{AB}|$ and $|\overrightarrow{BA}|$ with the help of Pythagoras's theorem as

$$|\overrightarrow{AB}| = |\overrightarrow{BA}| = \sqrt{|x_B - x_A|^2 + |y_B - y_A|^2} = \sqrt{3^2 + 4^2} = 5$$

(b) Similarly for $\overrightarrow{AB}$ we calculate

$$\overrightarrow{AB} = \begin{bmatrix} -1 \\ 4 \end{bmatrix} - \begin{bmatrix} 2 \\ 3 \end{bmatrix} = \begin{bmatrix} (-1-2) \\ (4-3) \end{bmatrix} = \begin{bmatrix} -3 \\ 1 \end{bmatrix}$$

while coordinates of $\overrightarrow{BA}$ are

$$\overrightarrow{BA} = \begin{bmatrix} 2 \\ 3 \end{bmatrix} - \begin{bmatrix} -1 \\ 4 \end{bmatrix} = \begin{bmatrix} (2-(-1)) \\ (3-4) \end{bmatrix} = \begin{bmatrix} 3 \\ -1 \end{bmatrix}$$

in both cases we calculate $|\overrightarrow{AB}|$ and $|\overrightarrow{BA}|$ with the help of Pythagoras's theorem as

$$|\overrightarrow{AB}| = |\overrightarrow{BA}| = \sqrt{|x_B - x_A|^2 + |y_B - y_A|^2} = \sqrt{3^2 + 1^2} = 2$$

(c)   For $\overrightarrow{AB}$ we calculate

$$\overrightarrow{AB} = \begin{bmatrix} 4 \\ 2 \end{bmatrix} - \begin{bmatrix} -1 \\ -3 \end{bmatrix} = \begin{bmatrix} (4-(-1)) \\ (2-(-3)) \end{bmatrix} = \begin{bmatrix} 5 \\ 5 \end{bmatrix}$$

while coordinates of $\overrightarrow{BA}$ are

$$\overrightarrow{BA} = \begin{bmatrix} -1 \\ -3 \end{bmatrix} - \begin{bmatrix} 4 \\ 2 \end{bmatrix} = \begin{bmatrix} (-1-4) \\ (-3-2) \end{bmatrix} = \begin{bmatrix} -5 \\ -5 \end{bmatrix}$$

in both cases we calculate $|\overrightarrow{AB}|$ and $|\overrightarrow{BA}|$ with the help of Pythagoras's theorem as

$$|\overrightarrow{AB}| = |\overrightarrow{BA}| = \sqrt{|x_B - x_A|^2 + |y_B - y_A|^2} = \sqrt{5^2 + 5^2} = 5\sqrt{2}$$

(d)   For $\overrightarrow{AB}$ we calculate

$$\overrightarrow{AB} = \begin{bmatrix} 3 \\ 2 \end{bmatrix} - \begin{bmatrix} 1 \\ -2 \end{bmatrix} = \begin{bmatrix} (3-1) \\ (2-(-2)) \end{bmatrix} = \begin{bmatrix} 2 \\ 4 \end{bmatrix}$$

while coordinates of $\overrightarrow{BA}$ are

$$\overrightarrow{BA} = \begin{bmatrix} 1 \\ -2 \end{bmatrix} - \begin{bmatrix} 3 \\ 2 \end{bmatrix} = \begin{bmatrix} (1-3) \\ (-2-2) \end{bmatrix} = \begin{bmatrix} -2 \\ -4 \end{bmatrix}$$

in both cases we calculate $|\overrightarrow{AB}|$ and $|\overrightarrow{BA}|$ with the help of Pythagoras's theorem as

$$|\overrightarrow{AB}| = |\overrightarrow{BA}| = \sqrt{|x_B - x_A|^2 + |y_B - y_A|^2} = \sqrt{2^2 + 2^2} = 2$$

Fig. 7.7 Example 7.1-7(a)

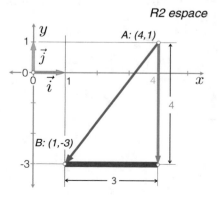

8. These vectors are placed in 3D space because there are three coordinates in each vector a_x, a_y, a_z.

(a)   Coordinates of $\overrightarrow{AB}$ we calculate as the end point minus the starting point,

$$\overrightarrow{AB} = \begin{bmatrix} 2 \\ 4 \\ -2 \end{bmatrix} - \begin{bmatrix} 4 \\ 1 \\ 6 \end{bmatrix} = \begin{bmatrix} (2-4) \\ (4-1) \\ (-2-6) \end{bmatrix} = \begin{bmatrix} -2 \\ 3 \\ -8 \end{bmatrix}$$

while coordinates of $\overrightarrow{BA}$ are

$$\overrightarrow{BA} = \begin{bmatrix} 4 \\ 1 \\ 6 \end{bmatrix} - \begin{bmatrix} 2 \\ 4 \\ -2 \end{bmatrix} = \begin{bmatrix} (4-2) \\ (1-4) \\ (6-(-2)) \end{bmatrix} = \begin{bmatrix} 2 \\ -3 \\ 8 \end{bmatrix}$$

in both cases, by inspection of graph, as for $\overrightarrow{AB}$ in Fig. 7.8, we find that vector is placed in diagonal of a cuboid, thus we calculate $|\overrightarrow{AB}|$ and $|\overrightarrow{BA}|$ with the help of Pythagoras's theorem as

$$|\overrightarrow{AB}| = |\overrightarrow{BA}| = \sqrt{|x_B - x_A|^2 + |y_B - y_A|^2 + |z_B - z_A|^2} = \sqrt{2^2 + 3^2 + 8^2} = \sqrt{77}$$

(b)   Relative to the point of origin, we write

$$\overrightarrow{AB} = \begin{bmatrix} 3 \\ 0 \\ -1 \end{bmatrix} - \begin{bmatrix} 2 \\ 3 \\ 4 \end{bmatrix} = \begin{bmatrix} (3-2) \\ (0-3) \\ (-1-4) \end{bmatrix} = \begin{bmatrix} 1 \\ -3 \\ -5 \end{bmatrix}$$

while coordinates of $\overrightarrow{BA}$ are

$$\overrightarrow{BA} = \begin{bmatrix} 2 \\ 3 \\ 4 \end{bmatrix} - \begin{bmatrix} 3 \\ 0 \\ -1 \end{bmatrix} = \begin{bmatrix} (2-3) \\ (3-0) \\ (4-(-1)) \end{bmatrix} = \begin{bmatrix} -1 \\ 3 \\ 5 \end{bmatrix}$$

in both cases, by inspection of graph, as for $\overrightarrow{AB}$ in Fig. 7.8, we find that vector is placed in diagonal of a cuboid, thus we calculate $|\overrightarrow{AB}|$ and $|\overrightarrow{BA}|$ with the help of Pythagoras's theorem as

$$|\overrightarrow{AB}| = |\overrightarrow{BA}| = \sqrt{|x_B - x_A|^2 + |y_B - y_A|^2 + |z_B - z_A|^2} = \sqrt{1^2 + 3^2 + 5^2} = \sqrt{35}$$

Fig. 7.8 Example 7.1-8(a)

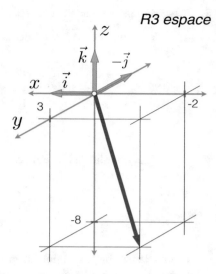

9. Given points and vectors, first we calculate the required vectors relative to the origin.

(a) $A(2, 3, 1)$, $B(3, -1, 0)$, $C = (-1, -2, 1)$, $D = (-3, 3, -2)$, therefore:

$$\overrightarrow{AB} = \begin{bmatrix} 3 \\ -1 \\ 0 \end{bmatrix} - \begin{bmatrix} 2 \\ 3 \\ 1 \end{bmatrix} = \begin{bmatrix} (3-2) \\ (-1-3) \\ (0-1) \end{bmatrix} = \begin{bmatrix} 1 \\ -4 \\ -1 \end{bmatrix}$$

$$\overrightarrow{CD} = \begin{bmatrix} -3 \\ 3 \\ -2 \end{bmatrix} - \begin{bmatrix} -1 \\ -2 \\ 1 \end{bmatrix} = \begin{bmatrix} (-3-(-1)) \\ (3-(-2)) \\ (-2-1) \end{bmatrix} = \begin{bmatrix} -2 \\ 5 \\ -3 \end{bmatrix}$$

Then, we calculate

$$\vec{m} = \overrightarrow{AB} + \overrightarrow{CD} = \begin{bmatrix} 1 \\ -4 \\ -1 \end{bmatrix} + \begin{bmatrix} -2 \\ 5 \\ -3 \end{bmatrix} = \begin{bmatrix} -1 \\ 1 \\ -4 \end{bmatrix}$$

$$\vec{n} = \overrightarrow{AB} - \overrightarrow{CD} = \begin{bmatrix} 1 \\ -4 \\ -1 \end{bmatrix} - \begin{bmatrix} -2 \\ 5 \\ -3 \end{bmatrix} = \begin{bmatrix} 3 \\ -9 \\ 2 \end{bmatrix}$$

(b) In this case $\vec{a} = (2, -3)$ and $\vec{b} = (3, -1)$ are already placed at the origin, thus

$$\vec{c} = \vec{a} + \vec{b} = \begin{bmatrix} 2 \\ -3 \end{bmatrix} + \begin{bmatrix} 3 \\ -1 \end{bmatrix} = \begin{bmatrix} 5 \\ -4 \end{bmatrix}$$

$$\vec{c} = \vec{a} - \vec{b} = \begin{bmatrix} 2 \\ -3 \end{bmatrix} - \begin{bmatrix} 3 \\ -1 \end{bmatrix} = \begin{bmatrix} -1 \\ -2 \end{bmatrix}$$

(c) Similarly,

$$\vec{p} = \vec{a} + \vec{b} = \begin{bmatrix} -1 \\ 2 \end{bmatrix} + \begin{bmatrix} 3 \\ 2 \end{bmatrix} = \begin{bmatrix} 2 \\ 4 \end{bmatrix}$$

$$\vec{m} = 2\vec{a} - \vec{b} = 2 \begin{bmatrix} -1 \\ 2 \end{bmatrix} - \begin{bmatrix} 3 \\ 2 \end{bmatrix} = \begin{bmatrix} (2 \times (-1) - 3) \\ (2 \times 2 - 2) \end{bmatrix} = \begin{bmatrix} -5 \\ 2 \end{bmatrix}$$

$$\vec{n} = 3\vec{a} - \vec{b} - \vec{c} = 3 \begin{bmatrix} -1 \\ 2 \end{bmatrix} - \begin{bmatrix} 3 \\ 2 \end{bmatrix} - \begin{bmatrix} -2 \\ -3 \end{bmatrix} = \begin{bmatrix} (3 \times (-1) - 3 - (-2)) \\ (3 \times 2 - 2 - (-3)) \end{bmatrix} = \begin{bmatrix} -4 \\ 7 \end{bmatrix}$$

$$\vec{r} = 2\vec{b} - \frac{1}{2}\vec{c} = 2 \begin{bmatrix} 3 \\ 2 \end{bmatrix} - \frac{1}{2} \begin{bmatrix} -2 \\ -3 \end{bmatrix} = \begin{bmatrix} (2 \times 3 - 1/2 \times (-2)) \\ (2 \times 2 - 1/2 \times (-3)) \end{bmatrix} = \begin{bmatrix} 7 \\ 11/2 \end{bmatrix}$$

10. By inspection of sketch in Fig. 7.9 (left) we write

$$\vec{a} = \vec{a_x} + \vec{a_y}$$
$$\vec{b} = \vec{b_x} + \vec{b_y}$$

where the horizontal and vertical projections are calculated as

$$|\vec{a_x}| = |\vec{a}| \cos \frac{\pi}{6} = \sqrt{3} \frac{\sqrt{3}}{2} = \frac{3}{2}$$

$$|\vec{a_y}| = |\vec{a}| \sin \frac{\pi}{6} = \sqrt{3} \frac{1}{2} = \frac{\sqrt{3}}{2}$$

Solution 1: in matrix form we write

$$\vec{a} = \begin{bmatrix} 3/2 \\ \sqrt{3}/2 \end{bmatrix} \qquad \vec{b} = \begin{bmatrix} 1 \\ 0 \end{bmatrix}$$

then, we calculate

$$\vec{p} = \vec{a} + \vec{b} \begin{bmatrix} 3/2 \\ \sqrt{3}/2 \end{bmatrix} + \begin{bmatrix} 1 \\ 0 \end{bmatrix} = \begin{bmatrix} 5/2 \\ \sqrt{3}/2 \end{bmatrix}$$

$$\vec{q} = \vec{a} - \vec{b} \begin{bmatrix} 3/2 \\ \sqrt{3}/2 \end{bmatrix} - \begin{bmatrix} 1 \\ 0 \end{bmatrix} = \begin{bmatrix} 1/2 \\ \sqrt{3}/2 \end{bmatrix}$$

Angle θ between $\vec{p}$ and $\vec{q}$ is found by inspection of Fig. 7.9 (right). We write,

$$\theta = \alpha - \beta$$

$$\tan \alpha = \frac{\sqrt{3}/2}{1/2} = \sqrt{3} \quad \therefore \quad \alpha = \arctan \sqrt{3} = \frac{\pi}{3} = 60°$$

$$\tan \beta = \frac{\sqrt{3}/2}{5/2} = \frac{\sqrt{3}}{5} \quad \therefore \quad \alpha = \arctan \frac{\sqrt{3}}{5} = 19.1°$$

$$\therefore$$

$$\theta = 40.89°$$

Fig. 7.9 Example 7.1-10

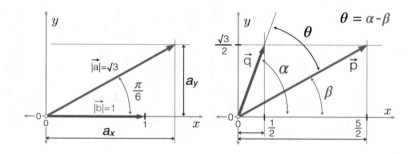

Solution 2: Alternatively, we can use for example Pythagoras's theorem and write

$$\vec{p} = \begin{bmatrix} 5/2 \\ \sqrt{3}/2 \end{bmatrix} \quad \therefore \quad |\vec{p}| \qquad = \sqrt{(5/2)^2 + (\sqrt{3}/2)^2} = \sqrt{7}$$

$$\vec{q} = \begin{bmatrix} 1/2 \\ \sqrt{3}/2 \end{bmatrix} \quad \therefore \quad |\vec{q}| \qquad = \sqrt{(1/2)^2 + (\sqrt{3}/2)^2} = 1$$

Then we can use the law of cosines for triangle formed by $\vec{p}$ and $\vec{q}$, see Fig. 7.10, and find that its third side is $a = 2$. Now all data is in place to calculate

$$a^2 = b^2 + c^2 - 2bc \cos \theta$$

$$\therefore$$

$$2^2 = (\sqrt{7})^2 + 1^2 - 2\sqrt{7} \cos \theta \quad \therefore \quad 2\sqrt{7}\theta = 4 \quad \therefore \quad \cos \theta = \frac{2}{\sqrt{7}}$$

$$\therefore$$

$$\theta = \arccos \frac{2}{\sqrt{7}} = 40.89°$$

Fig. 7.10 Example 7.1-10

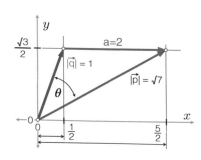

Exercise 7.2, page 132

1. Vector equations may be solved using the graphical representation.

(a) We resolve vector equation $\overrightarrow{AD} = 3\overrightarrow{AB} - {}^3/_2\overrightarrow{BC}$ by using, for example, the graphical method. One possible order of calculations is as follows.

Each pair of points in cartesian space $\mathscr{A} = (3, 5)$, $\mathscr{B} = (-1, 3)$, $\mathscr{C} = (1, -7)$ defines one vectors. Vectors $\overrightarrow{AB}$ and $\overrightarrow{BC}$, see Fig. 7.11, are translated from their initial positions to the origin 0, 0, Fig. 7.11.

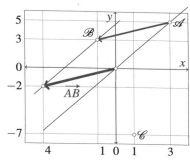 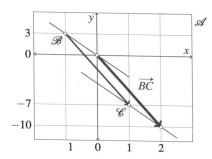

Fig. 7.11 Example 7.2-1(a)

We derive vector coordinates at the origin by separately calculating x and y vector projections. That is

$$\overrightarrow{AB} : (x_B - x_A, y_B - y_A) = (-1 - 3, 3 - 5) = (-4, -2)$$

$$\overrightarrow{BC} : (x_B - x_A, y_B - y_A) = (1 - (-1), -7 - 3) = (2, -10)$$

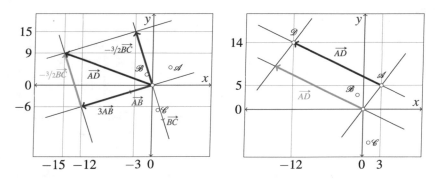

Fig. 7.12 Example 7.2-1(a)

Multiplication of each vector by a constant gives

$$3\overrightarrow{AB} : (3\overrightarrow{AB}_x, 3\overrightarrow{AB}_y) = (3 \times (-4), 3 \times (-2)) = (-12, -6)$$

$$-\frac{3}{2}\overrightarrow{BC} : \left(-\frac{3}{2}\overrightarrow{BC}_x, -\frac{3}{2}\overrightarrow{BC}_y\right) = \left(-\frac{3}{2} \times 2, -\frac{3}{2} \times (-10)\right) = (-3, 15)$$

Vector $\overrightarrow{AD} = 3\overrightarrow{AB} - \frac{3}{2}\overrightarrow{BC}$ as the sum of these two vectors, i.e. $3\overrightarrow{AB}$: $(-12, -6)$ and $-\frac{3}{2}\overrightarrow{BC}$: $(-3, 15)$, Fig. 7.12 (left), as

$$\overrightarrow{AD} : (-3 + (-12), 15 + (-6)) = (-15, 9)$$

In order to find coordinates of point $\mathcal{D}$, the origin of vector $\overrightarrow{AD}$ is translated to point $\mathcal{A}$: $(3, 5)$, Fig. 7.12 (right), and therefore we calculate

$$\mathcal{D} : (-15 + 3, 9 + 5) = (-12, 14)$$

(b)   Two lines

$$y_1 = m_1 x + b_1$$
$$y_2 = m_2 x + b_2$$

are orthogonal ($y_1 \perp y_2$) if

$$m_2 = -\frac{1}{m_1}$$

Slope of vector $\vec{n} = (1, 3)$ is found by definition (hint: $\tan x$ in a right angled triangle)

$$m_1 = \frac{\Delta y}{\Delta x} = \frac{3}{1} = 3$$

$$\therefore$$

$$m_2 = -\frac{1}{m_1} = -\frac{1}{3}$$

Thus, $y_2 = -1/3x + b_2$, where we calculate b_2 so that y_2 crosses point $\mathscr{A} : (x, y) = (2, 5)$. Therefore we write

$$y_2 = -\frac{1}{3}x + b_2$$

$$\therefore$$

$$5 = -\frac{1}{3} \times 2 + b_2 \quad \therefore \quad b_2 = \frac{17}{3}$$

$$\therefore$$

$$y_2 = -\frac{1}{3}x + \frac{17}{3}$$

as illustrated in Fig. 7.13.

Fig. 7.13 Example 7.2-1(b)

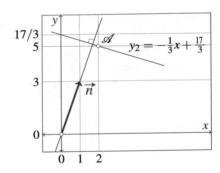

(c) Given two points in space $\mathscr{A} = (5, 4)$, $\mathscr{B} = (8, 6)$ slope m_1 of vector $\overrightarrow{AB}$ is found as

$$m_1 = \frac{\Delta y}{\Delta x} = \frac{2}{3} \quad \therefore \quad m_2 = -\frac{1}{m_1} = -\frac{3}{2}$$

We write line equation through point $\mathscr{A} : (0, 0)$, Fig. 7.14, where $b_2 = 0$ as

$$y_2 = -\frac{3}{2}x$$

Fig. 7.14 Example 7.2-1(c)

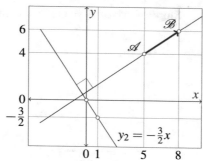

(d)   Two lines are parallel if their slopes are equal, i.e. $m_1 = m_2$. Given line equation $2x - 3y + 5 = 0$ and point in space $\mathscr{A}: (x, y) = (7, 6)$, Fig. 7.15, we derive equation of the parallel line as

$$2x - 3y + 5 = 0 \quad \therefore \quad y = \frac{2}{3}x + \frac{5}{3}$$

$$\therefore$$

$$y_2 = \frac{2}{3}x + b_2$$

$$\therefore$$

$$6 = \frac{2}{3} \times 7 + b_2 \quad \therefore \quad b_2 = \frac{4}{3}$$

$$\therefore$$

$$y_2 = \frac{2}{3}x + \frac{4}{3}$$

Fig. 7.15 Example 7.2-1(d)

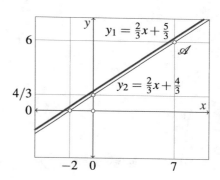

2. (a)   General equation of a circle is

$$(x - x_0)^2 + (y - y_0)^2 = r^2$$

where (x_0, y_0) are coordinates of the centre. We recall that this equation is yet another interpretation of Pythagoras's theorem Fig. 7.16.

Fig. 7.16 Example 7.2-2(a)

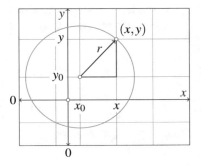

Given $C = (4, 0)$ and its radius equals $r = 3$, we derive the equation of a circle as follows

$$(x - 4)^2 + (y - 0)^2 = 3^2 \quad \therefore \quad (x - 4)^2 + y^2 - 9 = 0$$

$$\therefore$$

$$y = \pm\sqrt{9 - (x - 4)^2}$$

(b) Given equation is extended to fit into "binomial square" form as

$$x^2 + y^2 - 10x + 4y + 7 = 0$$

$$\underbrace{x^2 - 2 \times 5x + 25}_{} -25 + \underbrace{y^2 + 2 \times 2y + 4}_{} -4 + 7 = 0$$

$$\therefore$$

$$(x - 5)^2 + (y + 2)^2 = 22$$

In conclusion, centre coordinates are $(x_0, y_0) = (5, -2)$ and radius equals $r = \sqrt{22}$.

(c)   Given conditions, either point $\mathscr{A}$ or point $\mathscr{B}$ may serve as the circle centre, Fig. 7.17. Choosing $\mathscr{A}\colon (x_0, y_0) = (1, 2)$ while $\mathscr{B} = (-3, 4)$ we write

$$r = |\overrightarrow{AB}| = \sqrt{(1 - (-3))^2 + (4 - 2)^2} = \sqrt{20}$$

$$\therefore$$

$$(x - 1)^2 + (y - 2)^2 = 20$$

Fig. 7.17 Example 7.2-2(c)

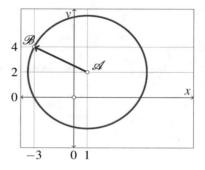

(d)   Given coordinates of circle centre $\mathscr{C} = (2, 0)$ and its radius $r = 5$ we write

$$(x - 2)^2 + y^2 = 25 \quad \therefore \quad x^2 - 4x + 4 + y^2 = 25$$

$$\therefore$$

$$y_1 = \pm\sqrt{21 + 4x - x^2}$$

At the same time, the line equation is

$$x - y - 3 = 0 \quad \therefore \quad y_2 = x - 3$$

The intersect point are found at the coordinates when the two curves are equal, i.e. $y_2 = y_1$, therefore we write the equality and square both sides as

$$x - 3 = \pm\sqrt{21 + 4x - x^2} \quad \Big|^2$$

$$\therefore$$

$$x^2 - 6x + 9 = 21 + 4x - x^2 \quad \therefore \quad 2x^2 - 10x - 12 = 0$$

$$\therefore$$

$$2x^2 + 2x - 12x - 12 = 0$$
$$2x(x + 1) - 12(x + 1) = 0$$
$$(x + 1)(2x - 12) = 0 \quad \therefore \quad 2(x + 1)(x - 6) = 0$$

$$\therefore$$

$$x_1 = -1, \quad x_2 = 6$$

Therefore, $y = x - 3 \quad \therefore \quad y_1 = -4, y_2 = 3$, Fig. 7.18.

Fig. 7.18 Example 7.2-2(d)

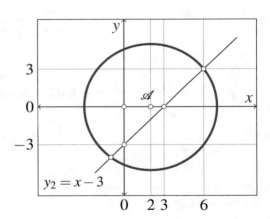

3. Non-right angle triangles are solved by the sin/cos laws.

(a)   Knowing "the cosine law" and "the sine law", i.e.

$$
\left.
\begin{aligned}
c^2 &= a^2 + b^2 - 2ab \cos \hat{C} \\
a^2 &= b^2 + c^2 - 2bc \cos \hat{A} \\
b^2 &= a^2 + c^2 - 2ac \cos \hat{B}
\end{aligned}
\right\} \quad \text{the cosine laws}
$$

$$
\left.
\frac{a}{\sin \hat{A}} = \frac{b}{\sin \hat{B}} = \frac{c}{\sin \hat{C}}
\right\} \quad \text{sine law}
$$

In general,

1. knowing all three sides of a triangle we use the cosine law;
2. knowing two sides and one angle we use the sine law if the angle is *not* in between the two sides, otherwise we use the cosine law;
3. knowing one side and two angles we use the sine law.

Given triangle whose sides are $AB = 4$, $AC = 6$, $BC = 8$, here $c = 4$, $b = 6$, $a = 8$, we write

$$
\cos \hat{A} = \frac{b^2 + c^2 - a^2}{2bc} = \frac{6^2 + 4^2 - 8^2}{2 \times 6 \times 4} = -\frac{1}{4}
$$

(b)   Given triangle as $AB = 3$, $AC = 4$, $\alpha = \pi/3$, i.e. $c = 3$, $b = 4$, $B\hat{A}C = 60°$, we write

$$
a^2 = b^2 + c^2 - 2bc \cos 60° = 25 - 24\frac{1}{2} \quad \therefore \quad a = \sqrt{13}
$$

Exercise 7.3, page 132

1. Vector addition is done by a simple chaining head of one vector to tail of the next one. Vector representing the total sum starts at tail of the first and ends at the head of last vector in the chain.

(a)   All vectors in 1D space are collinear (i.e. parallel), therefore their direction is already known. As a consequence, knowing only their magnitude is sufficient. In most general sense, numbers are viewed as the special case of vectors. As illustrated in Fig. 7.19 (left), sum of two numbers equals to sum of their magnitudes after accounting for their sign.

$$
2 + 3 = 5 \quad \text{and} \quad 2 + (-3) = -1
$$

(b) in 2D space not all vectors are collinear, which is to say that in general case, the sum of two vectors is not simply equal to the sum of their magnitudes. Instead, we must apply the rules of triangles. see Fig. 7.19 (right). In order to illustrate the practicality of matrix notification, we note that modules of vector projections add in the same manner as "regular" numbers. Formally, given that

$$\vec{c} = \vec{a} + \vec{b} \text{ if } \vec{a} = \begin{bmatrix} a_x \\ a_y \end{bmatrix} = \begin{bmatrix} 1 \\ 2 \end{bmatrix} \text{ and } \vec{b} = \begin{bmatrix} b_x \\ b_y \end{bmatrix} = \begin{bmatrix} 3 \\ -1 \end{bmatrix}$$

we write

$$\vec{c} = \vec{a} + \vec{b} = \begin{bmatrix} a_x \\ a_y \end{bmatrix} + \begin{bmatrix} b_x \\ b_y \end{bmatrix} = \begin{bmatrix} 1 \\ 2 \end{bmatrix} + \begin{bmatrix} 3 \\ -1 \end{bmatrix} = \begin{bmatrix} 1+3 \\ 2-1 \end{bmatrix} = \begin{bmatrix} 4 \\ 1 \end{bmatrix} = \begin{bmatrix} c_x \\ c_y \end{bmatrix}$$

Fig. 7.19 Example 7.3-1

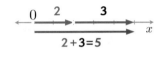

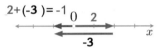

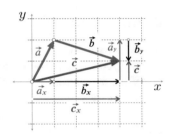

2. Multiplying a vector by a constant does not change the vector direction, only its magnitude, see Fig. 7.20. The constant multiplies all vector components, in this case both x_x and x_y.

(a) $\vec{a} = 2\vec{x} = 2\begin{bmatrix} x_x \\ x_y \end{bmatrix} = 2\begin{bmatrix} 3 \\ 2 \end{bmatrix} = \begin{bmatrix} 2\times 3 \\ 2\times 2 \end{bmatrix} = \begin{bmatrix} 6 \\ 4 \end{bmatrix}$

(b) $\vec{b} = \dfrac{2}{3}\vec{x} = \dfrac{2}{3}\begin{bmatrix} x_x \\ x_y \end{bmatrix} = \dfrac{2}{3}\begin{bmatrix} 3 \\ 2 \end{bmatrix} = \begin{bmatrix} 2/3\times 3 \\ 2/3\times 2 \end{bmatrix} = \begin{bmatrix} 2 \\ 4/3 \end{bmatrix}$

(c) $\vec{c} = -\dfrac{3}{4}\vec{x} = -\dfrac{3}{4}\begin{bmatrix} x_x \\ x_y \end{bmatrix} = -\dfrac{3}{4}\begin{bmatrix} 3 \\ 2 \end{bmatrix} = \begin{bmatrix} -3/4\times 3 \\ -3/4\times 2 \end{bmatrix} = \begin{bmatrix} -9/4 \\ -3/2 \end{bmatrix}$

Fig. 7.20 Example 7.3-2

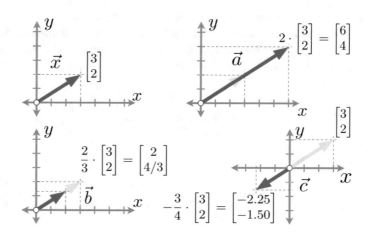

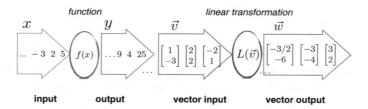

Exercise 7.4, page 133

1. In its basic form, a *function* is equivalent to a machine that for any given number input x performs an operation $f(x)$ and produces the corresponding output y. Similarly, for any given vector input $\vec{v}$ *linear operation* $L(\vec{v})$ performs an operation and produces the corresponding vector output $\vec{w}$, as illustrated in Fig. 7.21.

Fig. 7.21 Example 7.2-1

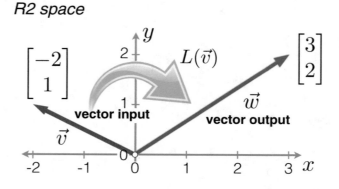

Another way to visualize operation of a linear transformation is to imagine that the input vector is displaced in space until it overlaps the output vector, see Fig. 7.22.

Fig. 7.22 Example 7.2-1

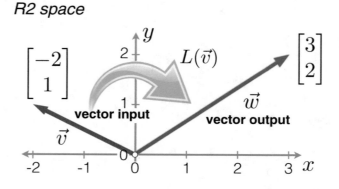

However, linear operation $L(\vec{v})$ displaces *all* vectors in R^2 vector space, that is to say it displaces *all the points on the surface*. For example, after linear transformation $L(\vec{v})$ displaces basis vector $\vec{i} = (1,0)$ into the new vector $L(\vec{i}) = (1,3)$, while at the same time basis vector $\vec{j} = (0,1)$ is displaced into the new vector $L(\vec{j}) = (2,-1)$ relative to the initial position, see Fig. 7.23. Since all points on the surface are also simultaneously displaced, we can imagine that the surface is "elastic" and it "stretched". In this analogy, now $L(\vec{i})$ and $L(\vec{j})$ represent basis vectors in the transformed space.

In summary, for a transformation to be *linear* it must obey the following rules:

1. after the transformation all grid lines must stay straight.
2. all the grid lines must stay parallel and equidistant because the distance between them must always be one basis vector.
3. the origin point must not move.

Due to these rules, the *numerical form* of a vector as written relative to the basis vectors $(\vec{i}, \vec{j})$ before, and $(L(\vec{i}), L(\vec{j}))$ after the transformation *stays the same*.

Fig. 7.23 Example 7.2-1

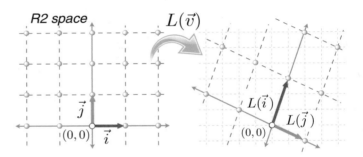

2. For example, we can create a graph as in Fig. 7.24, so that $\vec{v}$ is transformed into $L(\vec{v})$. We note the relationship between the original and transformed greed lines. In order to better visualize space transformation, the original grid is kept in the background. In addition, by following the transformation of basis vectors $(\vec{i}, \vec{j})$ we can visualize how the original R^2 space is rotated, flipped, and stretched into the new basis vectors $L(\vec{i}), L(\vec{j})$.

By inspection of grid in transformed space, we write the expression for transformed $L(\vec{v})$ within new coordinating system based on basis vectors $L(\vec{i}), L(\vec{j})$, and we compare it with the original vector $\vec{v}$ as

$$L(\vec{v}) = -2L(\vec{i}) + L(\vec{j}) \tag{7.3}$$

and

$$\vec{v} = -2\vec{i} + \vec{j} \tag{7.4}$$

In order to be truly linear transformation, the numerical forms of (7.3) and (7.4) are same. Consequently, coordinates of new basis vectors are, by definition, $L(\vec{i}) = (1,0)$ and $L(\vec{j}) = (0,1)$.

Fig. 7.24 Example 7.4-2

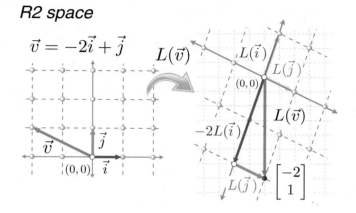

By inspection of graph, if expressed relative to the original basis vectors $\vec{i}$, $\vec{j}$, in this example we find that coordinates of

$$\vec{i} = (1, 0) \quad \therefore \quad L(\vec{i}) = (1, 3)$$

$$\vec{j} = (0, 1) \quad \therefore \quad L(\vec{j}) = (2, -1)$$

$$\therefore \quad L(\vec{v}) = (0, -7)$$

We can deduce matrix form of $L(\vec{v})$ as follows.

$$L(\vec{i}) = \begin{bmatrix} 1 \\ 3 \end{bmatrix} \quad L(\vec{j}) = \begin{bmatrix} 2 \\ -1 \end{bmatrix}$$

$$\therefore$$

$$L(\vec{v}) = -2L(\vec{i}) + L(\vec{j}) = -2\begin{bmatrix} 1 \\ 3 \end{bmatrix} + \begin{bmatrix} 2 \\ -1 \end{bmatrix} = \begin{bmatrix} -2 \\ -6 \end{bmatrix} + \begin{bmatrix} 2 \\ -1 \end{bmatrix} = \begin{bmatrix} 0 \\ -7 \end{bmatrix}$$

In compact matrix form, for an arbitrary $\vec{x}$ we prefer to write

$$\vec{x} = \begin{bmatrix} x_x \\ x_y \end{bmatrix} \quad \therefore \quad L(\vec{x}) = L\,\vec{x} = [L(\vec{i})\ L(\vec{j})]\,\vec{x} = \begin{bmatrix} 1 & 2 \\ 3 & -1 \end{bmatrix}\begin{bmatrix} x_x \\ x_y \end{bmatrix}$$

In this interpretation, all transformations of space can be formalized as the transformation matrix-vector product.

3. Given data, we write

$$L(\vec{i}) = \begin{bmatrix} 3 \\ 1 \end{bmatrix} \quad \text{and} \quad L(\vec{j}) = \begin{bmatrix} 1 \\ 2 \end{bmatrix}$$

$$\vec{x} = \begin{bmatrix} -1 \\ 2 \end{bmatrix} = (-1)\vec{i} + 2\vec{j}$$

Therefore,

$$L(\vec{x}) = (-1)L(\vec{i}) + 2L(\vec{j})$$

or, if measured in the original space:

$$= (-1)\begin{bmatrix} 3 \\ 1 \end{bmatrix} + 2\begin{bmatrix} 1 \\ 2 \end{bmatrix} = \begin{bmatrix} (-1) \times 3 + 2 \times 1 \\ (-1) \times 1 + 2 \times 2 \end{bmatrix}$$

$$= \begin{bmatrix} -1 \\ 3 \end{bmatrix}$$

The relation between the original and transformed space is illustrated in Fig. 7.25.

Fig. 7.25 Example 7.4-3

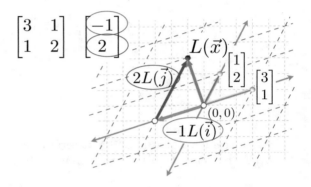

4. We follow movement of basis vectors $(\vec{i}, \vec{j})$ on graphs and read their new coordinates relative to

(a) Rotation of space is shown in Fig. 7.26. At the final position we find

$$L(\vec{i}) = \begin{bmatrix} 0 \\ 1 \end{bmatrix} \quad \text{and} \quad L(\vec{j}) = \begin{bmatrix} -1 \\ 0 \end{bmatrix} \quad \therefore \quad L(\vec{x}) = \begin{bmatrix} 0 & -1 \\ 1 & 0 \end{bmatrix}$$

which is matrix that performs $90°$ rotation of *all* points in R^2 space.

Fig. 7.26 Example 7.4-4(a)

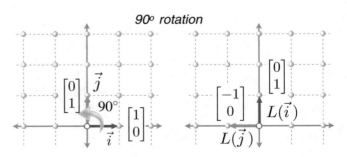

(b)   Operation of shear bends the space so that horizontal vectors are unchanged while vertical vectors bend $-45°$, as shown in Fig. 7.27. At the final position we find

$$L(\vec{i}\,) = \begin{bmatrix} 1 \\ 0 \end{bmatrix} \quad \text{and} \quad L(\vec{j}\,) = \begin{bmatrix} 1 \\ 1 \end{bmatrix} \quad \therefore \quad L(\vec{x}) = \begin{bmatrix} 1 & 1 \\ 0 & 1 \end{bmatrix}$$

which is matrix that performs shear of *all* points in R^2 space.

Fig. 7.27 Example 7.4-4(b)

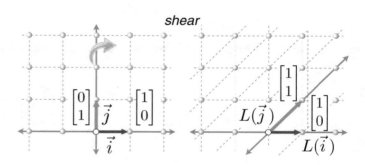

(c)   Linear transformations may be performed one after another. In the first step we do rotation $L_1(\vec{x})$ then sheer $L_2(\vec{x})$, as shown in Fig. 7.28. At the final position we find

$$L_1(\vec{x}) = \begin{bmatrix} 0 & -1 \\ 1 & 0 \end{bmatrix} \quad \text{and} \quad L_2(\vec{x}) = \begin{bmatrix} 1 & 1 \\ 0 & 1 \end{bmatrix} \quad \therefore \quad L(\vec{x}) = \begin{bmatrix} 1 & 1 \\ 0 & 1 \end{bmatrix}\begin{bmatrix} 0 & -1 \\ 1 & 0 \end{bmatrix} = \begin{bmatrix} 1 & -1 \\ 1 & 0 \end{bmatrix}$$

which is matrix that performs rotation plus shear of *all* points in R^2 space. We note that in this interpretation, $L(\vec{x})$ is matrix that does both transformations at the same time. Alternatively, multiple transformations are viewed as being equivalent to product of their respective matrix where transformation that is done first is written on the right side.

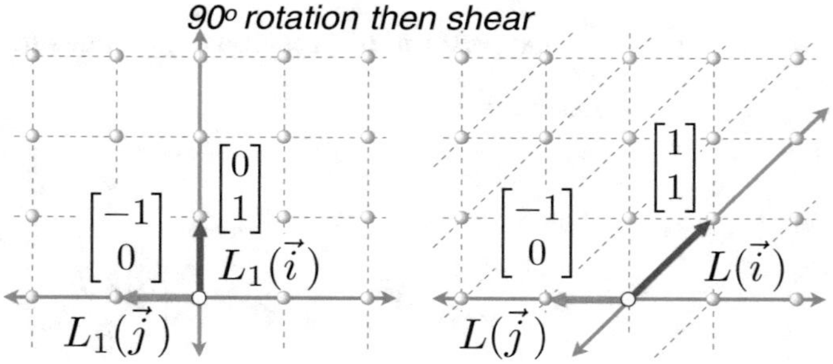

Fig. 7.28 Example 7.4-4(c)

90° rotation then shear

5. We follow movement of basis vectors $(\vec{i}, \vec{j})$ on graphs and read their final coordinates.

(a) This transformation simply enlarges the space by factor three in $\vec{i}$ direction and by factor two in $\vec{j}$ direction, see Fig. 7.29.

$$L(\vec{x}) = \begin{bmatrix} 3 & 0 \\ 0 & 2 \end{bmatrix} \quad \therefore \quad L(\vec{i}) = \begin{bmatrix} 3 \\ 0 \end{bmatrix} \quad \text{and} \quad L(\vec{j}) = \begin{bmatrix} 0 \\ 2 \end{bmatrix}$$

We note that relative position between basis vectors did not change, i.e. $\vec{j}$ is still on left relative to $\vec{i}$. In addition, the unity area increased by $3 \times 2 = 6$ factor. This area multiplication factor is geometrical interpretation of determinant, we use the syntax

$$|L(\vec{x})| = \begin{vmatrix} 3 & 0 \\ 0 & 2 \end{vmatrix} = 3 \times 2 - 0 \times 0 = 6$$

Fig. 7.29 Example 7.4-5(a)

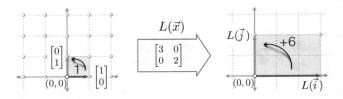

(b) This transformation performs multiple operations at the same time. By following basis vectors, we could imagine that 2D surface was rotated and then flipped over, see Fig. 7.30. In analogy to a sheet of a transparent paper, if the basis vectors are drawn on one side of the paper so that $\vec{j}$ is on the left side of $\vec{i}$, then if we flip the paper and look again at basis vectors, then from that perspective $\vec{j}$ is found on the right side of $\vec{i}$.

$$L(\vec{x}) = \begin{bmatrix} 1 & 2 \\ 1 & -1 \end{bmatrix} \quad \therefore \quad L(\vec{i}) = \begin{bmatrix} 1 \\ 1 \end{bmatrix} \quad \text{and} \quad L(\vec{j}) = \begin{bmatrix} 2 \\ -1 \end{bmatrix}$$

where determinant is calculated as

$$|L(\vec{x})| = \begin{vmatrix} 1 & 2 \\ 1 & -1 \end{vmatrix} = 1 \times (-1) - 1 \times 2 = -3$$

Geometrical interpretation is that the unity area is increased by factor of three; however, 2D surface is also "flipped over", which is indicated by the negative sign of "-3" multiplication factor.

Fig. 7.30 Example 7.4-5(b)

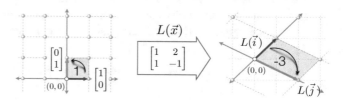

(c)   Linear transformations may be performed one after another. In the first step we do rotation $L_1(\vec{x})$ then sheer $L_2(\vec{x})$, as shown in Fig. 7.31. At the final position we find

$$L(\vec{x}) = \begin{bmatrix} 4 & 2 \\ 2 & 1 \end{bmatrix} \quad \therefore \quad L(\vec{i}) = \begin{bmatrix} 4 \\ 2 \end{bmatrix} \quad \text{and} \quad L(\vec{j}) = \begin{bmatrix} 2 \\ 1 \end{bmatrix}$$

where determinant is calculated as

$$|L(\vec{x})| = \begin{vmatrix} 4 & 2 \\ 2 & 1 \end{vmatrix} = 4 \times 1 - 2 \times 2 = 0$$

We note that $L(\vec{i})$ and $L(\vec{j})$ are *collinear*. After the transformation, 2D surface collapsed into 1D line. In the paper analogy, we can imagine that 2D surface is rotated so that we see its sideway projection, i.e. line. This reduction of space is geometrical interpretation of determinant equal zero. Alternative interpretation is that at least one of equation within system of equations in $L(\vec{x})$ is not independent.

Fig. 7.31 Example 7.4-5(c)

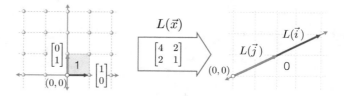

6. This type of problems we solve by both geometric and numerical methods.

(a)   Given three vectors, see Fig. 7.32, we find that $5\vec{b}$ to $\vec{a}$ closes the vector triangle, thus

$$\vec{m} = \vec{a} + 7\vec{b}$$

General form of linear combination takes the form of a sum, thus

$$\vec{m} = n\vec{a} + k\vec{b}$$

$$\therefore$$

$$\begin{bmatrix} 9 \\ 1 \end{bmatrix} = n \begin{bmatrix} 2 \\ 1 \end{bmatrix} + k \begin{bmatrix} 1 \\ 0 \end{bmatrix}$$

which is matrix form of the following system of equations

$$9 = 2n + 1k$$
$$1 = 1n + 0k$$

therefore, $n = 1$ and $k = 7$, which is to say $\vec{m} = \vec{a} + 7\vec{b}$.

(b)   Similarly,

$$\vec{m} = \vec{u} + \vec{v} + \vec{w} = \begin{bmatrix} 3 \\ -1 \end{bmatrix} + \begin{bmatrix} 1 \\ -2 \end{bmatrix} + \begin{bmatrix} -1 \\ 7 \end{bmatrix} = \begin{bmatrix} 3 \\ 4 \end{bmatrix}$$

Then,

$$\vec{m} = n\vec{u} + k\vec{v}$$

$$\therefore$$

$$\begin{bmatrix} 3 \\ 4 \end{bmatrix} = n \begin{bmatrix} 3 \\ -1 \end{bmatrix} + k \begin{bmatrix} 1 \\ -2 \end{bmatrix}$$

which is matrix form of the following system of equations

$$3 = 3n + 1k$$
$$4 = -1n - 2k$$

Cramer's rule gives,

$$\Delta \;=\; \begin{vmatrix} 3 & 1 \\ -1 & -2 \end{vmatrix} = -5 \neq 0$$

$$\Delta_n \;=\; \begin{vmatrix} 3 & 1 \\ 4 & -2 \end{vmatrix} = -10$$

$$\Delta_k \;=\; \begin{vmatrix} 3 & 3 \\ -1 & 4 \end{vmatrix} = 15$$

$$\therefore$$

$$n = \frac{\Delta_n}{\Delta} = \frac{-10}{-5} = 2$$

$$k = \frac{\Delta_k}{\Delta} = \frac{15}{-5} = -3$$

which is to say, $\vec{m} = 2\vec{u} - 3\vec{v}$

Fig. 7.32 Example 7.4-6(a)

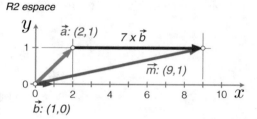

R2 espace

7. One method to prove that vectors are linearly independent is to confirm that the system's determinant is $\Delta \neq 0$. Alternatively, graphically the vectors should not be collinear (i.e. parallel)

(a)   Linear combination is

$$\vec{c} = k\vec{a} + n\vec{b}$$

$$\therefore$$

$$\begin{bmatrix} 7 \\ -4 \end{bmatrix} = k \begin{bmatrix} 3 \\ -2 \end{bmatrix} + n \begin{bmatrix} -2 \\ 1 \end{bmatrix}$$

which is matrix form of the following system of equations

$$7 = 3k - 2n$$

$$-4 = -2k + n$$

therefore, determinant is calculated as

$$\Delta = \begin{vmatrix} 3 & -2 \\ -2 & 1 \end{vmatrix} = -1 \neq 0$$

which is to say that equations (i.e. vectors) are indeed independent. Cramer's rule gives,

$$\Delta_k = \begin{vmatrix} 7 & -2 \\ -4 & 1 \end{vmatrix} = -1 \quad\Bigg\} \quad k = \frac{\Delta_k}{\Delta} = \frac{-1}{-1} = 1$$

$$\Delta_n = \begin{vmatrix} 3 & 7 \\ -2 & -4 \end{vmatrix} = 2 \quad\Bigg\} \therefore \quad n = \frac{\Delta_n}{\Delta} = \frac{2}{-1} = -2$$

which is to say, $\vec{c} = \vec{a} - 2\vec{b}$

(b) Linear combination is

$$\vec{c} = k\vec{a} + n\vec{b}$$

$$\therefore$$

$$\begin{bmatrix} 0 \\ -2 \end{bmatrix} = k \begin{bmatrix} 1 \\ 2 \end{bmatrix} + n \begin{bmatrix} 3 \\ 4 \end{bmatrix}$$

which is matrix form of the following system of equations

$$0 = k + 3n$$

$$-2 = 2k + 4n$$

therefore, determinant is calculated as

$$\Delta = \begin{vmatrix} 1 & 3 \\ 2 & 4 \end{vmatrix} = -2 \neq 0$$

which is to say that equations (i.e. vectors) are indeed independent. Cramer's rule gives,

$$\Delta_k = \begin{vmatrix} 0 & 3 \\ -2 & 4 \end{vmatrix} = 6 \quad\Bigg\} \quad k = \frac{\Delta_k}{\Delta} = \frac{6}{-2} = -3$$

$$\Delta_n = \begin{vmatrix} 1 & 0 \\ 2 & -2 \end{vmatrix} = -2 \quad\Bigg\} \therefore \quad n = \frac{\Delta_n}{\Delta} = \frac{-2}{-2} = 1$$

which is to say, $\vec{c} = -3\vec{a} + \vec{b}$

8. Main idea is that if vectors are linearly independent they are *not collinear*. Possible ways to formalize that statement is to write

$$\text{if } \vec{x} = k\vec{y} \quad \therefore \quad (\vec{x}, \vec{y}) \text{ are collinear}$$

where k is the multiplying constant. Alternatively, determinant of matrix that includes collinear vectors equals zero.

(a)   Stipulating that vectors are collinear, numerical method gives

$$\vec{b} = k\vec{a}$$

$$\therefore$$

$$\begin{bmatrix} 2 \\ 6 \end{bmatrix} = k \begin{bmatrix} 3 \\ \lambda \end{bmatrix}$$

which is matrix form of the following system of equations

$$\left. \begin{array}{l} 2 = 3k \\ 6 = \lambda k \end{array} \right\} \quad \therefore \quad \begin{array}{l} k = {}^3/2 \\ 6 = \lambda k \quad \therefore \quad \lambda = 9 \end{array}$$

which is to say, if $\lambda = 9$, then $(\vec{a}, \vec{b})$ are collinear, i.e. dependent and can be written as $\vec{b} = {}^3/2\vec{a}$, which is easily verified by performing the multiplication.
Alternatively, we calculate determinant as

$$\vec{b} = k\vec{a} \quad \therefore \quad \begin{bmatrix} 2 \\ 6 \end{bmatrix} = k \begin{bmatrix} 3 \\ \lambda \end{bmatrix}$$

Determinant is then calculated as

$$\Delta = \begin{vmatrix} 2 & 3 \\ 6 & \lambda \end{vmatrix} = 3 \times 6 - 2\lambda = 2\lambda - 18 = 0 \quad \therefore \quad \lambda = 9$$

which leads into the same conclusion as before.

(b)   Given vectors

$$\vec{a} = \begin{bmatrix} 6 \\ 8 \\ 4 \end{bmatrix} \qquad \vec{b} = \begin{bmatrix} 3 \\ 4 \\ 2 \end{bmatrix} \qquad \vec{c} = \begin{bmatrix} \lambda \\ 0 \\ 1 \end{bmatrix}$$

we see that $\vec{a} = 2\vec{b}$, which is to say that $(\vec{a}, \vec{b})$ are collinear (i.e. dependent) regardless of λ.
Alternatively, determinant is

$$\Delta = \begin{vmatrix} 6 & 3 & \lambda \\ 8 & 4 & 0 \\ 4 & 2 & 1 \end{vmatrix} \begin{matrix} 6 & 3 \\ 8 & 4 \\ 4 & 2 \end{matrix} = +6 \times 4 \times 1 + 3 \times 0 \times 4 + \lambda \times 8 \times 2$$

$$- 4 \times 4 \times \lambda - 2 \times 0 \times 6 - 1 \times 8 \times 3$$

$$= 24 + 16\lambda - 16\lambda - 24$$

$$= 0$$

that is to say, the determinant equals zero regardless of λ, same conclusion as already reached

Exercise 7.5, page 134

1. There are multiple methods for calculating determinants, here we illustrate two of the most common ones where either "$|A|$" or "Δ" annotates "determinant of matrix A".

(a) Second order determinants are calculated as the difference between products along their two diagonals.

$$|A_2| = \Delta_2 = \begin{vmatrix} 3 & 0 \\ 0 & 2 \end{vmatrix} = 3 \times 2 - 0 \times 0 = 6 \neq 0$$

(b) Similarly,

$$|A_2| = \begin{vmatrix} 1 & 2 \\ 1 & -1 \end{vmatrix} = 1 \times (-1) - 1 \times (-2) = -3 \neq 0$$

(c) Here, we note that the determinant equals zero, which is to say that two rows (i.e. equations) are not independent. Indeed, the two rows (i.e. equations) are related by a simple factor of two.

$$|A_2| = \begin{vmatrix} 4 & 2 \\ 2 & 1 \end{vmatrix} = 4 \times 1 - 2 \times 2 = 0$$

(d) Following the same idea of third and higher order determinants are calculated by calculating cross-products

$$\Delta_3 = \begin{vmatrix} 2 & 3 & 4 \\ 5 & -2 & 1 \\ 1 & 2 & 3 \end{vmatrix}$$

$$= +(2 \times (-2) \times 3) + (3 \times 1 \times 1) + (4 \times 5 \times 2)$$
$$-(1 \times (-2) \times 4) - (2 \times 1 \times 2) - (3 \times 5 \times 3)$$
$$= -12 + 3 + 40 + 8 - 4 - 45$$
$$= -10 \neq 0$$

Alternatively, we can use the method of cofactor expansions to calculate the same results as

$$\Delta_3 = \begin{bmatrix} 2 & 3 & 4 \\ 5 & -2 & 1 \\ 1 & 2 & 3 \end{bmatrix} + \begin{bmatrix} 2 & 3 & 4 \\ 5 & -2 & 1 \\ 1 & 2 & 3 \end{bmatrix} + \begin{bmatrix} 2 & 3 & 4 \\ 5 & -2 & 1 \\ 1 & 2 & 3 \end{bmatrix}$$

$$= (-1)^{1+1}\, 2 \begin{vmatrix} -2 & 1 \\ 2 & 3 \end{vmatrix} + (-1)^{1+2}\, 3 \begin{vmatrix} 5 & 1 \\ 1 & 3 \end{vmatrix} + (-1)^{1+3}\, 4 \begin{vmatrix} 5 & -2 \\ 1 & 2 \end{vmatrix}$$

$$= 2 \times (-2 \times 3 - 2 \times 1) - 3 \times (5 \times 3 - 1 \times 1) + 4 \times (5 \times 2 - 1 \times (-2))$$

$$= -10$$

2. Area of a surface formed by two vectors in R^2 space equals to the associated determinant. The same is valid in higher order spaces, thus volume of a cuboid formed by three vectors in R^3 space equals to the associated determinant, etc.

   (a) Given three vectors in R^3, we calculate the volume as

$$\Delta = \begin{vmatrix} 1 & 2 & 1 \\ -3 & 1 & 2 \\ 1 & -3 & 1 \end{vmatrix} \begin{matrix} 1 & 2 \\ -3 & 1 \\ 1 & -3 \end{matrix} = 1 + 4 + 9 - 1 + 6 + 6 = 25$$

   (b) Similarly,

$$\Delta = \begin{vmatrix} 1 & 0 & 3 \\ 0 & 1 & 4 \\ 3 & 2 & 0 \end{vmatrix} \begin{matrix} 1 & 0 \\ 0 & 1 \\ 3 & 2 \end{matrix} = -9 - 8 = -17$$

   We conclude that $V = 17$, albeit due to negative sign it is "inverted" in R^3 space.

3. Second order determinants are calculated by simple cross-product

   (a) $\begin{vmatrix} 2 & -1 \\ 1 & 0 \end{vmatrix} = 2 \times 0 - 1 \times (-1) = 1$

   (b) $\begin{vmatrix} \sqrt{3} & -3\sqrt{2} \\ \sqrt{3} & 2\sqrt{2} \end{vmatrix} = \sqrt{3} \times 2\sqrt{2} - \sqrt{3} \times (-3\sqrt{2}) = 2\sqrt{6} + 3\sqrt{6} = 5\sqrt{6}$

   (c) $\begin{vmatrix} 1 & i \\ -i & 1 \end{vmatrix} = 1 - (-i) \times i = 1 - 1 = 0$

   (d) $\begin{vmatrix} \sin\alpha & \cos\alpha \\ -\cos\alpha & \sin\alpha \end{vmatrix} = \sin\alpha \times \sin\alpha - (-\cos\alpha) \times \cos\alpha = \sin^2\alpha + \cos^2\alpha = 1$

4. Third order determinants may be calculated by cross-products.

   (a) $\begin{vmatrix} 3 & 2 & 1 \\ 4 & 5 & 6 \\ 8 & 9 & 7 \end{vmatrix} \begin{matrix} 3 & 2 \\ 4 & 5 \\ 8 & 9 \end{matrix} = -21$

   (b) $\begin{vmatrix} 2 & 3 & 1 \\ 3 & -1 & 2 \\ 1 & 1 & -3 \end{vmatrix} \begin{matrix} 2 & 3 \\ 3 & -1 \\ 1 & 1 \end{matrix} = 39$

(c) $\begin{vmatrix} 5 & 0 & 4 \\ 8 & 0 & -7 \\ 3 & 2 & 1 \end{vmatrix} \begin{matrix} 5 & 0 \\ 8 & 0 \\ 3 & 2 \end{matrix} = 134$

(d) $\begin{vmatrix} 1 & 3 & 5 \\ 7 & 9 & 11 \\ 13 & 15 & 17 \end{vmatrix} \begin{matrix} 1 & 3 \\ 7 & 9 \\ 13 & 15 \end{matrix} = 0$

Exercise 7.6, page 134

1. In order to keep track of matrix transformations, the three equations are numbered as (1), (2), (3) and their coefficients are placed in first, second, and third matrix row. In consequence, the vertical columns are where the coefficients of x, y, z variables are placed. In addition, the solutions of each equation are placed in the fourth column of expanded matrix.

$$(x) \quad (y) \quad (z)$$

$$5x - 5y - 15z = 40 \quad (1)$$

$$4x - 2y - 6z = 19 \quad (2)$$

$$3x - 6y - 17z = 41 \quad (3)$$

The objective of this technique is to transform the expanded matrix into its equivalent diagonalized form (i.e. all coefficient found at diagonal equal "1", while at the same time all the other coefficients equal to "0").

$$\begin{bmatrix} 5 & -5 & -15 & | & 40 \\ 4 & -2 & -6 & | & 19 \\ 3 & -6 & -17 & | & 41 \end{bmatrix} \Rightarrow \begin{bmatrix} 1 & 0 & 0 & | & a \\ 0 & 1 & 0 & | & b \\ 0 & 0 & 1 & | & c \end{bmatrix} \quad \therefore \quad \begin{matrix} 1 \times x + 0 \times y + 0 \times z = a \\ 0 \times x + 1 \times y + 0 \times z = b \\ 0 \times x + 0 \times y + 1 \times z = c \end{matrix} \quad \therefore \quad \begin{matrix} x = a \\ y = b \\ z = c \end{matrix}$$

We start by transforming coefficient found at first-row–first-column position into "1". In order to do so, obviously, it must be divided by five. Which is to say that all coefficients in the first row are also divided by five. Then systematically transform the other coefficients at and below diagonal. In order to annotate transformations, for example, the operation of division of each term in the first row equation (1) by five is written as "$\leftarrow (1) \div [5]$". Of course, there are multiple ways to transform the matrix that do not influence the final solution, it is matter of practice and preference which particular intermediate steps are taken. However, good strategy is to transform the diagonal and lower triangle coefficients first. Then, transform the upper triangle coefficients.

$$\begin{bmatrix} 5 & -5 & -15 & | & 40 \\ 4 & -2 & -6 & | & 19 \\ 3 & -6 & -17 & | & 41 \end{bmatrix} \quad \leftarrow (1) \div [5]$$

$$\begin{bmatrix} 1 & -1 & -3 & | & 8 \\ 4 & -2 & -6 & | & 19 \\ 3 & -6 & -17 & | & 41 \end{bmatrix} \quad \begin{matrix} \\ \leftarrow (2) - 4 \times (1) \\ \leftarrow (3) - 3 \times (1) \end{matrix}$$

$$\begin{bmatrix} 1 & -1 & -3 & | & 8 \\ 0 & 2 & 6 & | & -13 \\ 0 & -3 & -8 & | & 17 \end{bmatrix} \quad \leftarrow (2) \div [2]$$

$$\begin{bmatrix} 1 & -1 & -3 & | & 8 \\ 0 & 1 & 3 & | & -13/2 \\ 0 & -3 & -8 & | & 17 \end{bmatrix} \quad \leftarrow (3) + 3 \times (2)$$

$$\begin{bmatrix} 1 & -1 & -3 & | & 8 \\ 0 & 1 & 3 & | & -13/2 \\ 0 & 0 & 1 & | & -5/2 \end{bmatrix} \leftarrow (1) + (2)$$

$$\begin{bmatrix} 1 & 0 & 0 & | & 3/2 \\ 0 & 1 & 3 & | & -13/2 \\ 0 & 0 & 1 & | & -5/2 \end{bmatrix} \leftarrow (2) - 3 \times (3)$$

$$\begin{bmatrix} 1 & 0 & 0 & | & 3/2 \\ 0 & 1 & 0 & | & 1 \\ 0 & 0 & 1 & | & -5/2 \end{bmatrix} \quad \therefore \quad \{x, y, z\} = \left\{ \frac{3}{2}, 1, -\frac{5}{2} \right\}$$

2. Since all three points A, B, C must satisfy the quadratic equation $f(x)$ at the same time, we write the following system of equations

$$\therefore \quad ax^2 + bx + c = f(x)$$

$$\underline{(x, y) = (-2, 20)} \quad \therefore \quad a(-2)^2 + b(-2) + c = 20$$

$$\underline{(x, y) = (1, 5)} \quad \therefore \quad a(1)^2 + b(1) + c = 5$$

$$\underline{(x, y) = (3, 25)} \quad \therefore \quad a(3)^2 + b(3) + c = 25$$

This problem can be formalized in matrix form as

$$(a) \quad (b) \quad (c)$$

$$4a - 2b + c = 20 \quad (1)$$

$$a + b + c = 5 \quad (2)$$

$$9a + 3b + c = 25 \quad (3)$$

We start by taking advantage of the fact that the first coefficient of (2) already equals one, thus we exchange positions of (1) and (2). In any case, the initial order of equations is arbitrary.

$$\begin{bmatrix} 4 & -2 & 1 & | & 20 \\ 1 & 1 & 1 & | & 5 \\ 9 & 3 & 1 & | & 25 \end{bmatrix} \quad (1) \leftrightarrow (2)$$

$$\begin{bmatrix} 1 & 1 & 1 & | & 5 \\ 4 & -2 & 1 & | & 20 \\ 9 & 3 & 1 & | & 25 \end{bmatrix} \leftarrow (2) - 4 \times (1)$$

$$\begin{bmatrix} 1 & 1 & 1 & | & 5 \\ 0 & -6 & -3 & | & 0 \\ 9 & 3 & 1 & | & 25 \end{bmatrix} \leftarrow (2) \div [-6]$$

$$\begin{bmatrix} 1 & 1 & 1 & | & 5 \\ 0 & 1 & 1/2 & | & 0 \\ 9 & 3 & 1 & | & 25 \end{bmatrix} \leftarrow (3) - 9 \times (1)$$

$$\begin{bmatrix} 1 & 1 & 1 & 5 \\ 0 & 1 & 1/2 & 0 \\ 0 & -6 & -8 & -20 \end{bmatrix} \leftarrow (3) + 6 \times (2)$$

$$\begin{bmatrix} 1 & 1 & 1 & 5 \\ 0 & 1 & 1/2 & 0 \\ 0 & 0 & -5 & -20 \end{bmatrix} \leftarrow (3) \div [-5]$$

$$\begin{bmatrix} 1 & 1 & 1 & 5 \\ 0 & 1 & 1/2 & 0 \\ 0 & 0 & 1 & 4 \end{bmatrix} \leftarrow (1) - (2)$$

$$\begin{bmatrix} 1 & 0 & 1/2 & 5 \\ 0 & 1 & 1/2 & 0 \\ 0 & 0 & 1 & 4 \end{bmatrix} \leftarrow (2) - 1/2(3)$$

$$\begin{bmatrix} 1 & 0 & 1/2 & 5 \\ 0 & 1 & 0 & -2 \\ 0 & 0 & 1 & 4 \end{bmatrix} \leftarrow (1) - 1/2(3)$$

$$\begin{bmatrix} 1 & 0 & 0 & 3 \\ 0 & 1 & 0 & -2 \\ 0 & 0 & 1 & 4 \end{bmatrix} \quad \therefore \quad \{a, b, c\} = \{3, -2, 4\}$$

Which is to say that $f(x) = 3x^2 - 2x + 4$.

3. Cramer's rule is based on determinants of the extended matrix.

(a) Principal and two sub-determinants are

$$A = \begin{bmatrix} 4 & -3 & 0 \\ 2 & 3 & 18 \end{bmatrix} \quad \therefore \quad \Delta = \begin{vmatrix} 4 & -3 \\ 2 & 3 \end{vmatrix} = 4 \times 3 - 2 \times (-3) = 18 \neq 0$$

$$\Delta_{x_1} = \begin{vmatrix} 0 & -3 \\ 18 & 3 \end{vmatrix} = 0 \times 3 - 18 \times (-3) = 3 \times 18$$

$$\Delta_{x_2} = \begin{vmatrix} 4 & 0 \\ 2 & 18 \end{vmatrix} = 4 \times 18 - 2 \times (0) = 4 \times 18$$

By definition, we find

$$x_1 = \frac{\Delta_{x_1}}{\Delta} = \frac{3 \times 18}{18} = 3$$

$$x_2 = \frac{\Delta_{x_1}}{\Delta} = \frac{4 \times 18}{18} = 4$$

(b)   Principal and three sub-determinants are

$$A = \begin{bmatrix} 5 & -5 & -15 & 40 \\ 4 & -2 & -6 & 19 \\ 3 & -6 & -17 & 41 \end{bmatrix} \quad \therefore \quad \Delta = \begin{bmatrix} 5 & -5 & -15 \\ 4 & -2 & -6 \\ 3 & -6 & -17 \end{bmatrix} \begin{matrix} 5 & -5 \\ 4 & -2 \\ 3 & -6 \end{matrix} = 10 \neq 0$$

$$\Delta_x = \begin{bmatrix} 40 & -5 & -15 \\ 19 & -2 & -6 \\ 41 & -6 & -17 \end{bmatrix} \begin{matrix} 40 & -5 \\ 19 & -2 \\ 41 & -6 \end{matrix} = 15$$

$$\Delta_y = \begin{bmatrix} 5 & 40 & -15 \\ 4 & 19 & -6 \\ 3 & 41 & -17 \end{bmatrix} \begin{matrix} 5 & 40 \\ 4 & 19 \\ 3 & 41 \end{matrix} = 10$$

$$\Delta_z = \begin{bmatrix} 5 & -5 & 40 \\ 4 & -2 & 19 \\ 3 & -6 & 41 \end{bmatrix} \begin{matrix} 5 & -5 \\ 4 & -2 \\ 3 & -6 \end{matrix} = -25$$

By definition, we find

$$x = \frac{\Delta_x}{\Delta} = \frac{15}{10} = \frac{3}{2}$$

$$y = \frac{\Delta_y}{\Delta} = \frac{10}{10} = 1$$

$$z = \frac{\Delta_z}{\Delta} = \frac{-25}{10} = -\frac{5}{2}$$

4. Matrix form of this system is

$$A \vec{v} = \vec{b}$$

where

$$A = \begin{bmatrix} 1 & 1 & 0 \\ 0 & 1 & 1 \\ -1 & 0 & 2 \end{bmatrix} \quad \vec{v} = \begin{bmatrix} x \\ y \\ z \end{bmatrix} \quad \vec{b} = \begin{bmatrix} 7 \\ 8 \\ 7 \end{bmatrix}$$

That is to say,

$$\begin{bmatrix} 1 & 1 & 0 \\ 0 & 1 & 1 \\ -1 & 0 & 2 \end{bmatrix} \begin{bmatrix} x \\ y \\ z \end{bmatrix} = \begin{bmatrix} 7 \\ 8 \\ 7 \end{bmatrix}$$

Determinants are therefore,

$$\Delta = \begin{vmatrix} 1 & 1 & 0 \\ 0 & 1 & 1 \\ -1 & 0 & 2 \end{vmatrix} \begin{matrix} 1 & 1 \\ 0 & 1 \\ -1 & 0 \end{matrix} = 2 + (-1) + 0 - 0 - 0 - 0 = 1 \neq 0$$

$$\Delta_x = \begin{vmatrix} 7 & 1 & 0 \\ 8 & 1 & 1 \\ 7 & 0 & 2 \end{vmatrix} \begin{matrix} 7 & 1 \\ 8 & 1 \\ 7 & 0 \end{matrix} = 14 + 7 + 0 - 0 - 0 - 16 = 5$$

$$\Delta_y = \begin{vmatrix} 1 & 7 & 0 \\ 0 & 8 & 1 \\ -1 & 7 & 2 \end{vmatrix} \begin{matrix} 1 & 7 \\ 0 & 8 \\ -1 & 7 \end{matrix} = 16 + (-7) + 0 - 0 - 7 - 0 = 2$$

$$\Delta_z = \begin{vmatrix} 1 & 1 & 7 \\ 0 & 1 & 8 \\ -1 & 0 & 7 \end{vmatrix} \begin{matrix} 1 & 1 \\ 0 & 1 \\ -1 & 0 \end{matrix} = 7 + (-8) + 0 - (-7) - 0 - 0 = 6$$

Determinant $\Delta = 1 (\neq 0)$ therefore we calculate

$$x = \frac{\Delta_x}{\Delta} = \frac{5}{1} = 5 \quad y = \frac{\Delta_y}{\Delta} = \frac{2}{1} = 2 \quad z = \frac{\Delta_z}{\Delta} = \frac{6}{1} = 6$$

that is to say, $\vec{v} = \begin{bmatrix} 5 \\ 2 \\ 6 \end{bmatrix}$

or, $(x, y, z) = (5, 2, 6)$

Geometric interpretation is that transformation A stretches the space as necessary until $\vec{v}$ is equal to $\vec{b}$.

5. It is important to master multiple techniques for solving the system of linear equations.

(a) For example, we can use matrix form of the given system of equations and Cramer's rule to write

$$\begin{bmatrix} 1 & 2 \\ 3 & -1 \end{bmatrix} \begin{bmatrix} x \\ y \end{bmatrix} = \begin{bmatrix} -5 \\ 13 \end{bmatrix} \quad \therefore \quad \left.\begin{aligned} \Delta &= \begin{vmatrix} 1 & 2 \\ 3 & -1 \end{vmatrix} = -7 \\ \Delta_x &= \begin{vmatrix} -5 & 2 \\ 13 & -1 \end{vmatrix} = -21 \\ \Delta_y &= \begin{vmatrix} 1 & -5 \\ 3 & 13 \end{vmatrix} = 28 \end{aligned}\right\} \begin{aligned} x &= \frac{\Delta_x}{\Delta} = \frac{-21}{-7} = 3 \\ y &= \frac{\Delta_y}{\Delta} = \frac{28}{-7} = -4 \end{aligned}$$

(b)   For example we can use the elimination method

$$x + 2y = -8$$
$$2x - y = -1 \quad \times 2$$

$$x + 2y = -8$$
$$4x - 2y = -2 \quad \therefore$$

$$5x = -10 \quad \therefore \quad x = -2$$
$$-2 + 2y = -8 \quad \therefore \quad y = -3$$

(c)   For example we can use the matrix transformation method and given system

$$x + y + z = 36$$
$$2x \quad - z = -17$$
$$6x \quad - 5z = 7$$

we write,

$$\begin{bmatrix} 1 & 1 & 1 & | & 36 \\ 2 & 0 & -1 & | & -17 \\ 6 & 0 & -5 & | & 7 \end{bmatrix} \leftarrow (2) \div [2]$$

$$\begin{bmatrix} 1 & 1 & 1 & | & 36 \\ 1 & 0 & -1/2 & | & -17/2 \\ 6 & 0 & -5 & | & 7 \end{bmatrix} \begin{array}{l} \leftarrow (2) - (1) \\ \leftarrow (3) - 6 \times (1) \end{array}$$

$$\begin{bmatrix} 1 & 1 & 1 & | & 36 \\ 0 & -1 & -3/2 & | & -89/2 \\ 0 & -6 & -11 & | & -209 \end{bmatrix} \leftarrow (3) - 6 \times (2)$$

$$\begin{bmatrix} 1 & 1 & 1 & | & 36 \\ 0 & 1 & 3/2 & | & 89/2 \\ 0 & 0 & -2 & | & 58 \end{bmatrix} \leftarrow (3) \div [-2]$$

$$\begin{bmatrix} 1 & 1 & 1 & | & 36 \\ 0 & 1 & 3/2 & | & 89/2 \\ 0 & 0 & 1 & | & -29 \end{bmatrix} \leftarrow (1) - (2)$$

$$\begin{bmatrix} 1 & 0 & -1/2 & -17/2 \\ 0 & 1 & 3/2 & 89/2 \\ 0 & 0 & 1 & -29 \end{bmatrix} \begin{matrix} \leftarrow (1) + 1/2(3) \\ \leftarrow (2) - 3/2(3) \\ \end{matrix}$$

$$\begin{bmatrix} 1 & 0 & 0 & -23 \\ 0 & 1 & 0 & 88 \\ 0 & 0 & 1 & -29 \end{bmatrix}$$

$$\therefore \quad \{x, y, z\} = \{-23, 88, -29\}$$

(d)   For example, using Cramer's rule

$$\Delta = \begin{vmatrix} 2 & 3 & -1 \\ 1 & 1 & 2 \\ 2 & -1 & 1 \end{vmatrix} \begin{matrix} 2 & 3 \\ 1 & 1 \\ 2 & -1 \end{matrix} = 2 + 12 + 1 + 2 + 4 - 3 = 18$$

$$\Delta_x = \begin{vmatrix} 5 & 3 & -1 \\ 7 & 1 & 2 \\ 1 & -1 & 1 \end{vmatrix} \begin{matrix} 5 & 3 \\ 7 & 1 \\ 1 & -1 \end{matrix} = 5 + 6 + 7 + 1 + 10 - 21 = 8$$

$$\Delta_y = \begin{vmatrix} 2 & 5 & -1 \\ 1 & 7 & 2 \\ 2 & 1 & 1 \end{vmatrix} \begin{matrix} 2 & 5 \\ 1 & 7 \\ 2 & 1 \end{matrix} = 14 + 20 - 1 + 14 - 4 - 5 = 38$$

$$\Delta_z = \begin{vmatrix} 2 & 3 & 5 \\ 1 & 1 & 7 \\ 2 & -1 & 1 \end{vmatrix} \begin{matrix} 2 & 3 \\ 1 & 1 \\ 2 & -1 \end{matrix} = 2 + 42 - 5 - 10 + 14 - 3 = 40$$

Then,

$$x = \frac{\Delta_x}{\Delta} = \frac{4}{9} \qquad y = \frac{\Delta_y}{\Delta} = \frac{19}{9} \qquad z = \frac{\Delta_z}{\Delta} = \frac{20}{9}$$

Exercise 7.7, page 135

1. Matrix-vector product is calculated as linear combination.

(a) $\begin{bmatrix} 1 & 1 \\ 2 & -1 \end{bmatrix} \begin{bmatrix} 3 \\ 0 \end{bmatrix} = 3 \begin{bmatrix} 1 \\ 2 \end{bmatrix} + 0 \begin{bmatrix} 1 \\ -1 \end{bmatrix} = \begin{bmatrix} 3 \\ 6 \end{bmatrix}$

(b) $\begin{bmatrix} 2 & -1 \\ 3 & 2 \end{bmatrix} \begin{bmatrix} -5 \\ 3 \end{bmatrix} = -5 \begin{bmatrix} 2 \\ 3 \end{bmatrix} + 3 \begin{bmatrix} -1 \\ 2 \end{bmatrix} = \begin{bmatrix} -13 \\ -9 \end{bmatrix}$

(c) $\begin{bmatrix} 2 & -3 \\ 1 & 2 \end{bmatrix} \begin{bmatrix} 3 \\ -2 \end{bmatrix} = 3 \begin{bmatrix} 2 \\ 1 \end{bmatrix} + (-2) \begin{bmatrix} -3 \\ 2 \end{bmatrix} = \begin{bmatrix} 12 \\ -1 \end{bmatrix}$

(d) $\begin{bmatrix} 1 & 2 \\ 2 & -1 \end{bmatrix} \begin{bmatrix} -8 \\ -1 \end{bmatrix} = (-8) \begin{bmatrix} 1 \\ 2 \end{bmatrix} + (-1) \begin{bmatrix} 2 \\ -1 \end{bmatrix} = \begin{bmatrix} -10 \\ -15 \end{bmatrix}$

In geometric interpretation, we follow the basis vectors. Columns of matrix hold the coordinates of transformed basis vectors as measured in the original space.

For example, in (a) we look at matrix as linear transformation of vector $(3, 0)$. The vector coordinates stay numerically same in both original and transformed spaces, see Fig. 7.33. Note that the product solution $(3, 6)$ gives coordinates of $\vec{v}$ as measured in the original space. In order to create graphs for (b) to (d) we take the same approach.

Fig. 7.33 Example 7.7-1

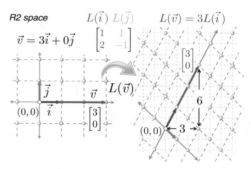

2. Geometrically, matrix–matrix product can be interpreted as "linear transformation of linear transformation". Thus, we can reason it out in two steps.

(a)   First matrix on the left side of product holds coordinates of the original basis vectors $(\vec{i}, \vec{j})$ whose magnitudes equal one.

$$\begin{bmatrix} 1 & 1 \\ 2 & -1 \end{bmatrix}\begin{bmatrix} 2 & -1 \\ 3 & 2 \end{bmatrix}$$

Result of this first transformation are transformed basis vectors whose coordinates (in the original space) are $(1, 2)$ and $(1, -1)$, see Fig. 7.34 (narrow line arrows). Second matrix contains coordinates of these two new basis vectors, as measured in the transformed space, see Fig. 7.34(wide line arrows).

$$\begin{bmatrix} 1 & 1 \\ 2 & -1 \end{bmatrix}\begin{bmatrix} 2 & -1 \\ 3 & 2 \end{bmatrix}$$

Therefore, we read the coordinates of two final vectors in the original units (grid in the background) and conclude that

$$\begin{bmatrix} 1 & 1 \\ 2 & -1 \end{bmatrix}\begin{bmatrix} 2 & -1 \\ 3 & 2 \end{bmatrix} = \begin{bmatrix} 5 & 1 \\ 1 & -4 \end{bmatrix} \tag{7.5}$$

Alternatively, we use numerical method to multiply two matrix formally and to search the product result in matrix form as follows

$$\begin{bmatrix} 1 & 1 \\ 2 & -1 \end{bmatrix}\begin{bmatrix} 2 & -1 \\ 3 & 2 \end{bmatrix} = \begin{bmatrix} i_x & j_x \\ i_y & j_x \end{bmatrix}$$

where matrix–matrix multiplication is reduced to two matrix-vector multiplications and each column of the resulting matrix (7.5) are calculated as follows

$$\vec{i}: \quad \begin{bmatrix} 1 & 1 \\ 2 & -1 \end{bmatrix}\begin{bmatrix} 2 \\ 3 \end{bmatrix} = 2\begin{bmatrix} 1 \\ 2 \end{bmatrix} + 3\begin{bmatrix} 1 \\ -1 \end{bmatrix} = \begin{bmatrix} (2+3) \\ (4-3) \end{bmatrix} = \begin{bmatrix} 5 \\ 1 \end{bmatrix}$$

$$\vec{j}: \quad \begin{bmatrix} 1 & 1 \\ 2 & -1 \end{bmatrix}\begin{bmatrix} -1 \\ 2 \end{bmatrix} = (-1)\begin{bmatrix} 1 \\ 2 \end{bmatrix} + 2\begin{bmatrix} 1 \\ -1 \end{bmatrix} = \begin{bmatrix} (-1+2) \\ (-2-2) \end{bmatrix} = \begin{bmatrix} 1 \\ -4 \end{bmatrix}$$

Both methods, naturally, produce the same result.

(b) Similarly, we multiply two matrix as follows

$$\begin{bmatrix} 2 & -1 \\ 3 & 2 \end{bmatrix}\begin{bmatrix} 2 & -3 \\ 1 & 2 \end{bmatrix} = \begin{bmatrix} i_x & j_x \\ i_y & j_x \end{bmatrix}$$

and, matrix–matrix multiplication is reduced to two matrix-vector multiplications as follows

$$\vec{i}: \quad \begin{bmatrix} 2 & -1 \\ 3 & 2 \end{bmatrix}\begin{bmatrix} 2 \\ 1 \end{bmatrix} = 2\begin{bmatrix} 2 \\ 3 \end{bmatrix} + \begin{bmatrix} -1 \\ 2 \end{bmatrix} = \begin{bmatrix} (4-1) \\ (6+2) \end{bmatrix} = \begin{bmatrix} 3 \\ 8 \end{bmatrix}$$

$$\vec{j}: \quad \begin{bmatrix} 2 & -1 \\ 3 & 2 \end{bmatrix}\begin{bmatrix} -3 \\ 2 \end{bmatrix} = (-3)\begin{bmatrix} 2 \\ 3 \end{bmatrix} + 2\begin{bmatrix} -1 \\ 2 \end{bmatrix} = \begin{bmatrix} (-6-2) \\ (-9+4) \end{bmatrix} = \begin{bmatrix} -8 \\ -5 \end{bmatrix}$$

Therefore,

$$\begin{bmatrix} 2 & -1 \\ 3 & 2 \end{bmatrix}\begin{bmatrix} 2 & -3 \\ 1 & 2 \end{bmatrix} = \begin{bmatrix} 3 & -8 \\ 8 & -5 \end{bmatrix}$$

(c) We multiply two matrix as follows

$$\begin{bmatrix} 2 & -3 \\ 1 & 2 \end{bmatrix}\begin{bmatrix} 1 & 1 \\ 2 & -1 \end{bmatrix} = \begin{bmatrix} i_x & j_x \\ i_y & j_x \end{bmatrix}$$

and, matrix–matrix multiplication is reduced to two matrix-vector multiplications as follows

$$\vec{i}: \quad \begin{bmatrix} 2 & -3 \\ 1 & 2 \end{bmatrix}\begin{bmatrix} 1 \\ 2 \end{bmatrix} = \begin{bmatrix} 2 \\ 1 \end{bmatrix} + 2\begin{bmatrix} -3 \\ 2 \end{bmatrix} = \begin{bmatrix} (2-6) \\ (1+4) \end{bmatrix} = \begin{bmatrix} -4 \\ 5 \end{bmatrix}$$

$$\vec{j}: \quad \begin{bmatrix} 2 & -3 \\ 1 & 2 \end{bmatrix}\begin{bmatrix} 1 \\ -1 \end{bmatrix} = \begin{bmatrix} 2 \\ 1 \end{bmatrix} + (-1)\begin{bmatrix} -3 \\ 2 \end{bmatrix} = \begin{bmatrix} (2+3) \\ (1-2) \end{bmatrix} = \begin{bmatrix} 5 \\ -1 \end{bmatrix}$$

Therefore,

$$\begin{bmatrix} 2 & -3 \\ 1 & 2 \end{bmatrix}\begin{bmatrix} 1 & 1 \\ 2 & -1 \end{bmatrix} = \begin{bmatrix} -4 & 5 \\ 5 & -1 \end{bmatrix}$$

Fig. 7.34 Example 7.7-2

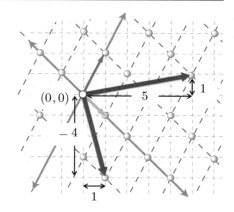

Exercise 7.8, page 136

1. Matrices with different sizes are multiplied as

$$(A \cdot B)_{ij} = \sum_{k=1}^{n} a_{ik} b_{kj} \tag{7.6}$$

where size of product matrix $(A \cdot B)_{ij}$ depends on the two matrices being multiplied. For example, given $A_{2\times3}$ (two rows and three columns) and $B_{3\times2}$ (three rows and two columns) the resulting matrix must have size $[2 \times 2]$ (two rows and two columns), which is to say that not all products are possible. That is because the number of columns in the first matrix must be equal to the number of rows in the second matrix, otherwise the product is not possible.

The sum of row–column products in (7.6) is calculated as follows.

$$(i, j) = (1, 1): \quad \begin{bmatrix} 3 & 0 & 5 \\ -2 & -1 & 4 \end{bmatrix} \begin{bmatrix} 1 & 2 \\ 5 & -1 \\ 0 & -6 \end{bmatrix} = 3 \times 1 + 0 \times 5 + 5 \times 0 = 3$$

$$(ab)_{11} = 3$$

$$(i, j) = (1, 2): \quad \begin{bmatrix} 3 & 0 & 5 \\ -2 & -1 & 4 \end{bmatrix} \begin{bmatrix} 1 & 2 \\ 5 & -1 \\ 0 & 6 \end{bmatrix} = 3 \times 2 + 0 \times (-1) + 5 \times (-6) = -24$$

$$(ab)_{12} = -24$$

$$(i, j) = (2, 1): \quad \begin{bmatrix} 3 & 0 & 5 \\ -2 & -1 & 4 \end{bmatrix} \begin{bmatrix} 1 & 2 \\ 5 & -1 \\ 0 & -6 \end{bmatrix} = (-2) \times 1 + (-1) \times 5 + 4 \times 0 = -7$$

$$(ab)_{21} = -7$$

$$(i, j) = (2, 2): \quad \begin{bmatrix} 3 & 0 & 5 \\ -2 & -1 & 4 \end{bmatrix} \begin{bmatrix} 1 & 2 \\ 5 & -1 \\ 0 & 6 \end{bmatrix} = (-2) \times 2 + (-1) \times (-1) + 4 \times (-6) = -27$$

$$(ab)_{22} = -27$$

Therefore:

$$A \cdot B = \begin{bmatrix} 3 & -24 \\ -7 & -27 \end{bmatrix}$$

Exercise 7.9, page 136

1. The matrix-vector products are calculated as follows.

   (a) In this case row–vector products are,

   $$A\,\vec{x_1} = \begin{bmatrix} -2 & -9 & -1 \\ 2 & -6 & -5 \\ -4 & -3 & 4 \end{bmatrix} \begin{bmatrix} 1 \\ 1 \\ 2 \end{bmatrix} = \begin{bmatrix} (-2 - 9 - 2) \\ (2 - 6 - 10) \\ (-4 - 3 + 8) \end{bmatrix} = \begin{bmatrix} -13 \\ -14 \\ 1 \end{bmatrix} \neq 0$$

   (b) However, the same matrix A and $\vec{x_2}$ product is,

   $$A \cdot \vec{x_2} = \begin{bmatrix} -2 & -9 & -1 \\ 2 & -6 & -5 \\ -4 & -3 & 4 \end{bmatrix} \begin{bmatrix} 13 \\ -4 \\ 10 \end{bmatrix} = \begin{bmatrix} (-26 + 36 - 10) \\ (26 + 24 - 50) \\ (-52 + 12 + 40) \end{bmatrix} = \begin{bmatrix} 0 \\ 0 \\ 0 \end{bmatrix}$$

2. By definition, eigenvector $\vec{v}$ has the property that when multiplied by a matrix A gives a scalar multiple of itself. The multiplying scalar is called eigenvalue,

   $$A \cdot \vec{v} = \lambda\,\vec{v} \tag{7.7}$$

   (a) Matrix A and $\vec{v_1}$,

   $$A \cdot \vec{v_1} = \begin{bmatrix} 3 & -3 \\ 2 & -4 \end{bmatrix} \begin{bmatrix} 3 \\ 1 \end{bmatrix} = \begin{bmatrix} (3 \times 3 - 1 \times 3) \\ (2 \times 3 - 4 \times 1) \end{bmatrix} = \begin{bmatrix} 6 \\ 2 \end{bmatrix} = 2 \begin{bmatrix} 3 \\ 1 \end{bmatrix} = \lambda\,\vec{v_1}$$

   We conclude that, indeed $\vec{v_1}$ is eigenvector of matrix A and eigenvalue is $\lambda = 2$.

   (b) However, the same matrix A multiplied by $\vec{v_2}$ gives

   $$A \cdot \vec{v_2} = \begin{bmatrix} 3 & -3 \\ 2 & -4 \end{bmatrix} \begin{bmatrix} 2 \\ 1 \end{bmatrix} = \begin{bmatrix} (2 \times 3 - 1 \times 3) \\ (2 \times 2 - 4 \times 1) \end{bmatrix} = \begin{bmatrix} 3 \\ 0 \end{bmatrix} \neq \lambda \begin{bmatrix} 2 \\ 1 \end{bmatrix}$$

   We conclude that $\vec{v_2}$ is not eigenvector of matrix A because there is no eigenvalue that is λ possible.

3. Calculating eigenvectors and eigenvalues involves calculation of the characteristic polynomial defined as

$$A \cdot \vec{v} = \lambda \vec{v} \quad \therefore \quad A \cdot \vec{v} - \lambda \vec{v} = 0$$

$$\therefore$$

$$(A - \lambda I)\, \vec{v} = 0 \tag{7.8}$$

where I is the *identity matrix* (the square matrix of the same size as A with ones on the main diagonal and zeros elsewhere). Note that we must use the identity matrix I (which is equivalent to number one for number operations), so that we can actually subtract a scalar from a matrix. In order to satisfy (7.8), since $\vec{v} \neq 0$ it must be that *determinant* of $(A - \lambda I)$ equals zero, thus

$$\det(A - \lambda I) = 0 \tag{7.9}$$

is the formal definition of the characteristic polynomial.
Therefore, we write,

$$\det(A - \lambda I) = \left\| \begin{bmatrix} 4 & 6 & 10 \\ 3 & 10 & 13 \\ -2 & -6 & -8 \end{bmatrix} - \lambda \begin{bmatrix} 1 & 0 & 0 \\ 0 & 1 & 0 \\ 0 & 0 & 1 \end{bmatrix} \right\| = \left\| \begin{bmatrix} 4 & 6 & 10 \\ 3 & 10 & 13 \\ -2 & -6 & -8 \end{bmatrix} - \begin{bmatrix} \lambda & 0 & 0 \\ 0 & \lambda & 0 \\ 0 & 0 & \lambda \end{bmatrix} \right\|$$

$$= \begin{vmatrix} 4 - \lambda & 6 & 10 \\ 3 & 10 - \lambda & 13 \\ -2 & -6 & -8 - \lambda \end{vmatrix} \quad \text{(use any method to calculate det)}$$

$$= (4 - \lambda) \begin{vmatrix} 10 - \lambda & 13 \\ -6 & -8 - \lambda \end{vmatrix} - 6 \begin{vmatrix} 3 & 13 \\ -2 & -8 - \lambda \end{vmatrix} + 10 \begin{vmatrix} 3 & 10 - \lambda \\ -2 & -6 \end{vmatrix}$$

$$= (4 - \lambda)\,[(10 - \lambda)(-8 - \lambda) - (-6) \times 13]$$

$$\quad - 6\,[3(-8 - \lambda) - (-2) \times 13]$$

$$\quad + 10\,[3 \times 6 - (-2)(10 - \lambda)]$$

$$= (-\lambda^3 - 6\lambda^2 - 6\lambda - 8) + (-12 + 18\lambda) + (20 - 20\lambda)$$

$$= \underline{-\lambda^3 + 6\lambda^2 - 8\lambda = 0} \tag{7.10}$$

Roots of the characteristic polynomial (7.10) are found as,

$$-\lambda^3 + 6\lambda^2 - 8\lambda = 0$$

$$\lambda^3 - 6\lambda^2 + 8\lambda = 0$$

$$\lambda(\lambda^2 - 6\lambda + 8) = 0$$

$$\lambda(\lambda^2 - 4\lambda - 2\lambda + 8) = 0$$

$$\lambda[\lambda(\lambda - 4) - 2(\lambda - 4)] = 0$$

$$\lambda(\lambda - 2)(\lambda - 4) = 0$$

$$\therefore$$

$$\lambda_1 = 0, \quad \lambda_2 = 2, \quad \lambda_3 = 4$$

For each of eigenvalues $\lambda_{1,2,3}$ we calculate eigenvectors $\vec{v} = (x, y, z)$ by solving (7.8) so that

$$(A - \lambda I)\, \vec{v} = \left[\begin{bmatrix} 4 & 6 & 10 \\ 3 & 10 & 13 \\ -2 & -6 & -8 \end{bmatrix} - \lambda \begin{bmatrix} 1 & 0 & 0 \\ 0 & 1 & 0 \\ 0 & 0 & 1 \end{bmatrix} \right] \begin{bmatrix} x \\ y \\ z \end{bmatrix}$$

$$= \begin{bmatrix} 4-\lambda & 6 & 10 \\ 3 & 10-\lambda & 13 \\ -2 & -6 & -8-\lambda \end{bmatrix} \begin{bmatrix} x \\ y \\ z \end{bmatrix} = \begin{bmatrix} 0 \\ 0 \\ 0 \end{bmatrix}$$

Case $\lambda = 0$: By using the matrix transformations method we write,

$$\left| \begin{array}{ccc|c} \mathbf{4} & 6 & 10 & 0 \\ 3 & \mathbf{10} & 13 & 0 \\ -2 & -6 & -\mathbf{8} & 0 \end{array} \right| \quad \leftarrow (1) \div [4]$$

$$\left| \begin{array}{ccc|c} 1 & 3/2 & 5/2 & 0 \\ 3 & 10 & 13 & 0 \\ -2 & -6 & -8 & 0 \end{array} \right| \quad \begin{array}{l} \leftarrow (2) - 3 \times (1) \\ \leftarrow (3) \div [-2] \end{array}$$

$$\left| \begin{array}{ccc|c} 1 & 3/2 & 5/2 & 0 \\ 0 & 11/2 & 11/2 & 0 \\ 0 & 3/2 & 3/2 & 0 \end{array} \right| \quad \begin{array}{l} \leftarrow (2) \times 2/11 \\ \leftarrow (3) \times 2/3 \end{array}$$

$$\left| \begin{array}{ccc|c} 1 & 3/2 & 5/2 & 0 \\ 0 & 1 & 1 & 0 \\ 0 & 1 & 1 & 0 \end{array} \right| \quad \leftarrow (3) - (2)$$

$$\left| \begin{array}{ccc|c} 1 & 3/2 & 5/2 & 0 \\ 0 & 1 & 1 & 0 \\ 0 & 0 & 0 & 0 \end{array} \right| \quad \begin{array}{l} \therefore \quad 1x + 3/2\,y + 3/2\,z = 0 \\ \therefore \quad 0x + 1y + 1z = 0 \\ \therefore \quad z: \text{ free variable !} \end{array}$$

In the z-column and the third row we reached the identity $0 \times z = 0$, which is to say that any value of z is valid choice, thus we choose a dummy value $\underline{z = t}$.
With this choice, from the second and first equations we calculate

$$y + z = 0 \quad \therefore \quad y = -z \quad \therefore \quad \underline{y = -t}$$

$$x + \frac{3}{2}y + \frac{5}{2}z = 0$$

$$\therefore$$

$$x - \frac{3}{2}t + \frac{5}{2}t = 0 \quad \therefore \quad x + t = 0 \quad \therefore \quad \underline{x = -t}$$

Which is to say that t has an arbitrary value except $t \neq 0$. The reason is that any multiple of eigenvector is still vector in the same direction, thus we choose a convenient scale.

$$\vec{v_1} = \begin{bmatrix} x \\ y \\ z \end{bmatrix} = \begin{bmatrix} -t \\ -t \\ t \end{bmatrix} = t \begin{bmatrix} -1 \\ -1 \\ 1 \end{bmatrix} = \begin{bmatrix} -1 \\ -1 \\ 1 \end{bmatrix} \quad \text{(since } t \neq 0, \text{ we choose } t = 1\text{)}$$

Case $\lambda = 2$: thus, we write,

$$\begin{vmatrix} \mathbf{2} & 6 & 10 & | & 0 \\ 3 & \mathbf{8} & 13 & | & 0 \\ -2 & -6 & \mathbf{-10} & | & 0 \end{vmatrix} \quad \leftarrow (1) \div [2]$$

$$\begin{vmatrix} 1 & 3 & 5 & | & 0 \\ 3 & 8 & 13 & | & 0 \\ -2 & -6 & -10 & | & 0 \end{vmatrix} \quad \begin{array}{l} \leftarrow (2) - 3 \times (1) \\ \leftarrow (3) + 2 \times (1) \end{array}$$

$$\begin{vmatrix} 1 & 3 & 5 & | & 0 \\ 0 & -1 & -2 & | & 0 \\ 0 & 0 & 0 & | & 0 \end{vmatrix} \quad \leftarrow (2) \div [-1]$$

$$\begin{vmatrix} 1 & 3 & 5 & | & 0 \\ 0 & 1 & 2 & | & 0 \\ 0 & 0 & 0 & | & 0 \end{vmatrix} \quad \begin{array}{l} \therefore \quad 1x + 3y + 5z = 0 \\ \therefore \quad 0x + 1y + 2z = 0 \\ \therefore \quad z: \text{free variable !} \quad \therefore \quad \underline{z = t} \end{array}$$

From the second and first equations we calculate

$$y + 2z = 0 \quad \therefore \quad y = -2z \quad \therefore \quad \underline{y = -2t}$$

$$x + 3y + 5z = 0$$

$$\therefore$$

$$x - 6t + 5t = 0 \quad \therefore \quad x - t = 0 \quad \therefore \quad \underline{x = t}$$

Thus we choose a convenient scale as

$$\vec{v_2} = \begin{bmatrix} x \\ y \\ z \end{bmatrix} = \begin{bmatrix} t \\ -2t \\ t \end{bmatrix} = t \begin{bmatrix} 1 \\ -2 \\ 1 \end{bmatrix} = \begin{bmatrix} 1 \\ -2 \\ 1 \end{bmatrix}$$

Case $\lambda = 4$: thus, we write,

$$\begin{vmatrix} \mathbf{0} & 6 & 10 & | & 0 \\ 3 & \mathbf{6} & 13 & | & 0 \\ -2 & -6 & \mathbf{-12} & | & 0 \end{vmatrix} \quad \leftarrow (1) \leftrightarrow (2)$$

$$\begin{vmatrix} 3 & 6 & 13 & | & 0 \\ 0 & 6 & 10 & | & 0 \\ -2 & -6 & -12 & | & 0 \end{vmatrix} \quad \begin{array}{l} \leftarrow (1) \div [3] \\ \\ \leftarrow (3) \div [-2] \end{array}$$

$$\begin{vmatrix} 1 & 2 & {}^{13}/_3 & | & 0 \\ 0 & 6 & 10 & | & 0 \\ 1 & 3 & 6 & | & 0 \end{vmatrix} \quad \begin{array}{l} \leftarrow (2) \div [6] \\ \leftarrow (3) - (1) \end{array}$$

$$\begin{vmatrix} 1 & 2 & {}^{13}/_3 & | & 0 \\ 0 & 1 & {}^5/_3 & | & 0 \\ 0 & 1 & {}^5/_3 & | & 0 \end{vmatrix} \quad \leftarrow (3) - (2)$$

$$
\begin{vmatrix} 1 & 2 & 13/3 & 0 \\ 0 & 1 & 5/3 & 0 \\ 0 & 0 & 0 & 0 \end{vmatrix}
\qquad
\begin{aligned}
&\therefore \quad 1x + 2y + {}^{13}/_3\, z = 0 \\
&\therefore \quad 0x + 1y + {}^{5}/_3\, z = 0 \\
&\therefore \quad z: \text{ free variable !} \quad \therefore \ \underline{z = t}
\end{aligned}
$$

From the second and first equations we calculate

$$
y + \frac{5}{3}z = 0 \quad \therefore \quad y = -\frac{5}{3}z \quad \therefore \quad \underline{y = -\frac{5}{3}t}
$$

$$
x + 2y + \frac{13}{3}z = 0
$$

$$
\therefore
$$

$$
x - \frac{10}{3}t + \frac{13}{3}t = 0 \quad \therefore \quad x + t = 0 \quad \therefore \quad \underline{x = -t}
$$

Thus we choose a convenient scale as

$$
\vec{v_3} = \begin{bmatrix} x \\ y \\ z \end{bmatrix} = \begin{bmatrix} -t \\ -5t/3 \\ t \end{bmatrix} = \begin{bmatrix} -3 \\ -5 \\ 3 \end{bmatrix} \quad \text{(we choose } t = 3\text{)}
$$

Good practice is to verify results, for example, in case of $\vec{v_2}$ and $\lambda = 2$ we calculate,

$$
A\,\vec{v_2} = \begin{bmatrix} 4 & 6 & 10 \\ 3 & 10 & 13 \\ -2 & -6 & -8 \end{bmatrix} \begin{bmatrix} 1 \\ -2 \\ 1 \end{bmatrix} = \begin{bmatrix} (4 - 12 + 10) \\ (3 - 20 + 13) \\ (-2 + 12 - 8) \end{bmatrix} = \begin{bmatrix} 2 \\ -4 \\ 2 \end{bmatrix} = 2 \begin{bmatrix} 1 \\ -2 \\ 1 \end{bmatrix} = \lambda_2\,\vec{v_2}
$$

Exercise 7.10, page 136

1. Calculation of a matrix inverse involves similar assumptions as to the product of a non-zero number and its inverse that equals one. For a matrix we write

$$
A\,A^{-1} = A^{-1}\,A = I, \quad \text{if} \ \ \Delta_A \neq 0 \tag{7.11}
$$

First, we verify if determinant equals zero or not, which answers the question of the inverse matrix existence.

$$
\Delta = \begin{vmatrix} 4 & -2 \\ 2 & 3 \end{vmatrix} = 4 \times 3 - 2 \times (-2) = 16 \neq 0 \quad \therefore \quad A^{-1} \text{ exists}
$$

Technique used to calculate the inverse matrix is based on "Gauss" form of (7.11). First, we start with the $A\,A^{-1}$ part with the objective to use matrix transformations and derive the $A^{-1}\,A$ side.

$$
\begin{bmatrix} 4 & -2 & 1 & 0 \\ 2 & 3 & 0 & 1 \end{bmatrix} \quad \leftarrow (1) \div [4]
$$

$$
\underbrace{}_{A} \ \underbrace{}_{I}
$$

$$\begin{bmatrix} 1 & -1/2 & | & 1/4 & 0 \\ 2 & 3 & | & 0 & 1 \end{bmatrix} \quad \leftarrow (2) - 2 \times (1)$$

$$\begin{bmatrix} 1 & -1/2 & | & 1/4 & 0 \\ 0 & 4 & | & -1/2 & 1 \end{bmatrix} \quad \leftarrow (2) \div [4]$$

$$\begin{bmatrix} 1 & -1/2 & | & 1/4 & 0 \\ 0 & 1 & | & -1/8 & 1/4 \end{bmatrix} \quad \leftarrow (1) + 1/2 \times (2)$$

$$\begin{bmatrix} 1 & 0 & | & 3/16 & 1/8 \\ 0 & 1 & | & -1/8 & 1/4 \end{bmatrix}$$

$$\underbrace{}_{I} \underbrace{}_{A^{-1}}$$

Which is the right side of (7.11), therefore,

$$A^{-1} = \begin{bmatrix} 3/16 & 1/8 \\ -1/8 & 1/4 \end{bmatrix}$$

To verify, we calculate

$$AA^{-1} = \begin{bmatrix} 4 & -2 \\ 2 & 3 \end{bmatrix} \begin{bmatrix} 3/16 & 1/8 \\ -1/8 & 1/4 \end{bmatrix} = \begin{bmatrix} 1 & 0 \\ 0 & 1 \end{bmatrix}$$

Because:

$$\begin{bmatrix} 4 & -2 \\ 2 & 3 \end{bmatrix} \begin{bmatrix} 3/16 \\ -1/8 \end{bmatrix} = 3/16 \begin{bmatrix} 4 \\ 2 \end{bmatrix} + {-1/8} \begin{bmatrix} -2 \\ 3 \end{bmatrix} = \begin{bmatrix} (12/16 + 1/4) \\ (6/16 - 3/8) \end{bmatrix} = \begin{bmatrix} 1 \\ 0 \end{bmatrix}$$

$$\begin{bmatrix} 4 & -2 \\ 2 & 3 \end{bmatrix} \begin{bmatrix} 1/8 \\ 1/4 \end{bmatrix} = 1/8 \begin{bmatrix} 4 \\ 2 \end{bmatrix} + 1/4 \begin{bmatrix} -2 \\ 3 \end{bmatrix} = \begin{bmatrix} (4/8 - 1/2) \\ (2/8 + 3/4) \end{bmatrix} = \begin{bmatrix} 0 \\ 1 \end{bmatrix}$$

Exercise 7.11, page 137

1. Calculation of a matrix powers involves several steps. By definition, a square matrix A is diagonalizable if it can be written as

$$A = PDP^{-1} \tag{7.12}$$

where:

1. P is a matrix that has its inverse, i.e. $PP^{-1} = I$ exists.
2. D is a diagonal matrix

In this form, for example, we write

$$A^5 = AAAAA = (PDP^{-1})(PDP^{-1})(PDP^{-1})(PDP^{-1})(PDP^{-1})$$
$$= PDDDDDP^{-1} = PD^5P^{-1}$$

That is to say, the problem of calculating A^5 is replaced with the problem of calculating D^5 and calculation of P^{-1} matrix. For higher order powers, this approach is faster relative to the trivial repeated multiplication. This is because the powers of a diagonal matrix are simple to calculate.

(a) Matrix D is diagonal, thus we write

$$D = \begin{bmatrix} 2 & 0 \\ 0 & 2 \end{bmatrix} \quad \therefore \quad D^2 = \begin{bmatrix} 2^2 & 0 \\ 0 & 2^2 \end{bmatrix} = \begin{bmatrix} 4 & 0 \\ 0 & 4 \end{bmatrix}$$

(b) Matrix D is diagonal, thus we write

$$D = \begin{bmatrix} 2 & 0 \\ 0 & 5 \end{bmatrix} \quad \therefore \quad D^5 = \begin{bmatrix} 2^5 & 0 \\ 0 & 5^5 \end{bmatrix} = \begin{bmatrix} 32 & 0 \\ 0 & 3125 \end{bmatrix}$$

(c) Powers of non-diagonal matrices may be done with the following technique.

1. Calculate determinant to verify if $\det A \neq 0$?

$$\det A = \begin{vmatrix} 2 & 0 \\ -1 & 3 \end{vmatrix} = 2 \times 3 - (-1) \times 0 = 6 \neq 0$$

2. Calculate the characteristic polynomial

$$\det(A - \lambda I) = \left| \begin{bmatrix} 2 & 0 \\ -1 & 3 \end{bmatrix} - \lambda \begin{bmatrix} 1 & 0 \\ 0 & 1 \end{bmatrix} \right| = \begin{vmatrix} 2 - \lambda & 0 \\ -1 & 3 - \lambda \end{vmatrix} = (2 - \lambda)(3 - \lambda) = 0$$

Therefore, $\lambda_1 = 2$ and $\lambda_2 = 3$.
3. Calculate eigenvectors,

$$\lambda_1 = 2: \quad A - \lambda_1 I = 0 \quad \therefore \quad \begin{bmatrix} 2 & 0 \\ -1 & 3 \end{bmatrix} - 2 \begin{bmatrix} 1 & 0 \\ 0 & 1 \end{bmatrix} = \begin{bmatrix} 2-2 & 0 \\ -1 & 3-2 \end{bmatrix} = \begin{bmatrix} 0 & 0 \\ -1 & 1 \end{bmatrix}$$

$$\therefore \quad \begin{bmatrix} 0 & 0 & 0 \\ -1 & 1 & 0 \end{bmatrix} \begin{matrix} \leftarrow x = t = 1 \\ \leftarrow -x + y = 0 \end{matrix} \quad \therefore \quad x = y = 1$$

$$\therefore \quad \vec{v_1} = \begin{bmatrix} 1 \\ 1 \end{bmatrix}$$

$$\lambda_1 = 3: \quad A - \lambda_1 I = 0 \quad \therefore \quad \begin{bmatrix} 2 & 0 \\ -1 & 3 \end{bmatrix} - 3 \begin{bmatrix} 1 & 0 \\ 0 & 1 \end{bmatrix} = \begin{bmatrix} 2-3 & 0 \\ -1 & 3-3 \end{bmatrix} = \begin{bmatrix} -1 & 0 \\ -1 & 0 \end{bmatrix}$$

$$\therefore \quad \begin{bmatrix} -1 & 0 & 0 \\ -1 & 0 & 0 \end{bmatrix} \begin{matrix} \leftarrow -x = 0 \\ \leftarrow y = t \end{matrix} \quad \begin{matrix} \therefore \quad x = 0 \\ \therefore \quad y = 1 \end{matrix}$$

$$\therefore \quad \vec{v_2} = \begin{bmatrix} 0 \\ 1 \end{bmatrix}$$

4. Write diagonal matrix, by definition we write

$$D = \begin{bmatrix} \lambda_1 & 0 \\ 0 & \lambda_2 \end{bmatrix} = \begin{bmatrix} 2 & 0 \\ 0 & 3 \end{bmatrix}$$

1. Calculate P and P^{-1}, by definition we write

$$P = [\vec{v_1}\ \vec{v_2}] = \begin{bmatrix} 1 & 0 \\ 1 & 1 \end{bmatrix}$$

and by using Gauss form we find

$$P\,I = \begin{bmatrix} 1 & 0 & 1 & 0 \\ 1 & 1 & 0 & 1 \end{bmatrix} \quad \leftarrow (2) - (1)$$

$$= \begin{bmatrix} 1 & 0 & 1 & 0 \\ 0 & 1 & -1 & 1 \end{bmatrix} = I\,P^{-1} \quad \therefore \quad P^{-1} = \begin{bmatrix} 1 & 0 \\ -1 & 1 \end{bmatrix}$$

(d) Calculate (7.12), as

$$A^5 = P\,D^5\,P^{-1} = \begin{bmatrix} 1 & 0 \\ 1 & 1 \end{bmatrix} \begin{bmatrix} 2^5 & 0 \\ 0 & 3^5 \end{bmatrix} \begin{bmatrix} 1 & 0 \\ -1 & 1 \end{bmatrix} = \begin{bmatrix} 1 & 0 \\ 1 & 1 \end{bmatrix} \begin{bmatrix} 32 & 0 \\ 0 & 243 \end{bmatrix} \begin{bmatrix} 1 & 0 \\ -1 & 1 \end{bmatrix}$$

$$= \begin{bmatrix} 32 & 0 \\ 32 & 243 \end{bmatrix} \begin{bmatrix} 1 & 0 \\ -1 & 1 \end{bmatrix} = \begin{bmatrix} 32 & 0 \\ -211 & 243 \end{bmatrix}$$

Limits

8

Important to Know

Basic rules for calculating limits	Infinite limits
$$\lim_{x \to x_0} \Big(C\, f(x) \Big) = C \lim_{x \to x_0} f(x)$$ $$\lim_{x \to x_0} \Big(f(x) \pm g(x) \Big) = \lim_{x \to x_0} f(x) \pm \lim_{x \to x_0} g(x)$$ $$\lim_{x \to x_0} \Big(f(x)\, g(x) \Big) = \lim_{x \to x_0} f(x) \lim_{x \to x_0} g(x)$$ $$\lim_{x \to x_0} \left(\frac{f(x)}{g(x)} \right) = \frac{\lim_{x \to x_0} f(x)}{\lim_{x \to x_0} g(x)}$$ $$\lim_{x \to x_0} \Big(f(x) \Big)^n = \left(\lim_{x \to x_0} f(x) \right)^n$$	$$\lim_{x \to +\infty} x^n = +\infty$$ $$\lim_{x \to -\infty} x^n = -\infty \quad n \text{ is odd}$$ $$\lim_{x \to -\infty} x^n = +\infty \quad n \text{ is even}$$ $$\lim_{x \to \pm\infty} \frac{1}{x^n} = 0$$ $$\lim_{x \to 0} \ln x = -\infty$$ $$\lim_{x \to \infty} \ln x = +\infty$$ $$\lim_{x \to -\infty} e^x = 0$$ $$\lim_{x \to \infty} e^x = +\infty$$

Some of the "famous" limits	
$$\lim_{x \to \pm\infty} \left(1 + \frac{1}{x} \right)^x = e$$	$$\lim_{x \to 0} \frac{e^x - 1}{x} = 1$$
$$\lim_{x \to 0} \frac{\sin x}{x} = 1$$	$$\lim_{x \to 0} \frac{\ln(1 + x)}{x} = 1$$

© Springer Nature Switzerland AG 2021
R. Sobot, *Engineering Mathematics by Example*,
https://doi.org/10.1007/978-3-030-79545-0_8

8.1   Exercises

8.1 *   Limits

1. Determine the following limits:

(a) $\lim\limits_{x \to 2} f(x), \quad f(x) = x$

(b) $\lim\limits_{x \to 2} f(x), \quad f(x) = \begin{cases} 3 & (x > 2) \\ 1 & (x < 2) \end{cases}$

(c) $\lim\limits_{x \to 2} f(x), \quad f(x) = \dfrac{|x - 2|}{x - 2}$

(d) $\lim\limits_{x \to 0} f(x), \quad f(x) = \dfrac{x^2 - 2x}{x}$

(e) $\lim\limits_{x \to -5} f(x), \quad f(x) = \dfrac{x^2 + 2x - 15}{x^2 + 8x + 15}$

(f) $\lim\limits_{x \to 1} f(x), \quad f(x) = \dfrac{\sqrt{x} - 1}{x - 1}$

8.2 **   Limits: Undefined Functions

1. Determine the following limits:

(a) $\lim\limits_{x \to 1} \dfrac{x^2 + 2x - 3}{x - 1}$

(b) $\lim\limits_{x \to \infty; x \to 0} f(x), \quad f(x) = \dfrac{\sin(x)}{x}$

(c) $\lim\limits_{x \to 0} f(x), \quad f(x) = \dfrac{x + \sin(x)}{x}$

(d) $\lim\limits_{x \to 0} f(x), \quad f(x) = \dfrac{\sin(2x)}{x}$

(e) $\lim\limits_{x \to 0} f(x), \quad f(x) = \dfrac{4 \sin(5x)}{\sin(4x)}$

(f) $\lim\limits_{x \to 0} f(x), \quad f(x) = \dfrac{\sin(2x)}{\sin(3x)}$

(g) $\lim\limits_{x \to \infty} f(x), \quad f(x) = \left(1 + \dfrac{1}{x}\right)^{x+5}$

(h) $\lim\limits_{x \to \infty} f(x), \quad f(x) = \left(1 + \dfrac{1}{x}\right)^{3x}$

(i) $\lim\limits_{x \to \infty} f(x), \quad f(x) = \left(1 + \dfrac{2}{x}\right)^{x}$

(j) $\lim\limits_{x \to \infty} f(x), \quad f(x) = \left(\dfrac{x}{1 + x}\right)^{x}$

(k) $\lim\limits_{x \to \infty} f(x), \quad f(x) = \left(\dfrac{x + 3}{x - 1}\right)^{x+1}$

8.3 ***   Asymptotes

1. Calculate limits of the following functions:

(a) $f(x) = \dfrac{x - 1}{x^2 - 5x + 6}$

(b) $f(x) = \dfrac{\sqrt{x + 1} - 3}{x - 8}$

(c) $f(x) = \dfrac{x^3}{x^2 - 4}$

Solutions

Exercise 8.1, page 188

1. There are various possibilities that determine whether the limiting value exists or not.

(a)   A continuous function is defined in all points, including for $x = 2$, and therefore calculating any limiting value is trivial. For example,

$$\lim_{x \to 2} f(x) = \lim_{x \to 2} x = 2$$

because both left side (i.e. $\lim_{x \to 2_-}$) and right side (i.e. $\lim_{x \to 2_+}$) limits point to the same value of function, i.e. $f(2) = 2$, Fig. 8.1.

Fig. 8.1 Example 8.1-1(a)

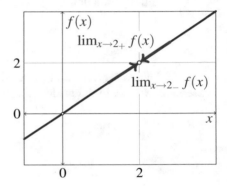

(b)   A non-continuous function may or may not have defined limit. For example, function $f(x)$ is *not* defined for $x = 2$. When calculating limits, we find

$$\lim_{x \to 2_-} f(x) = 1$$

$$\lim_{x \to 2_+} f(x) = 3$$

For the reason that the left side and right side limits are not equal, consequently, limit $\lim_{x \to 2} f(x)$ does not exist, Fig. 8.2.

Fig. 8.2 Example 8.1-1(b)

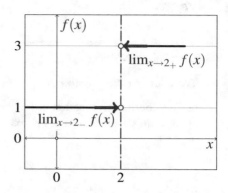

(c)   Similarly, function

$$f(x) = \frac{|x - 2|}{x - 2}$$

is not defined for $x = 2$, it equals 0/0. Consequently, the first limitation is $x \neq 2$. We analyse separately left and right sides of $x = 2$ point, Fig. 8.3.

1. if $x > 2$, then

$$|x - 2| = x - 2$$

$$\therefore \quad (x \neq 2)$$

$$\lim_{x \to 2_+} \frac{x - 2}{x - 2} = 1$$

2. if $x < 2$, then

$$|x - 2| = -(x - 2)$$

$$\therefore \quad (x \neq 2)$$

$$\lim_{x \to 2_-} \frac{-(x - 2)}{x - 2} = -1$$

Therefore, for the reason that the left and right side limits are not equal, the conclusion is that $\lim_{x \to 2} f(x)$ does not exist.

Fig. 8.3 Example 8.1-1(c)

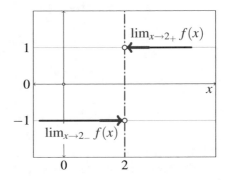

(d) For a special case of rational function, Fig. 8.4, after excluding division by zero, i.e. $x \neq 0$, we calculate the limit as

$$\lim_{x \to 0} f(x) = \frac{x^2 - 2x}{x}$$

$$= \lim_{x \to 0} \frac{\cancel{x}(x - 2)}{\cancel{x}}$$

$$= -2$$

which is equal from both sides of $x = 0$. In conclusion, even though $x \neq 0$, the limit exists in that point because both left and right side limits are equal (search for "Two Policemen and a Drunk" or "Sandwich" theorem).

Fig. 8.4 Example 8.1-1(d)

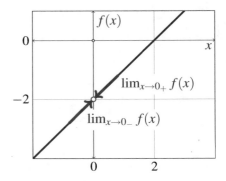

(e) Similar example of rational function where $x \neq -5$ also permits the existence of limit as

$$\lim_{x \to -5} f(x) = \lim_{x \to -5} \frac{x^2 + 2x - 15}{x^2 + 8x + 15} = \lim_{x \to -5} \frac{x^2 - 3x + 5x - 15}{x^2 + 3x + 5x + 15}$$

$$= \lim_{x \to -5} \frac{\cancel{x}(x - 3 + 5(x - 3)}{\cancel{x}(x + 3) + 5(x + 3)} \quad (x \neq 0)$$

$$= \lim_{x \to -5} \frac{(x - 3)\cancel{(x + 5)}}{(x + 3)\cancel{(x + 5)}} \quad (x \neq -5)$$

$$= \frac{-5 - 3}{-5 + 3} = 4$$

In conclusion, even though the function is not defined for $x = -5$, Fig. 8.5, limit $\lim_{x \to -5} f(x) = 4$. This case is also known as "pole–zero cancellation" because both numerator and denominator have one equal root (i.e. $x = -5$) that is cancelled, and therefore value of the function for $x = -5$ is neither zero nor infinity. We note that is not the case for $x = -3$ where the limit does not exist because

$$\lim_{x \to -3_-} f(x) = +\infty \quad \text{and} \quad \lim_{x \to -3_+} f(x) = -\infty$$

that is to say, they are not equal, and thus the limit does not exist.

Fig. 8.5 Example 8.1-1(e)

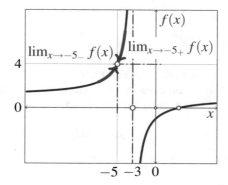

(f) Rational functions that contain radicals may also take advantage of pole–zero cancellation (and, with the help of $a^2 - b^2 = (a - b)(a + b)$ identity) as

$$\lim_{x \to 1} f(x) = \lim_{x \to 1} \frac{\sqrt{x} - 1}{x - 1} \quad (x \geq 0)$$

$$= \lim_{x \to 1} \frac{\cancel{\sqrt{x} - 1}}{(\cancel{\sqrt{1} - 1})(\sqrt{1} + 1)} \quad (x \neq 1)$$

$$= \lim_{x \to 1} \frac{1}{\sqrt{1} + 1} = \frac{1}{2}$$

In conclusion, even though the function is not defined for $x = 1$, due to pole–zero cancellation the limit exists, Fig. 8.6.

Fig. 8.6 Example 8.1-1(f)

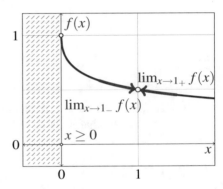

Exercise 8.2, page 188

1. Limits based on number e and sinc function are important.

(a) Knowing that sin function is limited to ± 1, we can deduce that

$$\lim_{x \to \infty} \text{sinc}(x) = \lim_{x \to \infty} \frac{\sin(x)}{x} = \frac{(\leq 1)}{\infty} = 0$$

(b) $\displaystyle \lim_{x \to 0} \frac{x + \sin(x)}{x} = \lim_{x \to 0} \left(1 + \overbrace{\frac{\sin(x)}{x}}^{1} \right) = 2$

(c) $\displaystyle \lim_{x \to 0} \frac{\sin(2x)}{x} = \lim_{x \to 0} \frac{2}{2} \frac{\sin(2x)}{x} = 2 \lim_{x \to 0} \frac{\sin(2x)}{2x} \quad (t = 2x) = 2 \lim_{x \to 0} \overbrace{\frac{\sin t}{t}}^{1} = 2$

(d) $\displaystyle \lim_{x \to 0} \frac{4 \sin(5x)}{\sin(4x)} = \lim_{x \to 0} \frac{\dfrac{\sin(5x)}{5x}}{\dfrac{\sin(4x)}{4} \dfrac{x}{x}} \cdot \frac{5x}{5x} = \lim_{x \to 0} \frac{5x}{x} \frac{\overbrace{\dfrac{\sin(5x)}{5x}}^{1}}{\underbrace{\dfrac{\sin(4x)}{4x}}_{1}} = 5$

(e) $\displaystyle\lim_{x\to 0}\frac{\sin(2x)}{\sin(3x)} = \lim_{x\to 0}\frac{\frac{2x}{2x}\sin(2x)}{\frac{3x}{3x}\sin(3x)} = \left\{\begin{array}{l} x \neq 0, \\[4pt] t = 2x, \\[4pt] p = 3x \end{array}\right\} = \lim_{x\to 0}\frac{2\cancel{x}}{3\cancel{x}}\frac{\cancel{\overset{1}{\frac{\sin(t)}{t}}}}{\cancel{\underset{1}{\frac{\sin(p)}{p}}}} = \frac{2}{3}$

(f) Forms that can be forced into limit of e,

$$\lim_{x\to\infty}\left(1+\frac{1}{x}\right)^{x+5} = \lim_{x\to\infty}\left[\left(1+\frac{1}{x}\right)^{x}\left(1+\frac{1}{x}\right)^{5}\right]$$

$$= \lim_{x\to\infty}\left(1+\frac{1}{x}\right)^{x}\lim_{x\to\infty}\left(1+\frac{1}{x}\right)^{5}$$

$$= \lim_{x\to\infty}\cancel{\left(1+\frac{1}{x}\right)^{x}}^{\,e}\left(1+\lim_{x\to\infty}\cancel{\frac{1}{x}}^{\,0}\right)^{5}$$

$$= e$$

(g) $\displaystyle\lim_{x\to\infty}\left(1+\frac{1}{x}\right)^{3x} = \lim_{x\to\infty}\left[\left(1+\frac{1}{x}\right)^{x}\right]^{3} = \left[\lim_{x\to\infty}\cancel{\left(1+\frac{1}{x}\right)^{x}}^{\,e}\right]^{3} = e^{3}$

(h) Similarly, a non-one numerator, for example, "2" in this example, may be temporarily "hidden" into another variable so that

$$\lim_{x\to\infty}\left(1+\frac{2}{x}\right)^{x} = \left\{\begin{array}{l} \dfrac{2}{x}=\dfrac{1}{t} \quad\therefore\quad x = 2t, \\[8pt] x\to\infty \quad\therefore\quad t\to\infty \end{array}\right\} = \lim_{t\to\infty}\left(1+\frac{1}{t}\right)^{2t}$$

$$= \left[\lim_{t\to\infty}\cancel{\left(1+\frac{1}{t}\right)^{t}}^{\,e}\right]^{2} = e^{2}$$

(i) Sometimes, rational forms may also be forced into

$$\lim_{x\to\infty}\left(\frac{x}{1+x}\right)^{x} = \lim_{x\to\infty}\left(\frac{1}{\frac{1+x}{x}}\right)^{x} = \lim_{x\to\infty}\frac{\cancel{1^{x}}^{\,1}}{\left(\dfrac{1+x}{x}\right)^{x}}$$

$$= \frac{1}{\lim_{x\to\infty}\cancel{\left(\dfrac{1+x}{x}\right)^{x}}^{\,e}} = \frac{1}{e}$$

(j)   Or, the combination of the above cases

$$\lim_{x \to \infty} \left(\frac{x+3}{x-1} \right)^{x+1} = \left\{ \begin{array}{l} x - 1 = t \quad \therefore \quad x = t + 1, \\[2mm] x \to \infty \quad \therefore \quad t \to \infty \end{array} \right\} = \lim_{t \to \infty} \left(\frac{t+4}{t} \right)^{t+2}$$

$$= \lim_{t \to \infty} \left(1 + \frac{4}{t} \right)^{t} \underbrace{\lim_{t \to \infty} (1 + \frac{4}{t})^2}_{0 \nearrow 1}$$

$$= \left\{ \begin{array}{l} \dfrac{4}{t} = \dfrac{1}{p} \quad \therefore \quad t = 4p, \\[3mm] t \to \infty \quad \therefore \quad p \to \infty \end{array} \right\} = \lim_{p \to \infty} \left(1 + \frac{1}{p} \right)^{4p}$$

$$= \left[\underbrace{\lim_{p \to \infty} \left(1 + \frac{1}{p} \right)^{p}}_{e} \right]^{4} = e^4$$

Exercise 8.3, page 188

1. Limits are systematically found around the domain extremes and the breaking points (i.e. vertical asymptotes).

(a)   Factorize $f(x)$

$$f(x) = \frac{x-1}{x^2 - 5x + 6} = \frac{x-1}{x^2 + 2x + 3x + 6} = \frac{x-1}{x(x+2) + 3(x+2)}$$

$$= \frac{x-1}{(x+2)(x+3)} = \frac{P(x)}{Q(x)}$$

Vertical asymptotes are found as roots of $Q(x) = 0$:
i.e. $(x+2)(x+3) = 0 \quad \therefore \quad x_1 = -2, \quad x_2 = -3$
The only zero of $f(x)$ is when $P(x) = 0 \quad \therefore \quad x - 1 - 0 \quad \therefore \quad x = 1$, which is different from the two poles, and therefore there is no pole–zero cancellation. Two vertical asymptotes define, in total, six limits to be calculated, see Fig. 8.7.

Fig. 8.7 Example 8.3-1(a)

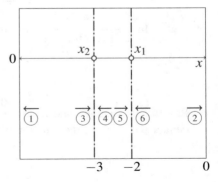

$$\text{(1)} \quad \lim_{x \to -\infty} \frac{x-1}{x^2 + 5x + 6} = \lim_{x \to -\infty} \frac{\cancel{x}\left(1 - \cancel{\frac{1}{x}}^{\,0}\right)}{\cancel{x}\left(x + 5 + \cancel{\frac{6}{x}}^{\,0}\right)} = \lim_{x \to -\infty} \frac{1}{x - 5} == \frac{1}{-\infty}$$

$$= 0_- \quad \text{(i.e. below the horizontal axis)}$$

$$\text{(2)} \quad \lim_{x \to +\infty} \frac{x-1}{x^2 + 5x + 6} = \lim_{x \to +\infty} \frac{\cancel{x}\left(1 - \cancel{\frac{1}{x}}^{\,0}\right)}{\cancel{x}\left(x + 5 + \cancel{\frac{6}{x}}^{\,0}\right)} = \lim_{x \to +\infty} \frac{1}{x - 5} == \frac{1}{+\infty}$$

$$= 0_+ \quad \text{(i.e. above the horizontal axis)}$$

$$\text{(3)} \quad \lim_{x \to -3_-} \frac{x-1}{(x+2)(x+3)} = \lim_{x \to -3_-} \frac{-3-1}{(-3+2)(x+3)} = \lim_{x \to -3_-} \frac{-4}{-(x+3)}$$

$$= \{\text{if } x < -3 \quad \therefore \quad (x+3) < 0\}$$

$$= \frac{-4}{+0} = -\infty$$

(a)

$$\text{(4)} \quad \lim_{x \to -3_+} \frac{x-1}{(x+2)(x+3)} = \lim_{x \to -3_+} \frac{-3-1}{(-3+2)(x+3)} = \lim_{x \to -3_+} \frac{-4}{-(x+3)}$$

$$= \{\text{if } x > -3 \quad \therefore \quad (x+3) > 0\}$$

$$= \frac{-4}{-0} = +\infty$$

$$\text{(5)} \quad \lim_{x \to -2_-} \frac{x-1}{(x+2)(x+3)} = \lim_{x \to -2_-} \frac{-2-1}{(x+2)(-2+3)} = \lim_{x \to -2_-} \frac{-3}{(x+2)}$$

$$= \{\text{if } x < -2 \quad \therefore \quad (x+2) < 0\}$$

$$= \frac{-3}{-0} = +\infty$$

$$\text{(6)} \quad \lim_{x \to -2_+} \frac{x-1}{(x+2)(x+3)} = \lim_{x \to -2_+} \frac{-2-1}{(x+2)(-2+3)} = \lim_{x \to -2_+} \frac{-3}{(x+2)}$$

$$= \{\text{if } x > -2 \quad \therefore \quad (x+2) > 0\}$$

$$= \frac{-3}{+0} = -\infty$$

Plot of this function shows (not to y-scale), see Fig. 8.8, that knowing limits we can already deduce the shape of the function (the complete function $f(x)$ is plot in the background).

Fig. 8.8 Example 8.3-1(a)

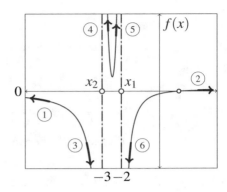

(b) $\quad f(x) = \dfrac{\sqrt{x+1}-3}{x-8} = \dfrac{P(x)}{Q(x)}$

Domain is defined for $x + 1 \geq 0 \;\; \therefore \;\; x \geq -1$. Furthermore, zero is found when $P(x) = 0 \;\; \therefore$
$\sqrt{x+1} - 3 = 0 \;\; \therefore \;\; x = 8$. However, vertical asymptote is found at $Q(x) = 0 \;\; \therefore \;\; x - 8 = 0 \;\; \therefore \;\; x = 8$. The conclusion is that the only pole–zero pair is cancelled, that is to say $f(8)$
equals neither zero nor infinity.
One break point $x = 8$ defines four limits to be calculated, see Fig. 8.9.

Fig. 8.9 Example 8.3-1(a)(b)

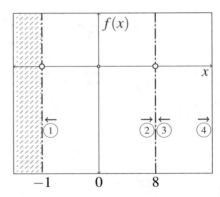

① $\displaystyle \lim_{x \to -1_+} \frac{\sqrt{x+1}-3}{x-8} = \frac{\sqrt{-1+1}-3}{-1-8} = \frac{-3}{-9} = \frac{1}{3}$

② $\displaystyle f(x) = \frac{\sqrt{x+1}-3}{x-8}\frac{\sqrt{x+1}}{\sqrt{x+1}} = \frac{(x+1)-9}{(x-8)\left(\sqrt{x+1}+3\right)} \quad (x \neq 8)$

$\displaystyle \qquad = \frac{\cancel{x-8}}{\cancel{(x-8)}\left(\sqrt{x+1}+3\right)}$

Therefore,

$$\lim_{x \to 8_-} \frac{1}{\sqrt{x+1}+3} = \frac{1}{\sqrt{8+1}+3} = \{x \ge -1\} = \frac{1}{6}$$

③ $$\lim_{x \to 8_+} \frac{1}{\sqrt{x+1}+3} = \frac{1}{\sqrt{8+1}+3} = \{x \ge -1\} = \frac{1}{6}$$

④ $$\lim_{x \to +\infty} \frac{1}{\sqrt{x+1}+3} = \frac{1}{\sqrt{+\infty+1}+3} = 0_+$$

(i.e. above the horizontal axes)

(c)   Vertical asymptotes are found as roots of $Q(x) = 0$:

$$Q(x) = 0 \quad \therefore \quad x^2 - 4 = 0 \quad \therefore \quad x_1 = -2, \quad x_2 = 2$$

Third order polynomial must have at least one real root; thus, real zeros of $f(x)$ are found as roots of $P(x) = 0$

$$P(x) = 0 \quad \therefore \quad x^3 = x\,x\,x = 0 \quad \therefore \quad x_{1,2,3} = 0$$

That is to say, all three roots of $P(x)$ are real and identical (i.e. "triple zero"). The zeros and poles of $f(x)$ are not equal, and thus there are no pole–zero cancellations. These two vertical asymptotes determine the following six limits to calculate:

① $$\lim_{x \to -\infty} \frac{x^3}{x^2 - 4} = \lim_{x \to -\infty} \frac{x\!\!\!/^2\, x}{x\!\!\!/^2 (1 - \underset{0}{\cancel{\frac{4}{x^2}}})} = -\infty$$

② $$\lim_{x \to -2_-} \frac{x^3}{x^2 - 4} = \lim_{x \to -2_-} \frac{x^3}{(x-2)(x+2)} = \lim_{x \to -2_-} \frac{(-2)^3}{(-2-2)(x+2)}$$

$$= \lim_{x \to -2_-} \frac{-8}{-4(x+2)} = \{\text{if } x < -2 \ \therefore \ (x+2) < 0\}$$

$$= \frac{-8}{-4(-0)} = -\infty$$

③ $$\lim_{x \to -2_+} \frac{x^3}{x^2 - 4} = \lim_{x \to -2_+} \frac{x^3}{(x-2)(x+2)} = \lim_{x \to -2_+} \frac{(-2)^3}{(-2-2)(x+2)}$$

$$= \lim_{x \to -2_+} \frac{-8}{-4(x+2)} = \{\text{if } x > -2 \ \therefore \ (x+2) > 0\}$$

$$= \frac{-8}{-4(+0)} = +\infty$$

④ $$\lim_{x \to 2_-} \frac{x^3}{x^2 - 4} = \lim_{x \to 2_-} \frac{x^3}{(x-2)(x+2)} = \lim_{x \to 2_-} \frac{(2)^3}{(x-2)(2+2)}$$

$$= \lim_{x \to 2_-} \frac{8}{4(x-2)} = \{\text{if } x < 2 \ \therefore \ (x-2) < 0\}$$

$$= \frac{8}{4(-0)} = -\infty$$

$$\lim_{x \to 2_+} \frac{x^3}{x^2 - 4} = \lim_{x \to 2_+} \frac{x^3}{(x - 2)(x + 2)} = \lim_{x \to 2_+} \frac{(2)^3}{(x + 2)(2 + 2)}$$

⑤
$$= \lim_{x \to 2_+} \frac{8}{4(x + 2)} = \left\{ \text{if } x > 2 \ \therefore \ (x - 2) > 0 \right\}$$

$$= \frac{8}{4(+0)} = +\infty$$

⑥ $$\lim_{x \to \infty} \frac{x^3}{x^2 - 4} = \lim_{x \to \infty} \frac{\cancel{x^2}\, x}{\cancel{x^2}\left(1 - \cancel{\frac{4}{x^2}}^{\ 0}\right)} = \infty$$

(d) In addition, we search for oblique asymptote(s) y_{aa} whose linear equation is found in form

$$y_{aa} = ax + b \quad \text{where,} \quad a = \lim_{x \to \infty} \frac{f(x)}{x} \quad \text{and} \quad b = \lim_{x \to \infty} (f(x) - ax)$$

Thus, we calculate

$$a = \lim_{x \to \infty} \frac{\dfrac{x^3}{x^2 - 4}}{x} = \lim_{x \to \infty} \frac{x^3}{x(x^2 - 4)} = \lim_{x \to \infty} \frac{\cancel{x^3}}{\cancel{x^3}\left(1 - \cancel{\frac{4}{x^2}}^{\ 0}\right)} = 1$$

$$b = \lim_{x \to \infty} \left(\frac{x^3}{x^2 - 4} - x \right) = \lim_{x \to \infty} \frac{x^3 - x(x^2 - 4)}{x^2 - 4} = \lim_{x \to \infty} \frac{\cancel{x^3} - \cancel{x^3} + 4x}{x^2 - 4}$$

$$= \lim_{x \to \infty} \frac{4\cancel{x}}{\cancel{x}\left(x - \cancel{\frac{4}{x}}^{\ 0}\right)} = \lim_{x \to \infty} \frac{4}{x} = 0$$

Therefore, there is one oblique asymptote $y_{aa} = ax + b = x$. Knowledge of the function's limits is sufficient to sketch its form, see Fig. 8.10.

Fig. 8.10 Example 8.3-1(c)

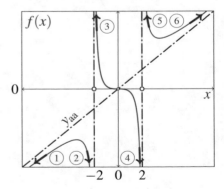

Derivatives

Important to Know

Basic tabular derivatives

$f(x)$	$f'(x)$
$a = \text{const.}$	0
x^n	$n\,x^{n-1}$
$\sqrt{x} = x^{1/2}$	$\dfrac{1}{2}\,x^{-1/2}$
$\ln x$	$\dfrac{1}{x}\quad (x \neq 0)$
e^x	e^x
a^x	$a^x\,\ln a$
$|x|$	$\text{sign}(t)\quad (x \neq 0)$
$\sin x$	$\cos x$
$\cos x$	$-\sin x$
$\tan x$	$\dfrac{1}{\cos^2 x} = 1 + \tan^2 x$
$\arcsin x$	$\dfrac{1}{\sqrt{1-x^2}}$
$\arccos x$	$-\dfrac{1}{\sqrt{1-x^2}}$
$\arctan x$	$\dfrac{1}{1+x^2}$

© Springer Nature Switzerland AG 2021
R. Sobot, *Engineering Mathematics by Example*,
https://doi.org/10.1007/978-3-030-79545-0_9

Basic rules of derivations

$$\left[C\,f(x)\right]' = C\,f'(x)$$

$$\left[f(x) \pm g(x)\right]' = f'(x) \pm g'(x)$$

$$\left[f(x)\,g(x)\right]' = f'(x)\,g(x) + f(x)\,g'(x)$$

$$\left[\frac{f(x)}{g(x)}\right]' = \frac{f'(x)\,g(x) - f(x)\,g'(x)}{g^2(x)}$$

$$\left[f(g(x))\right]' = f'(g(x))\,g'(x)$$

9.1   Exercises

9.1   Tabular Derivatives: Power Rule $(kx^n)' = k\,nx^{n-1}$

1. Calculate derivatives of the following functions.

(a) $f(x) = x^5$

(b) $h(t) = x^3 + 2x + 1$

(c) $g(x) = -\dfrac{1}{27}x^3 + \dfrac{1}{18}x^3$

(d) $h(x) = \dfrac{4}{x^3} - x^{-3}$

(e) $a(x) = \dfrac{1}{2}x^{\frac{1}{2}}$

(f) $h(x) = \dfrac{4}{3}x^{-\frac{3}{4}}$

2. Calculate derivatives of the following functions.

(a) $f(x) = \sqrt{x}$

(b) $h(x) = \sqrt[3]{x^2}$

(c) $g(x) = 5\sqrt[4]{x} - 2\sqrt[3]{x^3}$

(d) $b(x) = \dfrac{5}{\sqrt[4]{x}}$

(e) $t(x) = \dfrac{1}{\sqrt[3]{x^4}}$

(f) $g(t) = \dfrac{1}{3}t^3 - \dfrac{\sqrt{t}}{3} + 2$

9.2   Tabular Derivatives: Exponent $(a^x)' = a^x\,\ln a$

1. Calculate derivatives of the following functions.

(a) $f(x) = 1^x$

(b) $h(x) = 2^x$

(c) $g(x) = e^x$

(d) $b(x) = 10^x$

(e) $g(x) = a^x + x^a$

(f) $f(x) = e^x + x^3$

9.3 *   Tabular Derivatives: Logarithm $\left(\log_a x\right)' = \dfrac{1}{x\,\ln a}$

1. Calculate derivatives of the following functions.

(a) $f(x) = \ln x$

(b) $h(x) = \log x$

(c) $g(x) = \log_5 x$

9.4 **   Composite Functions: Chain Rule

Composite functions follow chain rule of derivation:

$$[h(x)]' = [f(g(x))]' = f'(g(x))\, g'(x)$$

That is to say, first we calculate derivative of the "external" function $f(x)$, then we multiply it with derivative of its argument function $g(x)$.

1. Calculate derivatives of simple composite functions.

    (a) $f(x) = e^{5x}$ (b) $h(x) = \sin\left(x^2\right)$ (c) $g(x) = \sin(abcx)$

2. Calculate derivatives of composite functions by using the power rule.

    (a) $f(x) = (1 - x)^5$ (b) $h(x) = \sqrt{x^2 + 2}$ (c) $g(x) = \sqrt[3]{\sqrt{2x - 5}}$

3. Calculate derivatives of composite functions by using the exponent rule.

    (a) $f(x) = e^{1-x}$ (b) $h(x) = 2^{x^2+2x-1}$ (c) $h(x) = e^{x^2}$

    (d) $g(x) = \sqrt[3]{e^{6x-3}}$ (e) $g(x) = \dfrac{1}{e^{-x^2/2}}$ (f) $g(x) = \dfrac{1}{e^{\sqrt{2-3x}}}$

4. Calculate derivatives of composite functions by using the logarithmic rule.

    (a) $f(x) = \ln(1 - x)$ (b) $h(x) = \log_2\left(x^2 + 2x + 1\right)$

    (c) $g(x) = \ln\sqrt{2 - x}$ (d) $g(t) = \dfrac{1}{\log(2t - 3)}$

5. Calculate derivatives of composite functions that include trigonometric functions.

    (a) $f(x) = \sin(2x - 1)$ (b) $h(x) = \sin\sqrt{x}$

    (c) $g(x) = \cos(\sin(1 - x))$ (d) $g(x) = \dfrac{1}{\ln(\sin x)}$

    (e) $g(x) = \sin\left(\dfrac{1}{x}\right)$ (f) $g(x) = \ln\left(\sin x + \sqrt{1 + \sin^2 x}\right)$

    (g) $g(x) = \sin^2 x + \sin^2\left(\dfrac{2\pi}{3} + x\right) + \sin^2\left(\dfrac{2\pi}{3} - x\right)$

9.5 **   Product Rule $(f(x)g(x))' = f'(x)g(x) + f(x)g'(x)$

1. Calculate derivatives by using the product rule.

    (a) $h(x) = x \sin x$ (b) $g(x) = e^x \cos x$

    (c) $f(x) = (2x - 1)(1 - x)$ (d) $f(x) = \sqrt{x} \ln x$

    (e) $f(x) = \sin(x - 1) \cos(1 - x)$ (f) $f(x) = \log(x - 1) \cos(1 - x)$

    (g) $h(t) = t^2\, 3^t$

9.6 **   Ratio Rule $\left(\dfrac{f(x)}{g(x)}\right)' = \dfrac{f'(x)g(x)-f(x)g'(x)}{g^2(x)}$

1. Calculate derivatives by using the ratio rule for the following functions in points where the derivatives exist.

(a) $h(x) = \dfrac{x}{\sin x}$

(b) $f(x) = \dfrac{\sin x}{e^x}$

(c) $f(x) = \dfrac{2x-1}{1-x}$

(d) $f(x) = \dfrac{\sqrt{x}}{\ln x}$

(e) $f(x) = \dfrac{\sin(x-1)}{\cos(1-x)}$

(f) $f(x) = \dfrac{\log(x-1)}{\cos(1-x)}$

9.7 *   Taylor Polynomial

1. Develop the following functions into its equivalent Tylor polynomials, calculate only the first four terms of the polynomials.

(a) $f(x) = e^x,\ x_0 = 0$

(b) $g(x) = \cos x,\ x_0 = 0$

(c) $f(x) = \sin x,\ x_0 = 0$

(d) $f(x) = \sqrt{x},\ x_0 = 1$

(e) $f(x) = \ln(1+x),\ x_0 = 0$

(f) $f(x) = e^{2x-x^2},\ x_0 = 0$

(g) $f(x) = \dfrac{x}{e^{x-1}},\ x_0 = 1$

9.8 *   L'Hôpital's Rule

1. Evaluate the following limits:

(a) $\lim\limits_{x\to 1} \dfrac{x^2+2x-3}{x-1}$

(b) $\lim\limits_{x\to 0} \dfrac{\sin(x)}{x}$

(c) $\lim\limits_{x\to 0} \dfrac{x+\sin(x)}{x}$

(d) $\lim\limits_{x\to 0} \dfrac{\sin(2x)}{x}$

(e) $\lim\limits_{x\to 0} \dfrac{4\sin(5x)}{\sin(4x)}$

(f) $\lim\limits_{x\to 0} \dfrac{\sin(2x)}{\sin(3x)}$

2. Evaluate the following limits:

(a) $\lim\limits_{x\to 0} x^x$

(b) $\lim\limits_{x\to\infty} x^{1/x}$

(c) $\lim\limits_{x\to\pi/2} \dfrac{\ln(\sin x)}{\cos x}$

(d) $\lim\limits_{x\to\pi/2} (\sin x)^{\tan x}$

(e) $\lim\limits_{x\to 0} \left(\dfrac{1}{x} - \dfrac{1}{\sin x}\right)$

(f) $\lim\limits_{x\to 0} x\,\ln x$

(g) $\lim\limits_{x\to-\infty} x\,e^x$

(h) $\lim\limits_{x\to\infty} \left(1+\dfrac{1}{x}\right)^{x+5}$

(i) $\lim\limits_{x\to\infty} \dfrac{e^x}{x^2}$

Solutions

Exercise 9.1, page 202

1. The power rule results in the following answers.

   (a) $f'(x) = (x^5)' = 5x^{5-1} = 5x^4$

   (b) $h'(t) = (x^3 + 2x + 1)' = 3x^2 + 2$

   (c) $g'(x) = \left(-\dfrac{1}{27}x^3 + \dfrac{1}{18}x^3\right)' = -\dfrac{\cancel{3}1}{27\,9}x^2 + \dfrac{\cancel{2}1}{18\,9}x^2 = 0$

   (d) $h'(x) = \left(\dfrac{4}{x^3} - x^{-3}\right)' = (4x^{-3} - x^{-3})' = -12x^{-4} + 3x^{-4} = -\dfrac{9}{x^4}$

   (e) $a'(x) = \left(\dfrac{1}{2}x^{\frac{1}{2}}\right)' = \dfrac{1}{2}\dfrac{1}{2}x^{(\frac{1}{2}-1)} = \dfrac{1}{4}x^{(-\frac{1}{2})} = \dfrac{1}{4x^{\frac{1}{2}}} = \dfrac{1}{4\sqrt{x}}$

   (f) $h'(x) = \left(\dfrac{4}{3}x^{-\frac{3}{4}}\right)' = \dfrac{\cancel{4}}{\cancel{3}}\left(-\dfrac{\cancel{3}}{\cancel{4}}\right)x^{-\frac{3}{4}-1} = -x^{-\frac{7}{4}} = -\dfrac{1}{x^{\frac{7}{4}}} = -\dfrac{1}{\sqrt[4]{x^7}}$

2. Radicals are equivalent to fractional powers.

   (a) $f'(x) = (\sqrt{x})' = \left(x^{\frac{1}{2}}\right)' = \dfrac{1}{2}x^{\frac{1}{2}-1} = \dfrac{1}{2}x^{-\frac{1}{2}} = \dfrac{1}{2\sqrt{x}}$

   (b) $h'(x) = \left(\sqrt[3]{x^2}\right)' = \left(x^{\frac{2}{3}}\right)' = \dfrac{2}{3}x^{\frac{2}{3}-1} = \dfrac{2}{3}x^{-\frac{1}{3}} = \dfrac{2}{3x^{\frac{1}{3}}} = \dfrac{2}{3\sqrt[3]{x}}$

   (c) $g'(x) = \left(5\sqrt[4]{x} - 2\sqrt[3]{x^3}\right)' = \left(5x^{\frac{1}{4}} - 2x^{\frac{3}{3}}\right)' = \dfrac{5}{4}x^{-\frac{3}{4}} - 2 = \dfrac{5}{4\sqrt[4]{x^3}} - 2$

   (d) $b'(x) = \left(\dfrac{5}{\sqrt[4]{x}}\right)' = \left(\dfrac{5}{x^{\frac{1}{4}}}\right)' = \left(5x^{-\frac{1}{4}}\right)' = 5\left(-\dfrac{1}{4}\right)x^{-\frac{5}{4}} = -\dfrac{5}{4\sqrt[4]{x^5}}$

   (e) $t'(x) = \left(\dfrac{1}{\sqrt[3]{x^4}}\right)' = \left(x^{-\frac{4}{3}}\right)' = -\dfrac{4}{3}x^{-\frac{7}{3}} = -\dfrac{4}{3\sqrt[3]{x^7}}$

   (f) $g'(t) = \left(\dfrac{1}{3}t^3 - \dfrac{\sqrt{t}}{3} + 2\right)' = t^2 - \dfrac{1}{6\sqrt{t}}$

Exercise 9.2, page 202

1. Derivatives of a^x functions include $\ln a$ term.

   (a) $f'(x) = (1^x)' = 1^x\,{}^{1}\ln(1)\,{}^{0} = 0$

   (b) $h'(x) = (2^x)' = 2^x \ln 2$

   (c) $g'(x) = (e^x)' = e^x \ln e\,{}^{1} = e^x$

   (d) $b'(x) = (10^x)' = 10^x \ln 10$

   (e) $g'(x) = (a^x + x^a)' = a^x \ln a + ax^{a-1}$

   (f) $f'(x) = (e^x + x^3)' = (e^x)' + (x^3)' = e^x + 3x^2$

Exercise 9.3, page 202

1. Derivatives of $\log_a x$ functions include $\ln a$ term.

(a) $f'(x) = (\ln x)' = \dfrac{1}{x \ln e} = \dfrac{1}{x}$

(b) $h'(x) = (\log x)' = \dfrac{1}{x \ln 10}$

(c) $g'(x) = \left(\log_5 x\right)' = \dfrac{1}{x \ln 5}$

Exercise 9.4, page 203

1. Derivatives of composite functions follow the "chain rule". First, we use derivative of the "external" function and keep its argument as is, then we multiply it with derivative of the argument function.

(a) $f'(x) = \left[e^{5x}\right]' = e^{5x} (5x)' = 5e^{5x}$

(b) $h'(x) = \left[\sin\left(x^2\right)\right]' = \cos\left(x^2\right)\left(x^2\right)' = 2x \cos\left(x^2\right)$

(c) $g'(x) = [\sin(abcx)]' = \cos(abcx)(abcx)' = abc \cos(abcx)$

2. Derivatives of composite functions follow the "chain rule".

(a) $f'(x) = \left[(1-x)^5\right]' = 5(1-x)^{5-1}(1-x)' = 5(1-x)^4(-1) = -5(1-x)^4$

(b) $h'(x) = \left(\sqrt{x^2+2}\right)' = \left[(x^2+2)^{1/2}\right]' = \dfrac{1}{2}(x^2+2)^{-1/2}(x^2+2)'$

$= \dfrac{2x}{2\sqrt{x^2+2}} = \dfrac{x}{\sqrt{x^2+2}}$

(c) $g'(x) = \left(\sqrt[3]{\sqrt{2x-5}}\right)' = \left[(2x-5)^{\frac{1}{2}\frac{1}{3}}\right]' = \left[(2x-5)^{\frac{1}{6}}\right]'$

$= \dfrac{1}{6}(2x-5)^{-\frac{5}{6}}(2x-5)' = \dfrac{2}{6}\dfrac{1}{3}(2x-5)^{-\frac{5}{6}} = \dfrac{1}{3\sqrt[6]{(2x-5)^5}}$

3. Derivatives of composite exponential functions. First we use the exponent rule, then multiply it with the derivative of its argument.

(a) $f'(x) = \left(e^{1-x}\right)' = e^{1-x} \ln e^{1} (1-x)' = -e^{1-x}$

(b) $h'(x) = \left(2^{x^2+2x-1}\right)' = 2^{x^2+2x-1} \ln(2)(x^2+2x-1)' = \ln(2)(2x+2)2^{x^2+2x-1}$

(c) $h'(x) = \left(e^{x^2}\right)' = e^{x^2} \ln e^{1} (x^2)' = 2x\,e^{x^2}$

(d) $g'(x) = \left(\sqrt[3]{e^{6x-3}}\right)' = \left[\left(e^{6x-3}\right)^{1/3}\right]' = \left(e^{2x-1}\right)' = 2\,e^{2x-1} \ln e^{1}$

(e) $g'(x) = \left(\dfrac{1}{e^{-x^2/2}}\right)' = \left(e^{x^2/2}\right)' = x\,e^{x^2/2}$

(f) $g'(x) = \left(\dfrac{1}{e^{\sqrt{2-3x}}}\right)' = \left(e^{-\sqrt{2-3x}}\right)' = \left(e^{-(2-3x)^{1/2}}\right)'$

$\qquad = \dfrac{1}{e^{\sqrt{2-3x}}} \ln e^{1} \left[-(2-3x)^{1/2}\right]'$

$\qquad = \dfrac{1}{e^{\sqrt{2-3x}}} \left(-\dfrac{1}{2}\right) (2-3x)^{-1/2}(2-3x)'$

$\qquad = \dfrac{3}{2\,e^{\sqrt{2-3x}}\sqrt{(2-3x)}}$

4. Derivatives of composite $\log_a x$ functions include $\ln a$ term, and derivative of its argument.

(a) $f'(x) = (\ln(1-x))' = \dfrac{1}{1-x} \ln e^{1} (1-x)' = -\dfrac{1}{1-x}$

(b) $h'(x) = \left(\log_2 (x^2+2x+1)\right)' = \dfrac{1}{(x^2+2x+1)\ln(2)}(x^2+2x+1)'$

$\qquad = \dfrac{2x+2}{\ln(2)(x^2+2x+1)} = \dfrac{2(x+1)}{\ln(2)(x+1)^2} = \dfrac{2}{\ln(2)(x+1)}$

(c) $g'(x) = \left(\ln\sqrt{2-x}\right)' = \dfrac{1}{\sqrt{2-x}\,\ln e^{1}}\left(\sqrt{2-x}\right)'$

$\qquad = \dfrac{1}{\sqrt{2-x}}\left[(2-x)^{1/2}\right]' = \dfrac{1}{\sqrt{2-x}}(2-x)^{-1/2}(2-x)'$

$\qquad = \dfrac{1}{2\sqrt{2-x}\sqrt{2-x}}(-1) = -\dfrac{1}{2(2-x)}$

(d) $g'(t) = \left(\dfrac{1}{\log(2t-3)}\right)' = \left[\left[\log(2t-3)\right]^{-1}\right]'$

$\qquad = -1\left[\log(2t-3)\right]^{-2}(\log(2t-3))'$

$\qquad = -\dfrac{1}{\log^2(2t-3)}\dfrac{1}{(2t-3)\ln 10}(2t-3)'$

$\qquad = -\dfrac{2}{\ln 10\,\log^2(2t-3)}\dfrac{1}{(2t-3)}$

5. Derivatives of composite trigonometric functions include derivative of its argument.

(a) $f'(x) = (\sin(2x - 1))' = \cos(2x - 1)\,(2x - 1)' = 2\,\cos(2x - 1)$

(b) $h'(x) = (\sin\sqrt{x})' = \cos(\sqrt{x})\,(\sqrt{x})' = \cos(\sqrt{x})\,(x^{1/2})' = \cos(\sqrt{x})\,\dfrac{1}{2}x^{1-1/2}$

$= \cos(\sqrt{x})\,\dfrac{1}{2}x^{-1/2} = \dfrac{\cos(\sqrt{x})}{2\sqrt{x}}$

(c) $g'(x) = [\cos(\sin(1 - x))]' = -\sin(\sin(1 - x))\,[\sin(1 - x)]'$

$= -\sin(\sin(1 - x))\,\cos(1 - x)\,(1 - x)' = \sin(\sin(1 - x))\,\cos(1 - x)$

(d) $g'(x) = \left(\dfrac{1}{\ln(\sin x)}\right)' = \left[(\ln(\sin x))^{-1}\right]' = (-1)\,(\ln(\sin x))^{-2}\,(\ln(\sin x))'$

$= -\dfrac{1}{\ln^2(\sin x)}\,\dfrac{1}{\sin x \ln e}\,(\sin x)' = -\dfrac{1}{\ln^2(\sin x)}\,\dfrac{\cos x}{\sin x} = -\dfrac{\cot x}{\ln^2(\sin x)}$

(e) $g'(x) = \left(\sin\left(\dfrac{1}{x}\right)\right)' = \cos\left(\dfrac{1}{x}\right)\left(\dfrac{1}{x}\right)'$

$= \cos\left(\dfrac{1}{x}\right)(x^{-1})' = \cos\left(\dfrac{1}{x}\right)(-1)\,x^{-2} = -\dfrac{\cos\left(\dfrac{1}{x}\right)}{x^2}$

(f) $g'(x) = \left[\ln\left(\sin x + \sqrt{1 + \sin^2 x}\right)\right]'$

$= \dfrac{1}{\sin x + \sqrt{1 + \sin^2 x}}\left(\sin x + \sqrt{1 + \sin^2 x}\right)'$

$= \dfrac{1}{\sin x + \sqrt{1 + \sin^2 x}}\left(\cos x + \dfrac{2\sin x \cos x}{2\sqrt{1 + \sin^2 x}}\right)$

$= \dfrac{\cos x}{\sin x + \sqrt{1 + \sin^2 x}}\,\dfrac{\sqrt{1 + \sin^2 x} + \sin x}{\sqrt{1 + \sin^2 x}} = \dfrac{\cos x}{\sqrt{1 + \sin^2 x}}$

(g)

$g'(x) = \left[\sin^2 x + \sin^2\left(\dfrac{2\pi}{3} + x\right) + \sin^2\left(\dfrac{2\pi}{3} - x\right)\right]'$

$\{\sin(x \pm y) = \sin x \cos y \pm \cos x \sin x\,\}$

$= \left[\sin^2 x + \left(\sin\dfrac{2\pi}{3}\cos x + \cos\dfrac{2\pi}{3}\sin x\right)^2\right.$

$\left. + \left(\sin\dfrac{2\pi}{3}\cos x - \cos\dfrac{2\pi}{3}\sin x\right)^2\right]'$

$$= \left[\sin^2 x + \left(\frac{\sqrt{3}}{2} \cos x - \frac{1}{2} \sin x \right)^2 + \left(\frac{\sqrt{3}}{2} \cos x + \frac{1}{2} \sin x \right)^2 \right]'$$

$$= 2 \sin x \cos x + 2 \left(\frac{\sqrt{3}}{2} \cos x - \frac{1}{2} \sin x \right) \left(\frac{\sqrt{3}}{2} \cos x - \frac{1}{2} \sin x \right)'$$

$$+ 2 \left(\frac{\sqrt{3}}{2} \cos x + \frac{1}{2} \sin x \right) \left(\frac{\sqrt{3}}{2} \cos x + \frac{1}{2} \sin x \right)'$$

$$= 2 \sin x \cos x + 2 \frac{1}{2} \frac{1}{2} (\sqrt{3} \cos x - \sin x)(-\sqrt{3} \sin x - \cos x)$$

$$+ 2 \frac{1}{2} \frac{1}{2} (\sqrt{3} \cos x + \sin x)(-\sqrt{3} \sin x + \cos x)$$

$$= 2 \sin x \cos x + \frac{1}{2}(-3 \cos x \sin x + \sqrt{3} \sin^2 x - \sqrt{3} \cos^2 x + \sin x \cos x$$

$$- 3 \cos x \sin x - \sqrt{3} \sin^2 x + \sqrt{3} \cos^2 x + \sin x \cos x)$$

$$= 2 \sin x \cos x + \frac{1}{2}(-4 \sin x \cos x) = 0$$

Exercise 9.5, page 203

1. Derivative of two functions product follows formula

$$(f(x)g(x))' = f'(x)g(x) + f(x)g'(x)$$

(a) $h'(x) = (x \sin x)' = x' \sin x + x(\sin x)' = \sin x + x \cos x$

(b) $g'(x) = (e^x \cos x)' = (e^x)' \cos x + e^x (\cos x)' = e^x \cos x - e^x \sin x$
$= e^x (\cos x - \sin x)$

(c) $f'(x) = [(2x - 1)(1 - x)]' = (2x - 1)'(1 - x) + (2x - 1)(1 - x)'$
$= 2(1 - x) - (2x - 1) = 2 - 2x - 2x + 1 = -4x + 3$

(d) $f'(x) = (\sqrt{x} \ln x)' = \dfrac{\ln x}{2\sqrt{x}} + \dfrac{\sqrt{x}}{x} = \dfrac{\ln x}{2\sqrt{x}} + \dfrac{1}{\sqrt{x}}$

(e) $f'(x) = [\sin(x - 1) \cos(1 - x)]'$
$= [\sin(x - 1)]' \cos(1 - x) + \sin(x - 1) [\cos(1 - x)]'$

$= \cos(x - 1)(x - 1)' \cos(1 - x) - \sin(x - 1) \sin(1 - x)(1 - x)'$

$= \cos(x - 1) \cos(1 - x) + \sin(x - 1) \sin(1 - x)$

$\{\cos(x - y) = \cos x \cos y + \sin x \sin y\}$

$= \cos(2x - 2) + \sin 0^{\,0}$

(f) $f'(x) = \left[\log(x-1)\cos(1-x)\right]'$

$$= \frac{1}{(x-1)\ln 10}\cos(1-x) + \log(x-1)\sin(1-x)$$

(g) $h'(t) = (t^2\, 3^t)' \quad = (t^2)'\, 3^t + t^2\, (3^t)' = 2t\, 3^t + t^2\, 3^t \ln 3 = 3^t t (2 + t \ln 3)$

Exercise 9.6, page 204

1. Derivative of two functions ratio follows formula

$$(f(x)g(x))' = \frac{f'(x)g(x) - f(x)g'(x)}{g^2(x)}$$

(a) $h'(x) = \left(\dfrac{x}{\sin x}\right)' = \dfrac{\sin x - x\cos x}{\sin^2 x}$

(b) $f'(x) = \left(\dfrac{\sin x}{e^x}\right)' = \dfrac{\cos x\, e^x - \sin x\, e^x}{e^{2x}} = e^x \dfrac{\cos x - \sin x}{e^{2x}} = \dfrac{\cos x - \sin x}{e^x}$

(c) $f'(x) = \left(\dfrac{2x-1}{1-x}\right)' = \dfrac{2(1-x) + (2x-1)}{(1-x)^2} = \dfrac{1}{(1-x)^2}$

(d) $f'(x) = \left(\dfrac{\sqrt{x}}{\ln x}\right)' = \dfrac{\frac{\ln x}{2\sqrt{2}} - \frac{\sqrt{x}}{x}}{\ln^2 x} = \dfrac{\frac{\ln x}{2\sqrt{2}} - \frac{1}{\sqrt{x}}}{\ln^2 x} = \dfrac{\ln x - 2}{2\sqrt{2}\ln^2 x}$

(e) $f'(x) = \left[\dfrac{\sin(x-1)}{\cos(1-x)}\right]' = \dfrac{\cos(x-1)\cos(1-x) - \sin(x-1)\sin(1-x)\,(-1)}{\cos^2(1-x)}$

$\{\cos(-x) = \cos x \quad \text{and} \quad \sin(-x) = -\sin(x)\}$

$$= \frac{\cos(x-1)\cos(x-1) + \sin(x-1)\sin(x-1)}{\cos^2(1-x)}$$

$$= \frac{\cos^2(x-1) + \sin^2(x-1)}{\cos^2(1-x)} = \frac{1}{\cos^2(1-x)}$$

(f) $f'(x) = \left(\dfrac{\log(x-1)}{\cos(1-x)}\right)' = \dfrac{\frac{\cos(1-x)}{(x-1)\ln 10} + \log(x-1)\sin(1-x)(-1)}{\cos^2(1-x)}$

$$= \frac{\cos(1-x) - \ln 10\,\log(x-1)\sin(1-x)\,(x-1)}{\ln 10\,(x-1)\cos^2(1-x)}$$

$\left\{\log a = \dfrac{\ln a}{\ln 10}\right\}$

$$= \frac{\cos(1-x) - \ln(x-1)\sin(1-x)\,(x-1)}{\ln 10\,(x-1)\cos^2(1-x)}$$

Exercise 9.7, page 204

1. Tylor formula gives an approximation of a function $f(x)$ in the form of polynomial of nth order as

$$f(x) = \sum_{k=0}^{n} \frac{f^{(k)}(x_0)}{k!}(x-x_0)^k + O(x-x_0)^n$$

where $f^{(k)}(x_0)$ is k^{th} derivative in $x = x_0$, and $O(x - x_0)^n$ is the reminder term so that $\lim_{x \to x_0} O(x - x_0)^n = 0$.

(a) First four terms of Tylor polynomial are found as

$$f(x) = e^x, \ x_0 = 0 \quad \therefore \quad f(0) = f'(0) = f''(0) = f'''(0) = e^0 = 1$$

$$\therefore$$

$$T(x) = \frac{f^{(0)}(0)}{0!}(x - 0)^0 + \frac{f'(0)}{1!}(x - 0)^1 + \frac{f''(0)}{2!}(x - 0)^2 + \frac{f'''(0)}{3!}(x - 0)^3$$

$$= 1 + x + \frac{x^2}{2} + \frac{x^3}{6}$$

Around the point $x = 0$, obviously, $f(x) = T(x) = 1$ and difference is relatively small in within $[-1, 1]$ interval, while further away $T(x)$ becomes less accurate approximation of $f(x)$, i.e. $\Delta = f(x) - T(x)$ becomes large, see Fig. 9.1.

Fig. 9.1 Example 9.7-1(a)

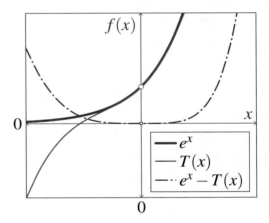

(b) Trigonometric functions are periodic, however, polynomial approximation within an interval inferior to the period of an even function is

$$g(x) = \cos x, \ x_0 = 0 \quad \therefore$$

$$f(0) = \cos 0 = 1, \quad f'(0) = -\sin 0 = 0,$$

$$f''(0) = -\cos 0 = -1, \quad f'''(0) = \sin 0 = 0,$$

therefore,

$$T(x) = \frac{f^{(0)}(0)}{0!}(x - 0)^0 + \frac{f'(0)}{1!}(x - 0)^1 + \frac{f''(0)}{2!}(x - 0)^2 + \frac{f'''(0)}{3!}(x - 0)^3$$

$$= 1 - \frac{x^2}{2}$$

Effectively, there are only two even order terms in $T(x)$ (odd order terms are cancelled) that reasonably well

(c)   approximate $\cos x$ within $[-1, 1]$ interval, see Fig. 9.2.

Fig. 9.2 Example 9.7-1(b)

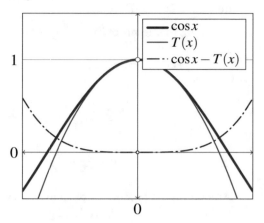

(d)   Approximation of an odd function is

$f(x) = \sin x$, $x_0 = 0$ $\therefore$

$f(0) = \sin 0 = 0$, $f'(0) = \cos 0 = 1$,

$f''(0) = -\sin 0 = 0$, $f'''(0) = -\cos 0 = -1$,

therefore,

$$T(x) = \frac{f^{(0)}(0)}{0!}(x-0)^0 + \frac{f'(0)}{1!}(x-0)^1 + \frac{f''(0)}{2!}(x-0)^2 + \frac{f'''(0)}{3!}(x-0)^3$$

$$= x - \frac{x^3}{6}$$

Effectively, there are only two odd order terms in $T(x)$ (even order terms are cancelled) that reasonably well approximate $\sin x$ within $[-1, 1]$ interval, see Fig. 9.3.

Fig. 9.3 Example 9.7-1(d)

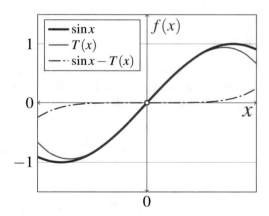

(e) We systematically calculate derivatives at $x_0 = 1$ as

$$f(x) = \sqrt{x} \quad \therefore \quad f(1) = 1$$

$$f'(x) = \frac{1}{2\sqrt{x}} \quad \therefore \quad f'(1) = \frac{1}{2}$$

$$f''(x) = -\frac{1}{4\sqrt{x^3}} \quad \therefore \quad f''(1) = -\frac{1}{4}$$

$$f'''(x) = \frac{3}{8\sqrt{x^5}} \quad \therefore \quad f'''(1) = \frac{3}{8}$$

Therefore,

$$T(x) = \frac{f^{(0)}(1)}{0!}(x-1)^0 + \frac{f'(1)}{1!}(x-1)^1 + \frac{f''(1)}{2!}(x-1)^2 + \frac{f'''(1)}{3!}(x-1)^3$$

$$= 1 + \frac{1}{2}(x-1) - \frac{1}{4 \times 2!}(x-1)^2 + \frac{3}{8 \times 3!}(x-1)^3$$

$$= 1 + \frac{x-1}{2} - \frac{(x-1)^2}{8} + \frac{(x-1)^3}{16}$$

Comparison between $f(x)$ and $T(x)$ is shown in Fig. 9.4.

Fig. 9.4 Example 9.7-1(e)

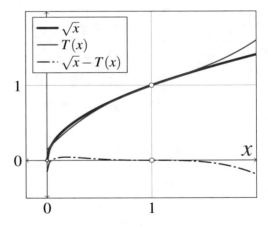

(f) $f(x) = \ln(1 + x)$, $x_0 = 0$

$$f(x) = \ln(1 + x) \quad \therefore \quad f(0) = 0$$

$$f'(x) = \frac{1}{1 + x} \quad \therefore \quad f'(0) = 1$$

$$f''(x) = -\frac{1}{(1 + x)^2} \quad \therefore \quad f''(0) = -1$$

$$f'''(x) = \frac{2}{(1 + x)^3} \quad \therefore \quad f'''(0) = 2$$

Therefore,

$$T(x) = \frac{f^{(0)}(0)}{0!}(x - 0)^0 + \frac{f'(0)}{1!}(x - 0)^1 + \frac{f''(0)}{2!}(x - 0)^2 + \frac{f'''(0)}{3!}(x - 0)^3$$

$$= x - \frac{1}{2}x^2 + \frac{1}{3}x^3$$

(g) $f(x) = e^{2x - x^2}$, $x_0 = 0$

$$f(x) = e^{2x - x^2} \quad \therefore \quad f(0) = 1$$

$$f'(x) = e^{2x - x^2}(2 - 2x) \quad \therefore \quad f'(0) = 2$$

$$f''(x) = e^{2x - x^2}(4x^2 - 8x + 2) \quad \therefore \quad f''(0) = 2$$

$$f'''(x) = e^{2x - x^2}(-8x^3 + 24x^2 - 12x - 4) \quad \therefore \quad f'''(0) = -4$$

Therefore,

$$T(x) = \frac{f^{(0)}(0)}{0!}(x - 0)^0 + \frac{f'(0)}{1!}(x - 0)^1 + \frac{f''(0)}{2!}(x - 0)^2 + \frac{f'''(0)}{3!}(x - 0)^3$$

$$= 1 + 2x + x^2 - \frac{2}{3}x^3$$

(h) $f(x) = \dfrac{x}{e^{x-1}}$, $x_0 = 1$

$$f(x) = \frac{x}{e^{x-1}} \quad \therefore \quad f(1) = 1$$

$$f'(x) = \frac{1-x}{e^{x-1}} \quad \therefore \quad f'(1) = 0$$

$$f''(x) = \frac{x-2}{e^{x-1}} \quad \therefore \quad f''(1) = -1$$

$$f'''(x) = \frac{3-x}{e^{x-1}} \quad \therefore \quad f'''(1) = 2$$

Therefore,

$$T(x) = \frac{f^{(0)}(1)}{0!}(x-1)^0 + \frac{f'(1)}{1!}(x-1)^1 + \frac{f''(1)}{2!}(x-1)^2 + \frac{f'''(1)}{3!}(x-1)^3$$

$$= 1 - \frac{1}{2}(x-1)^2 + \frac{1}{3}(x-1)^3$$

Exercise 9.8, page 204

1. L'Hôpital's rule is a very powerful technique to calculate limits that take the following indeterminate forms:

$$\frac{0}{0}, \quad \frac{\infty}{\infty}, \quad 0 \cdot \infty, \quad 0^0, \quad \infty^0, \quad \infty - \infty$$

Note that L'Hôpital's rule is applicable only in the case of rational functions that take the form of

$$\lim_{x \to x_0} \frac{f(x)}{g(x)} = \frac{f(x_0)}{g(x_0)} = \frac{0}{0} \quad \text{or,} \quad \frac{\infty}{\infty}$$

Then, the rational function's limit may be calculated after taking derivates of its numerator and denominator separately, that is to say

$$\text{if} \quad \lim_{x \to x_0} \frac{f(x)}{g(x)} = \frac{0}{0} \quad \text{or} \quad \frac{\infty}{\infty} \Rightarrow \lim_{x \to x_0} \frac{f(x)}{g(x)} = \lim_{x \to x_0} \frac{f'(x)}{g'(x)}$$

Some of the following examples may be solved by other methods (see Example 8.2-1), here we illustrate the use of L'Hôpital's rule.

(a) $\displaystyle \lim_{x \to 1} \frac{x^2 + 2x - 3}{x - 1} = \left(\frac{0}{0}\right) \overset{\text{L.H.}}{=} \lim_{x \to 1} \frac{(x^2 + 2x - 3)'}{(x - 1)'} = \lim_{x \to 1} \frac{2x + 2}{1} = 4$

(b) $\displaystyle \lim_{x \to 0} \frac{\sin(x)}{x} = \left(\frac{0}{0}\right) \overset{\text{L.H.}}{=} \lim_{x \to 0} \frac{(\sin(x))'}{(x)'} = \lim_{x \to 0} \frac{\cos(x)}{1} = 1$

(c) $\lim\limits_{x\to 0} \dfrac{x + \sin(x)}{x} = \left(\dfrac{0}{0}\right) \overset{\text{L.H.}}{=} \lim\limits_{x\to 0} \dfrac{(x + \sin(x))'}{(x)'} = \lim\limits_{x\to 0} \dfrac{1 + \cos(x)}{1} = 2$

(d) $\lim\limits_{x\to 0} \dfrac{\sin(2x)}{x} = \left(\dfrac{0}{0}\right) \overset{\text{L.H.}}{=} \lim\limits_{x\to 0} \dfrac{(\sin(2x))'}{(x)'} = \lim\limits_{x\to 0} \dfrac{2\cos(2x)}{1} = 2$

(e) $\lim\limits_{x\to 0} \dfrac{4\sin(5x)}{\sin(4x)} = \left(\dfrac{0}{0}\right) \overset{\text{L.H.}}{=} \lim\limits_{x\to 0} \dfrac{(4\sin(5x))'}{(\sin(4x))'} = \lim\limits_{x\to 0} \dfrac{4\times 5\cos(5x)}{4\cos(4x)} = 5$

(f) $\lim\limits_{x\to 0} \dfrac{\sin(2x)}{\sin(3x)} = \left(\dfrac{0}{0}\right) \overset{\text{L.H.}}{=} \lim\limits_{x\to 0} \dfrac{(\sin(2x))'}{(\sin(3x))'} = \lim\limits_{x\to 0} \dfrac{2\cos(2x)}{3\cos(3x)} = \dfrac{2}{3}$

2. Indeterminate forms (0^0), $(0\cdot\infty)$, $(\infty - \infty)$, etc. should be converted into the forms $(0/0)$ or (∞/∞), then L'Hôpital's rule applies. The first following example illustrates technique that helps us to resolve the exponential forms, which is used in the subsequent examples.

(a) $\lim\limits_{x\to 0_+} x^x = \left(0^0\right) = \lim\limits_{x\to 0_+} e^{\ln x^x} = \lim\limits_{x\to 0_+} e^{x\ln x} = e^{\lim_{x\to 0_+} x\ln x} = e^0 = 1$

(b) $\lim\limits_{x\to\infty} x^{1/x} = \left(\infty^0\right) = \left\{e^{\ln a} = a\right\} = \lim\limits_{x\to\infty} \exp\left(\ln x^{1/x}\right) = \left\{\ln a^b = b\ln a\right\}$

$\quad = \lim\limits_{x\to\infty} \exp\left(\dfrac{\ln x}{x}\right) = \exp\left(\lim\limits_{x\to\infty} \dfrac{\ln x}{x}\right) = \exp\left(\dfrac{\infty}{\infty}\right)$

$\quad \overset{\text{L.H.}}{=} \exp\left(\lim\limits_{x\to\infty} \dfrac{(\ln x)'}{x'}\right) = \exp\left(\lim\limits_{x\to\infty} \dfrac{1/x}{1}\right) = \exp\left(\dfrac{0}{1}\right) = e^0 = 1$

(c) $\lim\limits_{x\to\pi/2} \dfrac{\ln(\sin x)}{\cos x} = \left(\dfrac{0}{0}\right) \overset{\text{L.H.}}{=} \lim\limits_{x\to\pi/2} \dfrac{(\ln(\sin x))'}{(\cos x)'}$

$\quad = \left\{[f(g(x))]' = f'(g(x))\,g'(x)\right\}$

$\quad = \lim\limits_{x\to\pi/2} \dfrac{(\cos x/\sin x)}{(-\sin x)} = \lim\limits_{x\to\pi/2} \dfrac{\cos x}{-\sin^2 x} = \dfrac{0}{-1^2} = 0$

(d) $\lim\limits_{x\to\pi/2} (\sin x)^{\tan x} = \left(1^\infty\right) = \left\{e^{\ln a} = a, \quad \ln a^b = b\ln a\right\}$

$\quad = \lim\limits_{x\to\pi/2} \exp\left(\ln(\sin x)^{\tan x}\right) = \lim\limits_{x\to\pi/2} \exp\left(\tan x\,\ln(\sin x)\right)$

$\quad = \exp\left(\lim\limits_{x\to\pi/2} \dfrac{\sin x}{\cos x}^{1}\,\ln(\sin x)\right)$

$\quad = \exp\left(\lim\limits_{x\to\pi/2} \dfrac{\ln(\sin x)}{\cos x}\right) = \left\{\text{see Example 9.8-2(c)}\right\}$

$\quad = e^0 = 1$

(e) $\lim\limits_{x\to 0} \left(\dfrac{1}{x} - \dfrac{1}{\sin x}\right) = (\infty - \infty) = \lim\limits_{x\to 0} \left(\dfrac{\sin x - x}{x\sin x}\right) = \left(\dfrac{0}{0}\right)$

$\quad \overset{\text{L.H.}}{=} \lim\limits_{x\to 0} \left(\dfrac{(\sin x - x)'}{(x\sin x)'}\right)$

$\quad = \left\{[f(x)\,g(x)]' = f'(x)g(x) + f(x)g'(x)\right\}$

$\quad = \lim\limits_{x\to 0} \left(\dfrac{\cos x - 1}{\sin x + x\cos x}\right) = \left(\dfrac{0}{0}\right) \overset{\text{L.H.}}{=} \lim\limits_{x\to 0} \left(\dfrac{(\cos x - 1)'}{(\sin x + x\cos x)'}\right)$

$\quad = \lim\limits_{x\to 0} \left(\dfrac{-\sin x}{\cos x + \cos x - x\sin x}\right) = \dfrac{0}{2} = 0$

(f) $\displaystyle\lim_{x\to 0} x \ln x = (0\cdot\infty) = \lim_{x\to 0}\frac{\ln x}{1/x} = \left(\frac{\infty}{\infty}\right) \overset{\text{L.H.}}{=} \lim_{x\to 0}\frac{(\ln x)'}{(1/x)'} = \lim_{x\to 0}\frac{1/\!\!\!\;x}{-1/x^2}$

$\displaystyle\qquad = \lim_{x\to 0}(-x) = 0$

(g) $\displaystyle\lim_{x\to -\infty} x\, e^x = (\infty\cdot 0) = \lim_{x\to -\infty}\frac{x}{e^{-x}} = \left(\frac{\infty}{\infty}\right)\overset{\text{L.H.}}{=}\lim_{x\to -\infty}\frac{(x)'}{(e^{-x})'}$

$\displaystyle\qquad = \lim_{x\to -\infty}\frac{1}{-e^{-x}} = 0$

(h) $\displaystyle\lim_{x\to\infty}\left(1+\frac{1}{x}\right)^{x+5} = (1^\infty) = \lim_{x\to\infty}\exp\left[\ln\left(1+\frac{1}{x}\right)^{x+5}\right]$

$\displaystyle\qquad = \left\{\ln a^b = b\ln a,\ \ e^{\ln x} = x\right\}$

$\displaystyle\qquad = \exp\left[\lim_{x\to\infty}(x+5)\ln\left(1+\frac{1}{x}\right)\right]$

$\displaystyle\qquad = (\exp(\infty\cdot 0))$

$\displaystyle\qquad = \exp\left[\lim_{x\to\infty}\frac{\ln\left(1+\frac{1}{x}\right)}{\frac{1}{x+5}}\right] = \left(\exp\left(\frac{\infty}{\infty}\right)\right)$

$\displaystyle\qquad \overset{\text{L.H.}}{=}\exp\left[\lim_{x\to\infty}\frac{\left[\ln\left(1+\frac{1}{x}\right)\right]'}{\left[\frac{1}{x+5}\right]'}\right]$

$\displaystyle\qquad = \exp\left[\lim_{x\to\infty}\frac{\frac{-1/x^2}{1+1/x}}{\frac{1}{(x+5)^2}}\right] = \exp\left[\lim_{x\to\infty}\frac{(x+5)^2}{x^2+x}\right]$

$\displaystyle\qquad \overset{\text{L.H.}}{=}\exp\left[\lim_{x\to\infty}\frac{2(x+5)}{2x+1}\right]\overset{\text{L.H.}}{=}\exp\left[\lim_{x\to\infty}\frac{2}{2}\right]$

$\displaystyle\qquad = e^1 = e$

(i) $\displaystyle\lim_{x\to\infty}\frac{e^x}{x^2} = \left(\frac{\infty}{\infty}\right)\overset{\text{L.H.}}{=}\lim_{x\to\infty}\frac{(e^x)'}{(x^2)'} = \lim_{x\to\infty}\frac{e^x}{2x} = \left(\frac{\infty}{\infty}\right)\overset{\text{L.H.}}{=}\lim_{x\to\infty}\frac{(e^x)'}{(2x)'}$

$\displaystyle\qquad = \lim_{x\to\infty}\frac{e^x}{2} = \infty$

Function Analysis

Important to Know

Function analysis is usually done in the following steps

1. Domain: set of x where the function is defined,
2. Parity: if $f(x) = f(-x)$ (even), if $f(x) = -f(-x)$ (odd),
3. Zeros and sign: solve for $f(x) = 0$,
4. Vertical asymptotes: if around point x_0 it is true that $\lim_{x \to x_0} f(x) = \pm\infty$,
5. Horizontal asymptotes: if $\lim_{x \to \pm\infty} f(x) = a$, where $a = \text{const.}$,
6. Oblique asymptote $f_{bb}(x)$: the asymptote equitation $f_{bb}(x) = ax + b$, where $a = \lim_{x \to \infty}(f(x)/x)$ and $b = \lim_{x \to \infty}(f(x) - ax)$,
7. Critical points: solve for $f'(x) = 0$ and $f''(x) = 0$,
8. The function graph.

10.1 Exercises

10.1 Basic Functions: Review

---

1. Sketch graphs and list the main properties of the following basic functions.

(a) $f(x) = \dfrac{1}{x}$

(b) $f(x) = \sqrt[n]{x}$

(c) $f(x) = a^x$

(d) $f(x) = \log_a x$

(e) $f(x) = \sin x$ and $f(x) = \cos x$

(f) $f(x) = \tan x$

(g) $f(x) = \arctan x$

© Springer Nature Switzerland AG 2021
R. Sobot, *Engineering Mathematics by Example*,
https://doi.org/10.1007/978-3-030-79545-0_10

10.2   Composite Functions

1. Given that $x = g(t)$ and $y = f(x)$ derive composite function $y = f(g(t))$.

   (a) $x = t^2 + 1, \quad y = \sqrt{x}$

   (b) $x = \dfrac{1}{t+1}, \quad y = \ln x$

   (c) $x = -(t^2 + 1), \quad y = \sqrt{x}$

   (d) $x = -t^2, \quad y = \ln x$

2. Assuming a composite function is in the form $F(x) = f(g(x))$, determine $f(x)$ and $g(x)$ functions,

   (a) $F(x) = (2x + x^2)^4$

   (b) $F(x) = \cos^2 x$

   (c) $F(x) = \dfrac{\sqrt[3]{x}}{1 + \sqrt[3]{x}}$

   (d) $F(x) = \sqrt[3]{\dfrac{x}{1+x}}$

3. Assuming a composite function is in the form $F(x) = f(g(h(x)))$, determine $f(x)$, $g(x)$, and $h(x)$ functions,

   (a) $F(x) = 1 - 3^{x^2}$

   (b) $F(x) = \sqrt[4]{1 + |x|}$

4. Derive $f(x)$ given that:

   (a) $f(x+1) = 1 - x$

   (b) $f(2x - 1) = x^2 - 2x + 1$

   (c) $f\left(\dfrac{2x-1}{x}\right) = x$

   (d) $f(x^2) = \dfrac{9}{x}$

5. Given a graph of ln function, without using calculator or graphic tools, sketch the graph of $y = \ln(x - 2) - 1$ function.

10.3 ***   Analysis of Functions

1. Do complete study of the following function:

$$f(x) = \dfrac{x^3}{(x-1)^2}$$

2. Do complete study of the following function

$$f(x) = 2 - \dfrac{2(1-x)}{x^2+1}$$

3. Given function $f(x)$ studied in Example 10.3-2,

(a) Derive the equation of a tangent T_1 in point $x = 1$.

(b) Determine coordinates of point B so that tangent T_2 to $f(x)$ crossing B is parallel to line $y = -x$. What is the relationship between T_1 and T_2?

4. Do complete study of the following function

$$f(x) = \frac{x}{(x+1)(x-4)}$$

10.4 ***   Functions: Study

1. Analyse and plot graphs of the following functions.

(a) $f(x) = x^4 - 2x^2$

(b) $f(x) = \dfrac{4x}{4 - x^2}$

(c) $f(x) = e^{\frac{1}{x}}$

(d) $f(x) = (2 - x^2)e^x$

(e) $f(x) = x \ln^2 x$

(f) $f(x) = \left(x - \dfrac{1}{4}\right) \ln\left(1 - \dfrac{1}{x}\right)$

(g) $f(x) = \sqrt[3]{x^2 - x^3}$

(h) $f(x) = x^x$

Solutions

Exercise 10.1, page 219

1. Quick review of basic functions.

(a) Inverse, Fig. 10.1.

$$f(x) = \frac{\pm k}{x}$$

Domain: $x \in \mathbb{R}, \quad x \neq 0$

Asymptotes: (H): $y = 0$, (V): $x = 0$

Derivatives:

$$f'(x) = \mp \frac{1}{x^2}; \quad f''(x) = \pm \frac{2}{x^3}$$

Critical points: none

Fig. 10.1 Example 10.1-1(a)

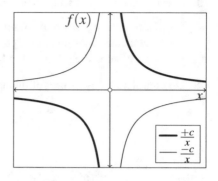

(b)   Radicals, Fig. 10.2.

$f(x) = \sqrt[n]{x}$

Domain:
$$x \geq 0 \ (n = 2k),$$
$$x \in \mathbb{R} \ (n = 2k + 1)$$
Derivatives:

$$f'(x) = \frac{\sqrt[n]{x}}{n\,x}; \quad f''(x) = \frac{1-n}{n^2}\frac{\sqrt[n]{x}}{x^2}$$

Critical points:
zero: $(0, 0)$, inflection: $(0, 0)$

Fig. 10.2 Example 10.1-1(b)

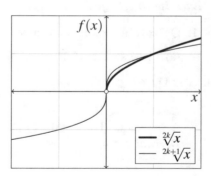

(c)   Exponent, Fig. 10.3.

$f(x) = a^x$

Domain: $x \in \mathbb{R}$
Asymptotes: (H): $y = 0$
Derivatives:

$$f'(x) = a^x \ln a; \quad f''(x) = a^x \ln^2 a$$

Critical points: none
Y intercept point: (0, 1)

Fig. 10.3 Example 10.1-1(c)

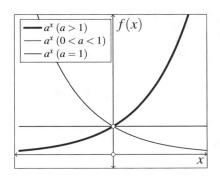

(d)   Logarithm, Fig. 10.4.

$f(x) = \log_a x$

Domain: $(x, a) > 0$, $a \neq 1$
Asymptotes: (V): $x = 0$
Derivatives:

$$f'(x) = \frac{1}{x \ln a}; \quad f''(x) = -\frac{1}{x^2 \ln a}$$

Critical points: zero: (0, 1)

Fig. 10.4 Example 10.1-1(c)

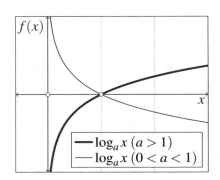

(e)   Sine and cosine, Fig. 10.5.

$f(x) = \sin x$ and $f(x) = \cos x$

Domain: $x \in \mathbb{R}$
Period: $T = 2\pi k, \; k \in \mathbb{Z}$
Derivatives:

$\sin x : \; f'(x) = \cos x; \; f''(x) = -\sin x$

$\cos x : \; f'(x) = -\sin x; \; f''(x) = -\cos x$

Critical points: zeros: $x_0 + k\pi$
($\sin x : x_0 = 0, \cos x : x_0 = {}^\pi/_2$)

Extreme points: min/max: $x_0 + k\pi$
($\sin x : x_0 = {}^\pi/_2, \cos x : x_0 = 0$)
$f(max/min) = \pm 1$

Fig. 10.5 Example 10.1-1(e)

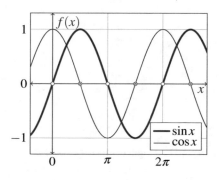

(f)   Tangent, Fig. 10.6.

$$f(x) = \tan x = \frac{\sin x}{\cos x}$$

Domain: $x \neq k\pi$
Period: $T = 2\pi k, \; k \in \mathbb{Z}$
Asymptotes: (V): $x = \pi/2 + k\pi$
Derivatives:

$$f'(x) = \frac{1}{\cos^2 x}; \; f''(x) = \frac{2\tan x}{\cos^2 x}$$

Critical points: zeros and inflection points: $x = k\pi$

Fig. 10.6 Example 10.1-1(f)

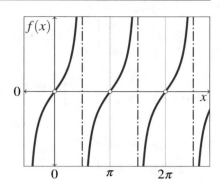

(g) Arctangent, Fig. 10.7.

$f(x) = \arctan x$

Domain: $x \in \mathbb{R}$

Asymptotes: (H): $y = \pm\pi/2$

Derivatives:

$$\sin x : \ f'(x) = \frac{1}{x^2 + 1};$$

$$f''(x) = -\frac{2x}{(x^2 + 1)^2}$$

Critical points: zero: $(0, 0)$

Extreme points: none

Fig. 10.7 Example 10.1-1(g)

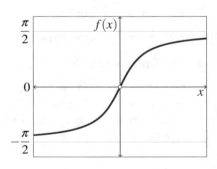

Exercise 10.2, page 220

1. Composite functions are created as follows.

(a) $\left.\begin{array}{l} x = t^2 + 1 \\ y = \sqrt{x} \end{array}\right\}$ $\therefore$ $y = f(g(t)) = f(t^2 + 1) = \sqrt{t^2 + 1}$

(b) $\left.\begin{array}{l} x = \frac{1}{t+1} \\ y = \ln x \end{array}\right\} \quad \therefore \quad y = f(g(t)) = f\left(\frac{1}{t+1}\right) = \ln\frac{1}{t+1}, \quad (t > |-1)$

(c) $\left.\begin{array}{l} x = -(t^2 + 1) \\ y = \sqrt{x} \end{array}\right\} \quad \therefore \quad y = f(g(t)) = f(-(t^2 + 1)) = \sqrt{-(t^2 + 1)}$

$$\therefore \quad f(t) \text{ not defined in } \mathbb{R}: \quad -(t^2 + 1) < 0 \ \forall x$$

(d) $\left.\begin{array}{l} x = -t^2 \\ y = \ln x \end{array}\right\} \quad \therefore \quad y = f(g(t)) = f(-t^2) = \ln(-t^2)$

$$\therefore \quad f(t) \text{ not defined in } \mathbb{R}: \quad -t^2 < 0 \ \forall x$$

2. By "decomposing" mathematical forms into its basic operations and clarifying the order of operations we can resolve more complicated structures.

   (a) $F(x) = f(g(x)) = (2x + x^2)^4$ where $f(x) = \underline{x^4}$ and $g(x) = \underline{2x + x^2}$

   (b) $F(x) = f(g(x)) = \cos^2 x$ where $f(x) = \underline{x^2}$ and $g(x) = \underline{\cos x}$

   (c) $F(x) = f(g(x)) = \underline{\dfrac{\sqrt[3]{x}}{1 + \sqrt[3]{x}}}$ where $f(x) = \underline{\dfrac{x}{1 + x}}$ and $g(x) = \underline{\sqrt[3]{x}}$

   (d) $F(x) = f(g(x)) = \underline{\sqrt[3]{\dfrac{x}{1 + x}}}$ where $f(x) = \underline{\sqrt[3]{x}}$ and $g(x) = \underline{\dfrac{x}{1 + x}}$

3. By inspection of operations and their order we write.

   (a) $F(x) = f(g(h(x))) = 1 - 3^{x^2}$ where $f(x) = 1 - x, \ g(x) = 3^x, \ h(x) = x^2$

   (b) $F(x) = f(g(h(x))) = \sqrt[4]{1 + |x|}$ where $f(x) = \sqrt[4]{x}, \ g(x) = 1 + x, \ h(x) = |x|$

4. Quick review of basic functions.

   (a) Given $f(x + 1) = 1 - x$, therefore the argument mapping is
   $(x + 1) \to x \ \therefore \ (x + 1) - 1 \to x - 1 \ \therefore \ x \to x - 1$. That is to say, if argument $x + 1$ is replaced with the argument x, then on the right side of the equation each occurrence of x must be replaced with $x - 1$, i.e.

   $$f(x) = 1 - (x - 1) = 2 - x$$

   (b) Similarly, $f(2x - 1) = x^2 - 2x + 1$, therefore
   $(2x - 1) \to x \ \therefore \ x \to (x + 1)/2$, i.e.

   $$f(x) = \left(\frac{x + 1}{2}\right)^2 - 2\left(\frac{x + 1}{2}\right) + 1 = \frac{1}{4}(x^2 - 2x + 1)$$

(c) Given $f\left(\dfrac{2x-1}{x}\right) = x$, therefore $x \to -\dfrac{1}{x-2}$, i.e.

$$f(x) = -\dfrac{1}{x-2}$$

(d) $f\left(x^2\right) = \dfrac{9}{x}$ $\therefore$ $f(x) = \dfrac{9}{\sqrt{x}}$, $(x > 0)$

5. Logarithmic function crosses the horizontal axis when its *argument* equals one. That is to say, the root of $\ln(1) = 0$ function, see Fig. 10.8. Equally, if the function's argument is an expression, again it has to equal one. Therefore, we write $x - 2 = 1$, that is $x = 3$ and simply translate horizontally the $\ln x$ graph, see Fig. 10.8. Adding a constant affects the y value *after* we account for the function's argument. That is to say, causes the vertical shift of $\ln(x - 2)$, see Fig. 10.8.

Fig. 10.8 Example 10.2-5

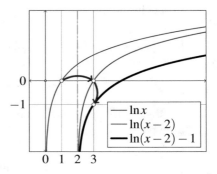

Exercise 10.3, page 220

1. Given rational function, a complete study may be systematically done as follows.

$$f(x) = \dfrac{x^3}{(x-1)^2} = \dfrac{P(x)}{Q(x)}$$

(a) *Domain of definition* $\mathbb{D}$: there is one double pole of $f(x)$ calculated as

$$Q(x) = 0 \quad \therefore \quad (x-1)^2 = 0 \quad \therefore \quad x_{1,2} = 1$$

therefore, $f(x)$ is defined for all real numbers except number one where the function becomes infinite (vertical asymptote), i.e. $\mathbb{D} : \mathbb{R}, x \neq 1$

(b) *Function's parity:* functions can be odd, even, or none.

$$\text{if:} \quad f(x) = f(-x) \quad \therefore \quad \text{even}$$

$$\text{if:} \quad f(x) = -f(-x) \quad \therefore \quad \text{odd}$$

We write,

$$f(-x) = \frac{(-x)^3}{((-x)-1)^2} = \frac{-x^3}{(-x-1)^2} = \frac{-x^3}{(-1)^2(x+1)^2} = -\frac{x^3}{(x+1)^2}$$

$$\neq f(x) \quad \therefore \quad \text{not even}$$

$$-f(-x) = -\left(-\frac{x^3}{(x+1)^2}\right) = \frac{x^3}{(x+1)^2} \neq f(x) \quad \therefore \quad \text{not odd}$$

In conclusion, there is no symmetry relative to neither y-axis (even) nor to the origin point (odd).

(c) *Function's sign:* There is one triple zero of $f(x)$ found when

$$f(x) = 0 \quad \therefore \quad P(x) = 0 \quad \therefore \quad x^3 = 0 \quad \therefore \quad x_{1,2,3} = 0$$

<u>Reminder:</u> in order to determine sign of an expression, we use rules for products and ratios, for example,

$$(neg) \times (neg) = (pos) \quad \text{or,} \quad \frac{(neg)}{(pos)} = (neg) \quad \text{etc.}$$

A function may change its sign (i.e. positivity/negativity) at zero and break points, which divide real domain $\mathbb{D}$ into the following intervals

x	$(-\infty; 0)$	0	$(0; 1)$	$(1; +\infty)$
x^3	$-$	0	$+$	$+$
$(x-1)^2$	$+$	$+$	$+$	$+$
$f(x)$	$-$	0	$+$	$+$

In conclusion, $f(x)$ is negative for $x < 0$, changes its sign at $x = 0$, and it is positive for $x > 0$.

(d) *Limits:* having one vertical asymptote at $x = 1$, there are four boundary limits to calculate. At far extremes, we find

$$\lim_{x \to -\infty} \frac{x^3}{x^2 - 2x + 1} = \lim_{x \to -\infty} \frac{\cancel{x^2}\, x}{\cancel{x^2} \left(1 - \overset{0}{\cancel{\frac{2}{x}}} + \overset{0}{\cancel{\frac{1}{x^2}}} \right)} = \lim_{x \to -\infty} x = -\infty$$

$$\lim_{x \to \infty} \frac{x^3}{x^2 - 2x + 1} = \lim_{x \to \infty} \frac{\cancel{x^2}\, x}{\cancel{x^2} \left(1 - \overset{0}{\cancel{\frac{2}{x}}} + \overset{0}{\cancel{\frac{1}{x^2}}} \right)} = \lim_{x \to \infty} x = \infty$$

Around vertical asymptote, we find

$$\lim_{x \to 1_-} \frac{x^3}{(x-1)^2} = \left\{ \text{for } x \neq 1 \quad \therefore \quad (x-1)^2 > 0 \ \text{regardless} \right\} = \frac{(1)^3}{+0} = +\infty$$

$$\lim_{x \to 1_+} \frac{x^3}{(x-1)^2} = \left\{ \text{for } x \neq 1 \quad \therefore \quad (x-1)^2 > 0 \ \text{regardless} \right\} = \frac{(1)^3}{+0} = +\infty$$

(e) *Oblique asymptote:* two limits are calculated to determine coefficients in

$$y_{aa} = ax + b$$

where

$$a = \lim_{x \to \infty} \frac{f(x)}{x} = \lim_{x \to \infty} \frac{x^3}{x(x-1)^2} = \left(\frac{\infty}{\infty} \right) \overset{\text{L.H.}}{=} \lim_{x \to \infty} \frac{(x^3)'}{(x^3 - 2x^2 + x)'}$$

$$= \lim_{x \to \infty} \frac{3x^2}{3x^2 - 4x + 1} = \left(\frac{\infty}{\infty} \right) \overset{\text{L.H.}}{=} \lim_{x \to \infty} \frac{6x}{6x - 4} = \left(\frac{\infty}{\infty} \right) \overset{\text{L.H.}}{=} \lim_{x \to \infty} \frac{6}{6}$$

$$= 1$$

$$b = \lim_{x \to \infty} (f(x) - x) = \lim_{x \to \infty} \left[\frac{x^3}{(x-1)^2} - x \right] = \lim_{x \to \infty} \left[\frac{x^3 - x(x-1)^2}{(x-1)^2} \right]$$

$$= \lim_{x \to \infty} \left[\frac{\cancel{x^3} - \cancel{x^3} + 2x^2 - 1}{x^2 - 2x + 1} \right] = \left(\frac{\infty}{\infty} \right) \overset{\text{L.H.}}{=} \lim_{x \to \infty} \frac{4x}{2x - 2} = \left(\frac{\infty}{\infty} \right) \overset{\text{L.H.}}{=} \lim_{x \to \infty} \frac{4}{2}$$

$$= 2$$

In conclusion, oblique asymptote is in form $y_{aa} = x + 2$.

(f)   *Critical points:* minimum/maxim and convex/concave points are found by solving $f'(x) = 0$ and $f''(x) = 0$ equations.

$$f'(x) = \left[\frac{x^3}{(x-1)^2}\right]' = \frac{3x^2(x-1)^2 - 2x^3(x-1)}{(x-1)^4} = \frac{x^2(x-1)[3(x-1) - 2x]}{(x-1)^3}$$

$$= \frac{x^2(x-3)}{(x-1)^3} = 0 \quad \therefore \quad x_{1,2} = 0, \quad x_3 = 3$$

We calculate $f(x)$ at these two points as, $f(0) = 0$ and $f(3) = 27/4$.

$$f''(x) = [f'(x)]' = \left[\frac{x^2(x-3)}{(x-1)^3}\right]' = \frac{[x^2(x-3)]'(x-1)^3 - x^2(x-3)\,[(x-1)^3]'}{(x-1)^6}$$

$$= \frac{(3x^2 - 6x)(x-1)^3 - 3x^2(x-3)(x-1)^2}{(x-1)^6}$$

$$= \frac{(x-1)^2[(3x^2 - 6x)(x-1) - 3x^2(x-3)]}{(x-1)^4}$$

$$= \frac{3x^3 - 9x^2 + 6x - 3x^3 + 9x^2}{(x-1)^4} = \frac{6x}{(x-1)^4} = 0 \quad \therefore \quad x = 0$$

Minimum/maxim and convex/concave points are determined by change of sign of critical points, we summarize as

x	$(-\infty; 0)$	0	$(0; 1)$	1	$(1; 3)$	3	$(1; +\infty)$
x^2	+	0	+		+	0	+
$x-3$	−	−	−		−	0	+
$(x-1)^3$	−	−	−		+	+	+
$f'(x)$	+	0	+		−	0	+
$f(x)$	↗	0	↗		↘	27/4	↗
						(min.)	

x	$(-\infty; 0)$	0	$(0; 1)$	1	$(1; +\infty)$
$6x$	−	0	+		+
$(x-1)^4$	+	+	+		+
$f''(x)$	−	0	+		+
$f(x)$	∩	0	U		U
$f(x)$		$(0,0)$			

In conclusion, there is one minimum at $min = (3, {}^{27}/_4)$ and on inflection point at $m = (0, 0)$.

(g) *Graphical representation:* the calculated results are sufficient to plot $f(x)$ graph, see Fig. 10.9.

Fig. 10.9 Example 10.3-1

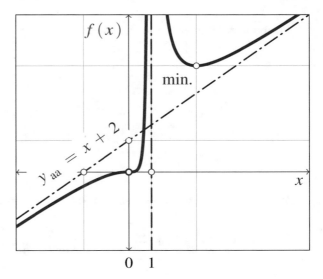

2. Following the same methodology as in Example 10.3-1, we analyse this function as follows,

$$f(x) = 2 - \frac{2(1-x)}{x^2+1} = \frac{2(x^2+1) - 2(1-x)}{x^2+1} = \frac{2x^2 + 2 - 2 + 2x)}{x^2+1} = \frac{2x(x+1)}{x^2+1}$$

(a) *Domain of definition* $\mathbb{D}$: there are no vertical asymptotes, $x \in [-\infty, +\infty]$

(b) *Function's parity:* We write,

$$f(-x) = \frac{2(-x)[(-x)+1]}{(-x)^2+1} = \frac{2x(1-x)}{x^2+1} \neq f(x) \quad \therefore \quad \text{not even}$$

$$-f(-x) = -\frac{2x(1-x)}{x^2+1} = \frac{2x(x-1)}{x^2+1} \neq f(x) \quad \therefore \quad \text{not odd}$$

(c) *Function's sign:* There are two zeros of $f(x)$ found as

$$f(x) = 0 \quad \therefore \quad P(x) = 0 \quad \therefore \quad 2x(x+1) = 0 \quad \therefore \quad x_1 = 0, \quad x_2 = -1$$

Denominator is $x^2 + 1 > 0$, therefore the sign of $f(x)$ is determined by numerator only.

x	$(-\infty; -1)$	-1	$(-1; 0)$	0	$(0; +\infty)$
$2x$	$-$		$-$		$+$
$x+1$	$-$		$+$		$+$
$f(x)$	$+$		$-$		$+$

(d) *Limits:* we find one horizontal asymptote at $y = 2$ as

$$\lim_{\pm\infty} \frac{2x^2 + 2x}{x^2 + 1} = \left(\frac{\infty}{\infty}\right) \overset{\text{L.H.}}{=} \lim_{\pm\infty} \frac{4x + 2}{2x} \overset{\text{L.H.}}{=} \lim_{\pm\infty} \frac{4}{2} = 2$$

(e) *Oblique asymptote:* by definition we write

$$y_{\text{aa}} = ax + b$$

where

$$a = \lim_{x \to \infty} \frac{f(x)}{x} = \lim_{x \to \infty} \frac{2x + 2}{x^2 + 1} = \left(\frac{\infty}{\infty}\right) \overset{\text{L.H.}}{=} \lim_{\pm\infty} \frac{2}{2x}^1 = 0$$

Therefore, there is no oblique asymptote.

(f) *Critical points:* minimum/maxim and convex/concave points are found by solving $f'(x) = 0$ and $f''(x) = 0$ equations.

$$f'(x) = \left\{ \left[\frac{P(x)}{Q(x)} \right]' = \frac{P'(x)Q(x) - P(x)Q'(x)}{Q^2(x)} \right\}$$

$$\therefore$$

$$= \frac{-2(x^2 - 2x - 1)}{(x^2 + 1)^2}$$

Therefore,

$$f'(x) = 0 \quad \therefore \quad x^2 - 2x + 1 = 0 \quad \therefore \quad x_{1,2} = \frac{2 \pm \sqrt{8}}{2} = 1 \pm \sqrt{2}$$

Therefore, we write the first derivative in its factorized form as

$$f'(x) = \frac{-2(x - x_1)(x - x_2)}{(x^2 + 1)^2}$$

$$= \frac{-2\left(x - (1 - \sqrt{2})\right)\left(x - (1 + \sqrt{2})\right)}{(x^2 + 1)^2}$$

Denominator of $f'(x)$ is greater than zero, thus we tabulate the corresponding critical points as follows.

x	$(-\infty; x_1)$	x_1	(x_1, x_2)	x_2	$(x_2; +\infty)$
-2	$-$		$-$		$-$
$x - x_1$	$-$	0	$+$		$+$
$x - x_2$	$-$		$-$	0	$+$
$f'(x)$	$-$	0	$+$	0	$-$
$f(x)$	$\searrow$	(min.)	$\nearrow$	(max.)	$\searrow$

Where coordinates of minimum and maximum points of $f(x)$ are calculated as

$$\text{(min.):} \quad x_1 = 1 - \sqrt{2} \quad \therefore \quad f(x_1) = \frac{2(1 - \sqrt{2})(1 - \sqrt{2} + 1)}{(1 - \sqrt{2})^2 + 1} = 1 - \sqrt{2}$$

$$\text{(max.):} \quad x_2 = 1 + \sqrt{2} \quad \therefore \quad f(x_1) = \frac{2(1 + \sqrt{2})(1 + \sqrt{2} + 1)}{(1 + \sqrt{2})^2 + 1} = 1 + \sqrt{2}$$

Similarly, convex/concave regions and coordinates and the corresponding inflection points are found as follows.

$$f''(x) = \frac{-4(-x^3 + 3x^2 + 3x - 1)}{(x^2 + 1)^3}$$

$$\therefore$$

$$f''(x) = 0 \quad \therefore \quad -x^3 + 3x^2 + 3x - 1 = 0$$

A third order polynomial must have at least one real zero. By factor theorem we search the first solution among the factors of the x^0 term, here it is ± 1. We find that $f''(-1) = 1 + 3 - 3 - 1 = 0$, therefore $f''(x)$ is divisible by $x + 1$, i.e. $x_1 = -1$. Consequently,

$$(-x^3 + 3x^2 + 3x - 1) \div (x + 1) = -x^2 + 4x - 1 = -(x - 2 + \sqrt{3})(x - 2 - \sqrt{3})$$

Therefore, we conclude that $x_1 = -1, x_2 = 2 - \sqrt{3}, x_3 = 2 + \sqrt{3}$ and we write the second derivative in its factorized form as

$$f''(x) = \frac{4(x + 1)(x - 2 + \sqrt{3})(x - 2 - \sqrt{3})}{(x^2 + 1)^3} \quad \text{where} \quad (x^2 + 1)^3 > 0 \ \forall x$$

Denominator of $f''(x)$ is greater than zero, thus we tabulate the corresponding critical points as follows.

x	$(-\infty; x_1)$	x_1	(x_1, x_2)	x_2	(x_2, x_3)	x_3	$(x_3; +\infty)$
$x - x_1$	$-$	0	$+$		$+$		$+$
$x - x_2$	$-$		$-$	0	$+$		$+$
$x - x_3$	$-$		$-$		$-$	0	$+$
$f''(x)$	$-$	0	$+$	0	$-$	0	$+$
$f(x)$	$\cap$	(y_{p1})	$\cup$	(y_{p2})	$\cap$	(y_{p3})	$\cup$

Where coordinates of inflection points of $f(x)$ are calculated as

$$(y_{p1}): \quad x_1 = -1 \quad \therefore \quad f(x_1) = \frac{2(-1)(-1+1)}{(-1)^2+1} = 0$$

$$(y_{p2}): \quad x_2 = 2 - \sqrt{3} \quad \therefore \quad f(x_2) = \frac{2(2-\sqrt{3})(2-\sqrt{3}+1)}{(2-\sqrt{3})^2+1} = \frac{3-\sqrt{3}}{2}$$

$$(y_{p3}): \quad x_2 = 2 + \sqrt{3} \quad \therefore \quad f(x_3) = \frac{2(2-\sqrt{3})(2-\sqrt{3}+1)}{(2-\sqrt{3})^2+1} = \frac{3+\sqrt{3}}{2}$$

(g) *Graphical representation:* in summary of function analysis, plot of $f(x)$ is shown in Fig. 10.10.

Fig. 10.10 Example 10.3-2

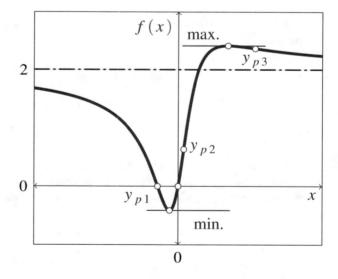

3. We use results of the analysis in Example 10.3-2.

(a)   Given condition that $x = 1$, we find that

$$f(1) = \frac{2(1+1)}{1^2 + 1} = 2 \quad \therefore \quad A = (1, 2)$$

Slope m_1 of tangent T_1 in $x = 1$ is calculated as

$$f'(1) = \frac{-2(1^2 - 2 - 1)}{(1^2 + 1)^2} = 1 \quad \therefore \quad m_1 = 1$$

Therefore, equation of tangent is $T_1 = m_1 x + b_1 = x + b_1$, where constant b_1 is found from the condition that T_1 crosses point $A = (1, 2)$. That is to say

$$\underline{x = 1:} \quad \therefore \quad T_1 = 2 \quad \therefore \quad 2 = 1 + b_1 \quad \therefore \quad b_1 = 1 \quad \therefore \quad T_1 = x + 1$$

We show this tangent in Fig. 10.11.

Fig. 10.11
Example 10.3-3(a)

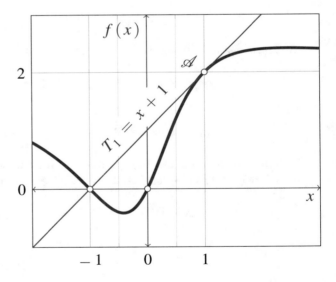

(b)   Given line $y = -x$ we know that its slope is $m_2 = -1$, therefore all parallel lines have the same slope. As a consequence

$$T_2 = -x + b_2$$

Now, the problem is to find at what x the slope (i.e. the first derivative) of $f(x)$ equals $f'(x) = m_2 = -1$? We write

$$f'(x) = \frac{-2(x^2 - 2x - 1)}{(x^2 + 1)^2} = -1$$

$$\therefore$$

$$-2(x^2 - 2x - 1) = -(x^4 + 2x^2 + 1) \quad \therefore \quad x^4 + 4x + 3 = 0$$

By exploiting the factor theorem, and factors of number 3, we find that

$$P(x) = x^4 + 4x + 3 \quad \therefore \quad P(-1) = (-1)^4 + 4(-1) + 3 = 0$$

Therefore, $P(x)$ is divisible by $x + 1$ binomial and we factorize $P(x)$ by divisions

$$(x^4 + 4x + 3) \div (x + 1) = x^3 - x^2 + x + 3$$

We repeat again the factor theorem to find

$$Q(x) = x^3 - x^2 + x + 3 \quad \therefore \quad Q(-1) = (-1)^3 - (-1)^2 + (-1) + 3 = 0$$

$$\therefore$$

$$(x^3 - x^2 + x + 3) \div (x + 1) = x^2 - 2x + 3$$

Therefore, after noting that this quadratic polynomial does not heave real zeros, we write: $x^4 + 4x + 3 = (x + 1)^2(x^2 - 2x + 3)$. Now we solve the fourth order equation as

$$(x + 1)^2(x^2 - 2x + 3) = 0 \quad \therefore \quad \underline{x_{1,2} = -1}$$

$$\therefore$$

$$f'(-1) = 0 \text{ and } \underline{f(-1) = 0}$$

We conclude the coordinates of $B = (-1, 0)$ because B point must be at $f(x)$ curve where the first derivative equals zero.

Therefore, equation of T_2 is derived as

$$T_2 = -x + b_2 \quad \therefore \quad -1 = 0x + b_2 \quad \therefore \quad b_2 = -1 \quad \therefore \quad T_2 = -x - 1$$

General condition for two lines to be orthogonal is that their slopes have relationship $m_1 = -1/m_2$, which is exactly the case with T_1 and T_2 (i.e. $1 = -1/(-1)$). Therefore they are normal to each other, as illustrated in Fig. 10.12.

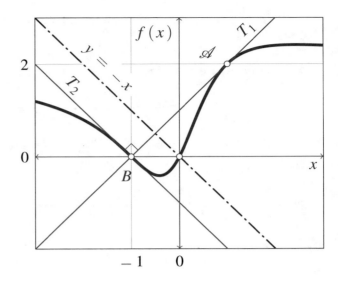

Fig. 10.12
Example 10.3-3(b)

4. Given rational function:

$$f(x) = \frac{x}{(x+1)(x-4)}$$

(a) *Domain of definition* $\mathbb{D}$:

$$x \in \Re : \quad (x+1) \neq 0, \Rightarrow (x \neq -1) \quad \text{and,}$$

$$x \in \Re : \quad (x-4) \neq 0, \Rightarrow (x \neq 4)$$

Therefore, there are two vertical asymptotes at $(x = -1)$ and $(x = 4)$.
Also, we note the product: $(x+1)(x-4) = x^2 - 3x - 4$

(b) *Function's parity:* functions can be odd, even, or none.

$$\text{if:} \quad f(x) = f(-x) \quad \therefore \quad \text{even}$$

$$\text{if:} \quad f(x) = -f(-x) \quad \therefore \quad \text{odd}$$

We write,

$$f(-x) = \frac{-x}{(-x+1)(-x-4)} = \frac{x}{(1-x)(4-x)} \neq f(x) \text{ not even}$$

$$-f(-x) = -\frac{x}{(1-x)(4-x)} \neq f(x) \text{ not odd}$$

In conclusion, there is no symmetry relative to neither y-axis (even) nor to the origin point (odd).

(c) *Limits:* we calculate $\lim_{x \to -\infty} f(x)$ and $\lim_{x \to +\infty} f(x)$

$$\lim_{x \to -\infty} f(x) = \lim_{x \to -\infty} \frac{x}{(x+1)(x-4)} = \lim_{x \to -\infty} \frac{x}{x^2 - 3x - 4} = \left(\frac{\infty}{\infty}\right)$$

$$\overset{\text{L.H.}}{=} \lim_{x \to -\infty} \frac{1}{2x - 3} = \frac{1}{-\infty} = 0_-$$

$$\lim_{x \to +\infty} f(x) = \lim_{x \to +\infty} \frac{x}{(x+1)(x-4)} = \lim_{x \to +\infty} \frac{x}{x^2 - 3x - 4} = \left(\frac{\infty}{\infty}\right)$$

$$\overset{\text{L.H.}}{=} \lim_{x \to +\infty} \frac{1}{2x - 3} = \frac{1}{+\infty} = 0_+$$

(d) *Limits at* $\underline{x = -1}$: we calculate $\lim_{x \to -1_-} f(x)$ et $\lim_{x \to -1_+} f(x)$

$$\lim_{x \to -1_-} f(x) = \lim_{x \to -1_-} \frac{x}{(x+1)(x-4)} = \lim_{x \to -1_-} \frac{-1}{\underbrace{(x+1)}_{<0}(-5)} = \frac{(-)}{(-)(-)} = -\infty$$

$$\lim_{x \to -1_+} f(x) = \lim_{x \to -1_+} \frac{x}{(x+1)(x-4)} = \lim_{x \to -1_+} \frac{-1}{\underbrace{(x+1)}_{>0}(-5)} = \frac{(-)}{(+)(-)} = +\infty$$

(e) *Limits at* $\underline{x = 4}$: we calculate $\lim_{x \to 4_-} f(x)$ et $\lim_{x \to 4_+} f(x)$

$$\lim_{x \to 4_-} f(x) = \lim_{x \to 4_-} \frac{x}{(x+1)(x-4)} = \lim_{x \to 4_-} \frac{4}{(5)\underbrace{(x-4)}_{<0}} = \frac{(+)}{(+)(-)} = -\infty$$

$$\lim_{x \to 4_+} f(x) = \lim_{x \to 4_+} \frac{x}{(x+1)(x-4)} = \lim_{x \to 4_+} \frac{4}{(5)\underbrace{(x-4)}_{>0}} = \frac{(+)}{(+)(+)} = +\infty$$

(f) *Function's zeros:* we calculate

$$f(x) = 0 \Rightarrow \frac{x}{(x+1)(x-4)} = 0 \quad \therefore \quad \underline{x = 0}$$

(g) *Derivatives and critical points:*

$$f(x) = \frac{x}{(x+1)(x-4)}$$

$$f'(x) = \left(\frac{x}{x^2 - 3x - 4}\right)' = \frac{(x^2 - 3x - 4) - x(2x-3)}{(x^2 - 3x - 4)^2} = \frac{x^2 - 3x - 4 - 2x^2 + 3x}{(x^2 - 3x - 4)^2}$$

$$= -\frac{x^2 + 4}{(x^2 - 3x - 4)^2} \neq 0 = -\frac{(+)}{(+)} \quad \therefore \quad f'(x) < 0; \quad (x \in \Re)$$

$$f''(x) = \left[-\frac{x^2 + 4}{(x^2 - 3x - 4)^2}\right]' = -\frac{2x(x^2 - 3x - 4)^2 - (x^2 + 4) \, 2 \, (x^2 - 3x - 4)(2x - 3)}{(x^2 - 3x - 4)^4}$$

$$= -\frac{2(x^2 - 3x - 4)\left[x(x^2 - 3x - 4) - (x^2 + 4)(2x - 3)\right]}{(x^2 - 3x - 4)^{4\,3}}$$

$$= -\frac{2\left[x^3 - 3x^2 - 4x - 2x^3 - 8x + 3x^2 + 12\right]}{(x^2 - 3x - 4)^3}$$

$$= 2\frac{x^3 + 12x - 12}{(x+1)^3(x-4)^3}$$

Without doing the numerical calculations, we can only estimate that $f''(x) = 0 \Rightarrow N(x) = x^3 + 12x - 12 = 0$ as: $N(0) = -12 < 0$ and $N(1) = 1 > 0$, Therefore, there must be a real zero $x = x_0$ found within the interval $[0, 1]$ very close to $x_0 \approx 1$. Therefore,

$$f(x) = 0; \quad \Rightarrow \quad x = 0, y = 0$$

$$f'(x) \neq 0; \quad \text{and} \quad f'(x) < 0$$

$$f''(x) = 0; \quad \Rightarrow \quad x_0 \approx 1; \quad \therefore \quad x_0 \approx 1, y_0 \approx -\frac{1}{6}$$

(h) *Summary of variations:*

x	$(-\infty, -1)$	(-1)	$(-1, 0)$	(0)	$(0, x_0)$	(x_0)	$(x_0, 4)$	(4)	$(4, +\infty)$
$f(x)$	$\frac{(-)}{(-)(-)}$		$\frac{(-)}{(+)(-)}$	0	$\frac{(+)}{(+)(-)}$	$\approx -1/6$	$\frac{(+)}{(+)(-)}$		$\frac{(+)}{(+)(+)}$
$f'(x)$	$(-)$		$(-)$	$-1/4$	$(-)$	$(-)$	$(-)$		$(-)$
$f(x)$	$\searrow$		$\searrow$	0	$\searrow$	$\searrow$	$\searrow$		$\searrow$
$f''(x)$	$\frac{(-)}{(-)(-)}$		$\frac{(-)}{(+)(-)}$	$+3/8$	$\frac{(-)}{(+)(-)}$	0	$\frac{(+)}{(+)(-)}$		$\frac{(+)}{(+)(+)}$
$f(x)$	$\cap$		$\cup$	$\cup$	$\cup$	(x_0, y_0)	$\cap$		$\cup$

(i)   *Oblique asymptote:* two limits are calculated to determine coefficients in

$$y_{aa} = ax + b$$

where

$$a = \lim_{x \to \pm\infty} \frac{f(x)}{x} = \lim_{x \to \pm\infty} \frac{\frac{1}{(x+1)(x-4)}}{x} = \lim_{x \to \pm\infty} \frac{1}{(x+1)(x-4)} = \frac{1}{\pm\infty} = 0$$

$$b = \lim_{x \to \pm\infty} \left(f(x) - ax^{\,0} \right) = \lim_{x \to \pm\infty} f(x) = (\text{already calculated}) = 0$$

therefore, because ($a = b = 0$), there is no oblique asymptote. Instead, there is one horizontal asymptote at $y = 0$.

(j)   *Graphical representation:* the complete function graph is shown in Fig. 10.13.

Fig. 10.13 Example 10.3-4

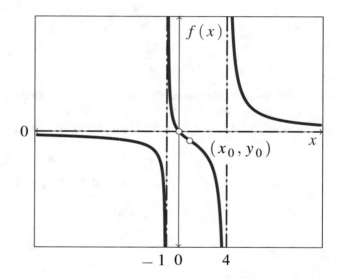

Exercise 10.4, page 221

1. Some of the typical function forms to study.

(a) $f(x) = x^4 - 2x^2$,   Fig. 10.14.
Domain: $x \in \mathbb{R}$
Derivatives:

$$f'(x) = 4x^3 - 4x;$$
$$f''(x) = 12x^2 - 4$$

Critical points:
zeros: $(-\sqrt{2}, 0)$, $(0, 0)$, $(0, 0)$, $(\sqrt{2}, 0)$
inflection points:
$(-1/\sqrt{3}, -5/9)$, $(1/\sqrt{3}, -5/9)$
extreme points:
min: $(-1, -1)$, $(1, -1)$ max: $(0, 0)$

Fig. 10.14
Example 10.4-1(a)

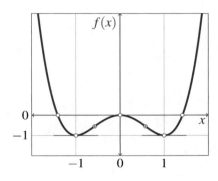

(b) $f(x) = \dfrac{4x}{4 - x^2}$,   Fig. 10.15.

Domain: $x \neq \pm 2$

Asymptotes:
(V): $x = -2$, $x = 2$, (H): $y = 0$

Derivatives:

$$f'(x) = \frac{4(4 + x^2)^2}{(4 - x^2)^2};$$

$$f''(x) = \frac{8x(x^2 + 12)^2}{(4 - x^2)^3}$$

Critical points:
zeros: $(0, 0)$
inflection point: $(0, 0)$

Fig. 10.15
Example 10.4-1(b)

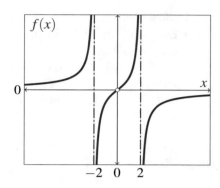

(c) $f(x) = e^{\frac{1}{x}}$,   Fig. 10.16.
Domain: $x \neq 0$
Asymptotes:
(V): $x = 0$, $x = 2$, (H): $y = 1$
Derivatives:

$$f'(x) = -\frac{e^{\frac{1}{x}}}{x^2};$$

$$f''(x) = \frac{e^{\frac{1}{x}}}{x^4}(2x + 1)$$

Limits: $\lim_{x \to 0_-} f(x) = 0$
Critical points:
inflection point: $(-1/2, 1/e^2)$

Fig. 10.16
Example 10.4-1(c)

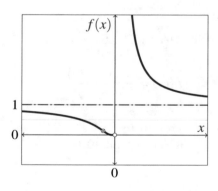

(d) $f(x) = (2 - x^2)e^x$, Fig. 10.17.

Domain: $x \in \mathbb{R}$

Derivatives:

$$f'(x) = -(x^2 + 2x - 2)e^x;$$

$$f''(x) = -x(x + 4)e^x$$

Critical points:

zeros: $(-\sqrt{2}, 0)$, $(\sqrt{2}, 0)$

extreme points:

min: $\left(-1 - \sqrt{3}, -\dfrac{2(1 + \sqrt{3})}{e^{1+\sqrt{3}}}\right)$

max: $\left(-1 + \sqrt{3}, -\dfrac{2(1 - \sqrt{3})}{e^{1-\sqrt{3}}}\right)$

inflection points: $(0, 2)$, $(-4, -14/e^4)$

Fig. 10.17
Example 10.4-1(d)

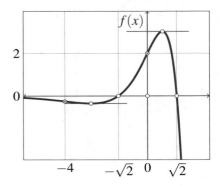

(e) $f(x) = x \ln^2 x$, Fig. 10.18.

Domain: $x > 0$

Derivatives:

$$f'(x) = (2 + \ln x) \ln x;$$

$$f''(x) = \frac{2(1 + \ln x)}{x}$$

Critical points:

zeros: $(0, 0)$, $(1, 0)$, $(1, 0)$

extreme points:

min: $(1, 0)$

max: $(1/e^2, 4/e^2)$

inflection points: $(1/e, 1/e)$

Fig. 10.18
Example 10.4-1(e)

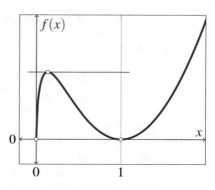

(f) $f(x) = \left(x - \dfrac{1}{4}\right) \ln \left(1 - \dfrac{1}{x}\right)$,   Fig. 10.19.

Domain: $x < 0$ and $x > 1$
Asymptotes:
(V): $x = 0$, $x = 1$, (H): $y = -1$
Derivatives:

$$f'(x) = \ln \left(1 - \dfrac{1}{x}\right) + \dfrac{4x - 1}{4x(x - 1)}$$

$$f''(x) = -\dfrac{2x + 1}{4x^2(x - 1)^2}$$

Critical point:
inflection points: $(-1/2, -3/4 \ln 3))$

Fig. 10.19
Example 10.4-1(f)

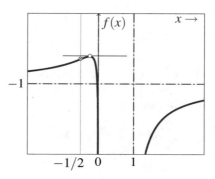

In order to find the extreme points, i.e. min and max, we must solve $f'(x) = 0$ equation. However, in this case the required equation is transcendental because it has the form of "$\ln h(x) = g(x)$", i.e.

$$f'(x) = \ln \left(1 - \dfrac{1}{x}\right) + \dfrac{4x - 1}{4x(x - 1)} = 0$$

$$\ln \left(1 - \dfrac{1}{x}\right) = -\dfrac{4x - 1}{4x(x - 1)}$$

where, in general case, there is no closed form analytical solution. We must use, for example, graphic method or numerical method to approximate the solution. Another possible method is to develop Taylor polynomial of $f'(x)$, then search for its roots. Note that some functions are very sensitive to the rounding numerical errors.

By inspection of $f(x)$, we already found that $f(x)$ is defined for $x < 0$, that the inflection point is at $x = -1/2$, that horizontal asymptote is at $y = -1$, where $f(x) < 0$ for $x > 1$. Thus, first good guess for position of the extreme point could be in the middle of $]0, -1/2[$ interval, for example, $x_0 = -1/4$. After simplification, second order Tylor polynomial around $x_0 = -1/4$ is

$$f'(x) \approx T(x) \approx -8.7x^2 - 5.6325x - 0.8546$$

$$\therefore$$

$$x_0 \approx \frac{5.6325 \pm \sqrt{5.6325^2 - 4 \times 8.7 \times 0.8546}}{-2 \times 8.7}$$

$$\therefore$$

$$x_0 \approx 0.243 \quad (\text{max.})$$

(the second root of quadratic equation is too far from the initial guess, where $T(x)$ is not correct anymore).

(g) $f(x) = \sqrt[3]{x^2 - x^3} = \sqrt[3]{x^2(1 - x)}$,   Fig. 10.20.
Domain: $x \in \mathbb{R}$
Asymptotes: (affine): $y = 1/3 - x$
Derivatives:

$$f'(x) = -\frac{x(3x - 2)}{3\sqrt[3]{x^2(1 - x)}}; \quad (x \neq 0)$$

$$f''(x) = -\frac{2\sqrt[3]{[x^2(1 - x)]^2}}{9x^2(x - 1)^2}$$

Critical points:
zeros: $(0, 0)$, $(1, 0)$
extreme points:

min: $(0, 0))$ max: $\left(\dfrac{2}{3}, -\dfrac{\sqrt[3]{4}}{3}\right)$

inflection points: $(0, 0)$, $(1, 0)$

Fig. 10.20
Example 10.4-1(g)

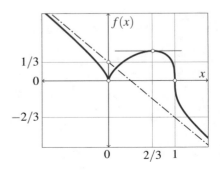

Note that $f'(0)$ is not defined, nevertheless there is change of sign of $f'(x)$, thus the *min.* point.

Hint: two limit calculations for the affine asymptote $y = ax + b$ reduce to

$$a = \lim_{x \to \infty} \frac{f(x)}{x} = \lim_{x \to \infty} \frac{\sqrt[3]{x^2 - x^3}}{x} = \lim_{x \to \infty} \sqrt[3]{\frac{1}{x} - 1} = -1$$

$$\therefore$$

$$b = \lim_{x \to \infty} f(x) - ax = \lim_{x \to \infty} \sqrt[3]{x^2 - x^3} + x = \lim_{x \to \infty} x\sqrt[3]{\frac{1}{x} - 1} + x$$

$$= \lim_{x \to \infty} x\left(\sqrt[3]{\frac{1}{x} - 1} + 1\right) = (\infty \cdot 0) = \lim_{x \to \infty} \frac{\left(\sqrt[3]{\frac{1}{x} - 1} + 1\right)}{\frac{1}{x}} = \left(\frac{0}{0}\right)$$

$$\overset{\text{L.H.}}{=} \lim_{x \to \infty} \frac{\frac{\sqrt[3]{\frac{1}{x} - 1}}{3x^2 - 3x}}{-\frac{1}{x^2}} = \lim_{x \to \infty} -x^2 \cdot \frac{\sqrt[3]{\frac{0}{\frac{1}{x}} - 1}}{3x^2\left(1 - \frac{\frac{1}{x}}{\frac{0}{x}}\right)} = \frac{1}{3}$$

Note that $\sqrt[3]{-1} = -1$, therefore $b = (-1)(-1)(1/3)$ is positive.

(h) $f(x) = x^x$,   Fig. 10.21.
Domain: $x > 0$
Derivatives:

$f'(x) = x^x(1 + \ln x); \;\; (x \neq 0)$

$f''(x) = x^x (1 + \ln x)^2 + x^{x-1}$

Critical points:
extreme point:
min: $\left(1/e, (1/e)^{1/e}\right)$

Fig. 10.21
Example 10.4-1(h)

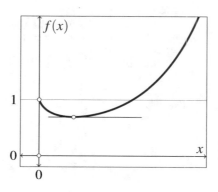

Integrals

11

Important to Know

Basic tabular integrals

$$
\begin{array}{cc}
f(x) & \int f(x)\, dx \\[2mm]
x^n & n\,x^{n-1} + C, \quad (n \neq -1) \\[2mm]
\dfrac{1}{x} & \ln|x| + C \quad (x \neq 0) \\[2mm]
e^{ax} & \dfrac{1}{a} e^{ax} C \\[2mm]
a^x & \dfrac{a^x}{\ln a} + C \\[2mm]
\sin x & -\cos x + C \\[2mm]
\cos x & \sin x + C \\[2mm]
\dfrac{1}{\sqrt{1 - x^2}} & \arcsin x + C \\[2mm]
\dfrac{1}{1 + x^2} & \arctan x + C
\end{array}
$$

Basic techniques of integration

$$
\int C\, f(x)\, dx = C \int f(x)\, dx
$$

$$
\int \left[f(x) \pm g(x) \right] dx = \int f(x)\, dx \pm \int g(x)\, dx
$$

$$
\int f(x)\, g'(x)\, dx = f(x)\, g(x) - \int f'(x)\, g(x)\, dx \quad \text{(partial integration)}
$$

© Springer Nature Switzerland AG 2021
R. Sobot, *Engineering Mathematics by Example*,
https://doi.org/10.1007/978-3-030-79545-0_11

How to choose the integration method? In general, try the possible methods in the following order:

1. check if the function is found in the list of tabular integrals (i.e. trivial solution);
2. simplify or modify the function's form by algebra methods, so that it is reduced to the tabulated forms;
3. one or more subsequent changes of variable that transform a given function to the tabulated forms;
4. partial integration, or
5. combination of change of variable and partial integration techniques.

11.1 Exercises

11.1   Basic Integrals

1. Calculate integrals of the following basic functions:

(a) $f(x) = x^3 + 2x - 1$

(b) $f(x) = x^{-3} + x$

(c) $f(x) = \dfrac{x+1}{\sqrt{x}}$

(d) $f(x) = (x+1)(2x+1)$

(e) $f(x) = 3\cos x + \sin x$

(f) $f(x) = e^x + \cos^3 x(1 + \tan^2 x)$

11.2 *   Integration by Substitution

1. Calculate integrals of the following functions by substitution of variables:

(a) $f(x) = x^2(x^3 + 1)^5$

(b) $f(x) = x\sqrt{x^2 + 1}$

(c) $f(x) = \dfrac{x^3}{\sqrt{1 - x^4}}$

(d) $f(x) = \dfrac{\sin x}{\cos^4 x}$

(e) $f(x) = \tan x$

(f) $f(x) = x\cos(x^2 + 2)$

(g) $f(x) = 1 + \tan^2 x$

11.3 *   Integration by Parts

1. Calculate integrals of the following functions by integration by parts:

(a) $f(x) = xe^x$

(b) $f(x) = x^2 e^{-x}$

(c) $f(x) = \ln x$

(d) $f(x) = x\sin x$

(e) $f(x) = x^2 \ln x$

(f) $f(x) = x^3 e^{x^2}$

(g) $f(x) = (x^2 - 1)e^x$

(h) $f(x) = e^x \sin x$

(i) $f(x) = \dfrac{\ln x}{x^2}$

11.4 **   Important Integrals

1. Calculate integrals of the following functions:

   (a) $\int \sin(2x)\,dx$

   (b) $\int \sin^2(x)\,dx$

   (c) $\int_0^1 f(x)^2\,dx$ given that $f(x)$ is a piecewise linear function as defined by its graph, see Fig. 11.1.

Fig. 11.1 Example 11.4-1(c)

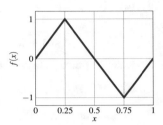

11.5 ***   Gaussian Integral

1. Evaluate integral

$$\int_{-\infty}^{\infty} e^{-x^2}\,dx$$

11.6 *   Integration of Rational Functions

1. Calculate the following rational function by using the method of integration by partial fractions. When possible, reuse derivations in Example 2.14-1.

   (a) $f(x) = \dfrac{x-1}{x^2+x}$ (b) $f(x) = \dfrac{x}{(x+1)(x-4)}$ (c) $f(x) = \dfrac{x+2}{x^3-2x^2}$

   (d) $f(x) = \dfrac{2x^2}{x^4-1}$ (e) $f(x) = \dfrac{1}{x^3-1}$

11.7 *   Trigonometric Substitution

1. Calculate integrals of the following functions by using trigonometric substitutions:

(a) $f(x) = \sqrt{9 - x^2}$ (b) $f(x) = \dfrac{x^2}{\sqrt{36 - x^2}}$ (c) $f(x) = \dfrac{\sqrt{25x^2 + 4}}{x^4}$

11.8 *   Definite Integrals

1. In the given closed interval (i.e. including the end points), calculate area under the function.

(a) $f(x) = x^3, \quad x \in [2, 3]$ (b) $f(x) = x(1 + x^3) \quad x \in [-1, 2]$

(c) $f(x) = \cos x \quad x \in [-\pi/4, \pi/4]$ (d) $f(x) = \cos x \quad x \in [0, 2\pi]$

(e) $f(x) = \dfrac{x}{\sqrt{1 + 3x^2}} \quad x \in [0, 2]$ (f) $f(x) = x^2 \cos x \quad x \in [0, \pi]$

2. Calculate area *between* two functions.

(a) $f(x) = x^2, \quad g(x) = 8 - x^2$

3. Given interval $x \in [5, 6]$, calculate area *under* function
$$f(x) = \frac{x}{(x + 1)(x - 4)}$$

4. Using the definite integrals, calculate

(a) area of right-angled triangle, where the two catheti equal $a = 2$ and $b = 1$;

(b) area of the circle whose radius equals $r = 1$; and

(c) volume of cuboid whose sides equal $a = 4, b = 3$, and $c = 2$.

5. In the given closed interval (i.e. including the end points), calculate the averages of the following functions:

(a) $f(x) = \sin x \quad x \in [0, 2\pi]$ (b) $f(x) = \cos x \quad x \in [\pi/2, 3\pi/2]$

(c) $f(x) = 1 + \cos x \quad x \in [0, 2\pi]$ (d) $f(x) = \cos^2 x \quad x \in [0, 2\pi]$

6. Given closed intervals (i.e. including the end points) and function graph, Fig. 11.2, calculate the averages

(a) $x \in [0, 1]$

(b) $x \in [0, 2]$

(c) $x \in [0, 3]$

(d) $x \in [0, 4]$

(e) $x \in [1, 2]$

(f) $x \in [2, 3]$

(g) $x \in [2, 4]$

(h) $x \in [0.5, 3]$

(i) $x \in [1.5, 3.5]$

Fig. 11.2 Example 11.8-6

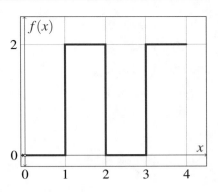

7. Given closed intervals (i.e. including the end points) and functions, Fig. 11.3, calculate the averages of $f(x)$ and $g(x)$ within each interval.

  (a) $x \in [0, 1]$

  (b) $x \in [0, 2]$

  (c) $x \in [0, 3]$

  (d) $x \in [0, 4]$

  (e) $x \in [0, 10]$

  (f) $x \in [5, 9]$

  (g) $x \in [1, 8]$

  (h) $x \in [1, 7]$

  (i) $x \in [2, 6]$

Fig. 11.3 Example 11.8-7

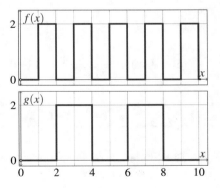

8. Given closed intervals (i.e. including the end points) and functions, Fig. 11.4, calculate the averages of $f(x)$ and $g(x)$ within each interval.

  (a) $x \in [0, 1]$

  (b) $x \in [0, 2]$

  (c) $x \in [0, 3]$

  (d) $x \in [0, 4]$

(e) $x \in [0, 10]$

(f) $x \in [5, 9]$

(g) $x \in [1, 8]$

(h) $x \in [1, 7]$

(i) $x \in [2, 6]$

Fig. 11.4 Example 11.8-8

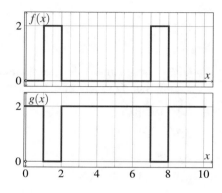

11.9 ***   Improper Integral

1. If possible, calculate area under the following functions:

(a) $f(x) = \dfrac{1}{x^2}, \quad x \in [1, \infty]$

(b) $f(x) = \dfrac{1}{x}, \quad x \in [1, \infty]$

(c) $f(x) = \dfrac{1}{\sqrt[3]{x^2}}, \quad x \in [-1, 1]$

(d) $f(x) = xe^{-x^2}, \quad x \in [-\infty, \infty]$

(e) $f(x) = \dfrac{1}{\sqrt{1-x}}, \quad x \in [-\infty, 0]$

(f) $f(x) = \dfrac{1}{1+x^2}, \quad x \in [0, \infty]$

11.10 ***   Improper Integral: Discussion

1. The limitation of blind implementation of the Newton–Leibniz formula is illustrated by the following functions:

(a) $f(x) = \dfrac{1}{x}, \quad x \in [-1, 1]$

(b) $f(x) = \sqrt{1 - \cos 2x}, \quad x \in [0, 100\pi]$

Solutions

Exercise 11.1, page 248

1. By using the list of basic integrals, we write

(a) $\displaystyle\int (x^3 + 2x - 1)\, dx = \left\{ \int x^n = \frac{x^{n+1}}{n+1} + C \right\}$

$\displaystyle = \frac{x^4}{4} + 2\frac{x^2}{2} - x + C = \frac{x^4}{4} + x^2 - x + C$

(b) $\displaystyle\int (x^{-3} + x)\, dx = \frac{x^{-2}}{-2} + \frac{x^2}{2} + C = -\frac{1}{2x^2} + \frac{x^2}{2} + C$

(c) $\displaystyle\int \frac{x+1}{\sqrt{x}}\, dx = \int \frac{x}{\sqrt{x}}\, dx + \int \frac{1}{\sqrt{x}}\, dx = \int x^{1/2}\, dx + \int x^{-1/2}\, dx$

$\displaystyle = \frac{x^{3/2}}{3/2} + \frac{x^{1/2}}{1/2} + C = 2\frac{x\sqrt{x}}{3} + 2\sqrt{x} + C = 2\sqrt{x}\left(\frac{x}{3} + 1\right) + C$

(d) $\displaystyle\int (x+1)(2x+1)\, dx = \int (2x^2 + 3x + 1)\, dx = \frac{2x^3}{3} + \frac{3x^2}{2} + x + C$

(e) $\displaystyle\int (3\cos x + \sin x)\, dx = \left\{ \int \cos x\, dx = \sin x, \quad \int \sin x\, dx = -\cos x \right\}$

$\displaystyle = 3\sin x - \cos x + C$

(f) $\displaystyle\int \left(e^x + \cos^3 x(1 + \tan^2 x) \right) dx = \int e^x\, dx + \int \left(\cos^3 x(1 + \tan^2 x) \right) dx$

$\displaystyle = e^x + \int \left[\cos^3 x \left(1 + \frac{\sin^2 x}{\cos^2 x} \right) \right] dx$

$\displaystyle = e^x + \int \left[\cos^3 x \left(\frac{\overbrace{\sin^2 x + \cos^2 x}^{1}}{\cos^2 x} \right) \right] dx$

$\displaystyle = e^x + \int \cos x\, dx = e^x + \sin x + C$

Exercise 11.2, page 248

1. Change of variables reduces integral to some of the basic forms.

(a) $\displaystyle\int x^2 (x^3 + 1)^5\, dx = \left\{ x^3 + 1 = t \quad \therefore \quad 3x^2\, dx = dt \quad \therefore \quad x^2\, dx = \frac{1}{3}\, dt \right\}$

$\displaystyle = \frac{1}{3} \int t^5\, dt = \frac{1}{3}\frac{t^6}{6} + C = \frac{1}{18}(x^3 + 1)^6 + C$

(b) $\displaystyle\int x\sqrt{x^2 + 1}\, dx = \left\{ t = x^2 + 1 \quad \therefore \quad 2x\, dx = dt \quad \therefore \quad x\, dx = \frac{1}{2}\, dt \right\}$

$\displaystyle = \frac{1}{2} \int \sqrt{t}\, dt = \frac{1}{2} \int t^{1/2}\, dt = \frac{1}{2}\frac{t^{3/2}}{3/2} = \frac{1}{3}\sqrt{(x^2 + 1)^3} + C$

(c) $\displaystyle\int \frac{x^3}{\sqrt{1-x^4}}\,dx = \left\{1-x^4=t \quad \therefore \quad -4x^3\,dx=dt \quad \therefore \quad x^3\,dx=-\frac{1}{4}\,dt\right\}$

$$= -\frac{1}{4}\int \frac{dt}{\sqrt{t}} = -\frac{1}{4}\int t^{-1/2}\,dt = -\frac{1}{2}\sqrt{1-x^4}+C$$

(d) $\displaystyle\int \frac{\sin x}{\cos^4 x}\,dx = \left\{\cos x=t \quad \therefore \quad -\sin x\,dx=dt \quad \therefore \quad \sin x\,dx=-\,dt\right\}$

$$= -\int \frac{dt}{t^4}\,dt = -\int t^{-4}\,dt = -\frac{t^{-3}}{-3}+C = \frac{1}{3\cos^3 x}+C$$

(e) $\displaystyle\int \tan x\,dx = \int \frac{\sin x}{\cos x}\,dx = \left\{\cos x=t \quad \therefore \quad -\sin x\,dx=dt\right\}$

$$= -\int \frac{dt}{t} = -\ln|t|+C = -\ln|\cos x|+C$$

(f) $\displaystyle\int x\cos(x^2+2) = \left\{x^2+2=t \quad \therefore \quad 2x\,dx=dt \quad \therefore \quad x\,dx=\frac{1}{2}\,dt\right\}$

$$= \frac{1}{2}\int \cos t\,dt = \frac{1}{2}\sin(x^2+2)+C$$

(g) $\displaystyle\int 1+\tan^2 x\,dx = \int \frac{\sin^2+\cos^2}{\cos^2}\,dx = \int \frac{1}{\cos^2}\,dx$

$$\left\{\begin{aligned} &t=\tan x=\frac{\sin x}{\cos x}\\[2mm] &dt=\frac{\cos x\cos x-\sin x(-\sin x)}{\cos^2 x}\,dx = \frac{\cos^2 x+\sin^2 x}{\cos^2 x}\,dx = \frac{1}{\cos^2 x}\,dx \end{aligned}\right\}$$

$$\therefore$$

$$= \int dt = t+C = \tan x+C$$

Exercise 11.3, page 248

1. The idea of integration by parts method is to choose one function so that its integral is simpler, while the other one has simpler derivative.

(a) $\displaystyle\int xe^x\,dx = \left\{\begin{aligned} &u=x &&\therefore\quad du=dx\\[2mm] &dv=e^x\,dx &&\therefore\quad v=\int e^x\,dx=e^x \end{aligned}\right\}$

$$= xe^x - \int e^x\,dx = xe^x - e^x + C = e^x(x-1)+C$$

(b) $\int x^2 e^{-x}\, dx =$

$$\left\{ \begin{array}{ll} u = x^2 & \therefore \quad du = 2x\, dx \\[2mm] dv = e^{-x}\, dx & \therefore \quad v = \int e^{-x}\, dx = \{t = -x\} = -e^{-x} \end{array} \right\}$$

$$= -x^2 e^{-x} - \int (-2x e^{-x})\, dx = -x^2 e^{-x} + 2\int x e^{-x}\, dx$$

$$\left\{ \begin{array}{ll} u = x & \therefore \quad du = dx \\[2mm] dv = e^{-x}\, dx & \therefore \quad v = \int e^{-x}\, dx = \{t = -x\} = -e^{-x} \end{array} \right\}$$

$$= -x^2 e^{-x} + 2\left(-x e^{-x} - \int (-e^{-x})\, dx \right)$$

$$= -x^2 e^{-x} - 2x e^{-x} - 2e^{-x} + C$$

$$= -e^{-x}\left(x^2 + 2x + 2 \right) + C$$

(c) $\int \ln x\, dx = \left\{ \begin{array}{ll} u = \ln x & \therefore \quad du = \dfrac{dx}{x} \\[2mm] dv = dx & \therefore \quad v = x \end{array} \right\}$

$$= x \ln x - \int \not{x} \frac{dx}{\not{x}} = x \ln x - x + C = x(\ln x - 1) + C$$

(d) $\int x \sin x\, dx = \left\{ \begin{array}{ll} u = x & \therefore \quad du = dx \\[2mm] dv = \sin x\, dx & \therefore \quad v = \int \sin x\, dx = -\cos x \end{array} \right\}$

$$= -x \cos x + \int \cos x\, dx = -x \cos x + \sin x + C$$

(e) $\int x^2 \ln x\, dx = \left\{ \begin{array}{ll} u = \ln x & \therefore \quad du = \dfrac{dx}{x} \\[2mm] dv = x^2\, dx & \therefore \quad v = \int x^2\, dx = \dfrac{x^3}{3} \end{array} \right\}$

$$= \frac{x^3}{3} \ln x - \int \frac{x^{\not{3}2}}{3} \frac{dx}{\not{x}1} = \frac{x^3}{3} \ln x - \frac{1}{3} \frac{x^3}{3} + C = \frac{x^3}{3} \ln x - \frac{x^3}{9} + C$$

(f) $\int x^3 e^{x^2}\, dx = \int x^2\, x\, e^{x^2}\, dx$

$$\left\{ \begin{array}{ll} u = x^2 & \therefore \quad du = 2x\, dx \\[2mm] dv = x\, e^{x^2}\, dx & \therefore \quad v = \int x\, e^{x^2}\, dx = \dfrac{1}{2} e^{x^2} \end{array} \right\}$$

$$\left\{ \text{where, } t = x^2 \quad \therefore \quad dt = 2x\, dx \quad \therefore \quad \int x\, e^{x^2}\, dx = \frac{1}{2} e^{x^2} \right\}$$

$$= \frac{1}{2} x^2 e^{x^2} - \int \frac{1}{2} e^{x^2}\, 2x\, dx = \frac{1}{2} x^2 e^{x^2} - \int x\, e^{x^2}\, dx$$

$$= \frac{1}{2} x^2 e^{x^2} - \frac{1}{2} e^{x^2} + C = \frac{e^{x^2}}{2} \left(x^2 - 1 \right) + C$$

(g) $\int (x^2 - 1)e^x \, dx = \int x^2 e^x \, dx - \int e^x \, dx$

$$\left\{ \int x^2 e^x \, dx = (\text{ by parts }) = e^x \left(x^2 - 2x + 2 \right) \right\}$$

$$= e^x \left(x^2 - 2x + 2 \right) - e^x + C$$

$$= e^x \left(x^2 - 2x + 1 \right) + C$$

(h) $\underbrace{\int e^x \sin x \, dx}_{\text{I}} = \begin{cases} u = e^x & \therefore \quad du = e^x \, dx \\ \\ dv = \sin x \, dx & \therefore \quad v = \int \sin x \, dx = -\cos x \end{cases}$

$$= -e^x \cos x + \int e^x \cos x \, dx$$

$$\begin{cases} u = e^x & \therefore \quad du = e^x \, dx \\ \\ dv = \cos x \, dx & \therefore \quad v = \int \cos x \, dx = \sin x \end{cases}$$

$$= -e^x \cos x + e^x \sin x - \underbrace{\int e^x \sin x \, dx}_{\text{I}}$$

This integral is an example of a "circular" form, i.e. after applying the integration by parts technique two times, we must again solve the original integral. However, we write

$$\text{I} = -e^x \cos x + e^x \sin x - \text{I} \quad \therefore \quad 2\text{I} = e^x \sin x - e^x \cos x$$

$$\therefore$$

$$\text{I} = \int e^x \sin x \, dx = \frac{e^x}{2} \left(\sin x - \cos x \right) + C$$

(i) $\int \frac{\ln x}{x^2} \, dx = \begin{cases} u = \ln x & \therefore \quad du = \dfrac{1}{x} \, dx \\ \\ dv = \dfrac{1}{x^2} \, dx & \therefore \quad v = \int \dfrac{1}{x^2} \, dx = -\dfrac{1}{x} \end{cases}$

$$= -\frac{\ln x}{x} + \int \frac{1}{x} \frac{1}{x} \, dx = -\frac{\ln x}{x} - \frac{1}{x} + C = -\frac{1}{x} (\ln x + 1) + C$$

Exercise 11.4, page 249

1. Three integrals in this exercise are typical forms encountered in engineering.

(a) Trigonometric functions of multiple angles are solved by change of variable technique.

$$t = 2x \quad \therefore \quad \frac{dt}{dx} = 2 \quad \therefore \quad dx = \frac{1}{2} dt$$

$$\therefore$$

$$\int \sin(2x)\, dx = \frac{1}{2} \int \sin t\, dt = -\frac{1}{2} \cos t = -\frac{1}{2} \cos(2x) + C$$

(b) Squares of sine/cosine functions are replaced by their trigonometric identities for double angles, followed by change of variable technique.

$$\sin x \sin x = \frac{1}{2} [\cos(x - x) - \cos(x + x)]$$

$$\sin^2(x) = \frac{1}{2} [\cos(0) - \cos(2x)] = \frac{1}{2} [1 - \cos(2x)]$$

Then,

$$\int \sin^2(x)\, dx = \int \frac{1}{2} [1 - \cos(2x)]\, dx = \frac{1}{2} \int dx - \frac{1}{2} \int \cos(2x)\, dx$$

$$= \frac{1}{2} \left(x - \frac{1}{2} \sin(2x) \right) + C$$

(c) Piecewise linear functions are among most often used approximations in engineering. Given their graphical representation, it is necessary to derive analytical forms for *each* of the linear sections separately. The main idea is that the operation of "integration" itself is fundamentally an operation of addition. Also, a linear function is the easiest function to integrate. Therefore, knowing a priori the analytical form of each linear piece, the overall integral is reduced to the sum of simple integrals of linear functions.

A linear section $y = ax + b$ is determined by two points in the plane, that is to say that knowing coordinates (x, y) of these two points on the line two constants (a, b) are determined with a simple algebra.

Here we derive three linear sections, as emphasized by different colors in graph and points $(\mathcal{A}, \mathcal{B}, \mathcal{C}, \mathcal{D})$, see Fig. 11.5.

1. $\overline{\mathcal{A}\mathcal{B}}$: at coordinates of the two end points (x, y), we write $y = ax + b$ as

$$(0, 0) \quad \therefore \quad 0 = a \times 0 + b \Rightarrow b = 0$$

$$(0.25, 1) \quad \therefore \quad 1 = a \times \frac{1}{4} + b \Rightarrow a = 4$$

Therefore, analytical form of $\overline{\mathscr{A}\mathscr{B}}$ linear segment is $f(x) = 4x$, which is valid in the interval $(0, 0.25)$. Therefore, in this interval, we calculate

$$I_1 = \int_0^{\frac{1}{4}} f(x)^2 \, dx = \int_0^{\frac{1}{4}} (4x)^2 \, dx = 16 \left. \frac{x^3}{3} \right|_0^{\frac{1}{4}} = \frac{16}{3} \left(\frac{1}{4^3} - 0 \right) = \frac{1}{12}$$

2. $\overline{\mathscr{B}\mathscr{C}}$: at coordinates of the two end points (x, y), we write $y = ax + b$ as

$$(0.25, 1) \quad \therefore \quad 1 = a \times \frac{1}{4} + b$$

$$(0.75, -1) \quad \therefore \quad -1 = a \times \frac{3}{4} + b$$

The solution of this system of two linear equations is $(a, b) = (-4, 2)$. Therefore, analytical form of $\overline{\mathscr{C}\mathscr{D}}$ linear segment is $f(x) = -4x + 2$, which is valid in the interval $(0.25, 0.75)$. Therefore, in this interval, we calculate

$$I_2 = \int_{\frac{1}{4}}^{\frac{3}{4}} f(x)^2 \, dx = \int_{\frac{1}{4}}^{\frac{3}{4}} (-4x + 2)^2 \, dx = \int_{\frac{1}{4}}^{\frac{3}{4}} (16x^2 - 16x + 4) \, dx$$

$$= 4 \left(4 \left. \frac{x^3}{3} \right|_{\frac{1}{4}}^{\frac{3}{4}} - 4 \left. \frac{x^2}{2} \right|_{\frac{1}{4}}^{\frac{3}{4}} + x \left. \right|_{\frac{1}{4}}^{\frac{3}{4}} \right)$$

$$= \left(16 \frac{1}{3} \frac{26}{64} - 16 \frac{1}{2} \frac{8}{16} + 4 \frac{2}{4} \right) = \left(\frac{13}{6} - 2 \right) = \frac{1}{6}$$

3. $\overline{\mathscr{C}\mathscr{D}}$: at coordinates of the two end points (x, y), we write $y = ax + b$ as

$$(0.75, -1) \quad \therefore \quad -1 = a \times \frac{3}{4} + b$$

$$(1, 0) \quad \therefore \quad 0 = a \times 1 + b$$

The solution of this system of two linear equations is $(a, b) = (4, -4)$. Therefore, analytical form of $\overline{\mathscr{C}\mathscr{D}}$ linear segment is $f(x) = 4x - 4$, which is valid in the interval $(0.75, 1)$. Therefore, in this interval, we calculate

$$I_3 = \int_{\frac{3}{4}}^{1} f(x)^2 \, dx = \int_{\frac{3}{4}}^{1} (4x - 4)^2 \, dx = \int_{\frac{3}{4}}^{1} (16x^2 - 32x + 16) \, dx$$

$$= 16 \left(\left. \frac{x^3}{3} \right|_{\frac{3}{4}}^{1} - 2 \left. \frac{x^2}{2} \right|_{\frac{3}{4}}^{1} + x \left. \right|_{\frac{3}{4}}^{1} \right)$$

$$= \left(16 \frac{1}{3} \frac{37}{64} - 16 \frac{7}{16} + 16 \frac{1}{4} \right) = \left(\frac{37}{12} - 3 \right) = \frac{1}{12}$$

(c) *(cont.)* After finding these three integrals, we return to the original questions to conclude

$$I = \int_0^1 f(x)^2 \, dx = I_1 + I_2 + I_3 = \frac{1}{12} + \frac{1}{6} + \frac{1}{12} = \frac{1}{3}$$

Geometrical interpretation is that integral I equals the total area underneath quadratic function $f(x)^2$ that consists of three distinct regions. These three quadratic sub-functions and their respective areas are shown by colored sections in Fig. 11.5.

Alternative Solution Due to symmetry of given $f(x)^2$, area I_1 in Fig. 11.5 is identical to areas of the other three segments, i.e. $(0.25, 0.5)$, $(0.5, 0.75)$, and $(0.75, 1.0)$. Thus, it is sufficient to solve only the first integral within $(0.0, 0.25)$ interval and keep in mind that it accounts for one-quarter of the total surface area I; that is to say,

$$I_1 = \int_0^{\frac{1}{4}} f(x)^2 \, dx = \int_0^{\frac{1}{4}} (4x)^2 \, dx = 16 \left. \frac{x^3}{3} \right|_0^{\frac{1}{4}} = \frac{16}{3} \left(\frac{1}{4^3} - 0 \right) = \frac{1}{12}$$

$$\therefore$$

$$I = 4 \, I_1 = 4 \times \frac{1}{12} = \frac{1}{3}$$

As a comparison, a simple integral of $f(x)$

$$A = \int_0^1 f(x) = \int_0^{\frac{1}{4}} 4x \, dx + \int_{\frac{1}{4}}^{\frac{3}{4}} (-4x + 2) \, dx + \int_{\frac{3}{4}}^1 (4x - 4) \, dx = 0$$

which may also be concluded by graph inspection; the area of triangle above zero (i.e. positive) equals the area of triangle below zero (i.e. negative), and therefore their sum equals zero. Another way to say is that average of this function equals zero.

Fig. 11.5 Example 11.4-1(c)

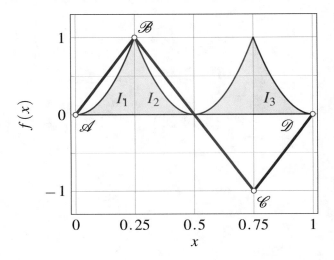

Exercise 11.5, page 249

1. Gaussian integral is one of the most important forms in science and engineering. There are multiple methods to solve it, one possible method exploits the symmetry of multivariable integral form by artificially adding second variable y as follows:

$$\int_{-\infty}^{\infty} e^{-x^2}\, dx = \int_{-\infty}^{\infty} \int_{-\infty}^{\infty} e^{-x^2-y^2}\, dx\, dy = \int_{-\infty}^{\infty} \int_{-\infty}^{\infty} e^{-x^2} e^{-y^2}\, dx\, dy$$

$$= \underbrace{\int_{-\infty}^{\infty} e^{-y^2}\, dy}_{I} \underbrace{\int_{-\infty}^{\infty} e^{-x^2}\, dx}_{I} = I^2$$

Even though there are two variables involved x and y, the two integrals have the same form, therefore the same form of their respective solutions. Therefore, we write

$$I^2 = \int_{-\infty}^{\infty} \int_{-\infty}^{\infty} e^{-(x^2+y^2)}\, dx\, dy$$

An illustration of how Cartesian coordinates (x, y) are transformed into polar coordinates (r, θ) is shown in Fig. 11.6. We observe the position of point $\mathscr{A}$ in Cartesian system and note the relations

$$x = r\cos\theta \quad y = r\sin\theta \quad \therefore \quad x^2 + y^2 = r \quad \text{and} \quad dx\, dy = r\, dr\, d\theta$$

The geometrical interpretation shows that the mapping between unity differential surfaces $dx\, dy = r\, dr\, d\theta$ is evident. In order to cover all available surfaces, the new variables are bound as

$$\left. \begin{array}{l} x \in [-\infty, \infty] \\ y \in [-\infty, \infty] \end{array} \right\} \quad \therefore \quad \left\{ \begin{array}{l} r \in [-\infty, \infty] \\ \theta \in [0, 2\pi] \end{array} \right.$$

We continue with the derivation of Gaussian integral in polar coordinates as

$$I^2 = \int_0^{2\pi} \int_0^{\infty} e^{-r^2} r\, dr\, d\theta = \int_0^{2\pi} d\theta \int_0^{\infty} e^{-r^2} r\, dr = 2\pi \int_0^{\infty} e^{-r^2} r\, dr$$

$$\left\{ t = -r^2 \quad \therefore \quad dt = -2r\, dr \quad \therefore \quad \int e^{-r^2} r\, dr = -\frac{1}{2} e^{-r^2} \right\}$$

$$= 2\pi \left(-\frac{1}{2}\right) e^{-r^2} \Big|_0^{\infty} = -\pi \left(e^{-\infty}{}^{0} - e^{0}{}^{1} \right) = \pi$$

$$\therefore$$

$$I = \int_{-\infty}^{\infty} e^{-x^2}\, dx = \sqrt{\pi}$$

Fig. 11.6 Example 11.5-1

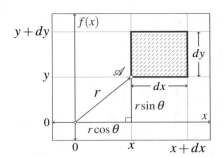
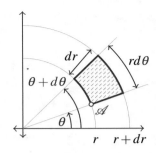

Exercise 11.6, page 249

1. Rational functions are decomposed by the partial fraction transformations.

(a)
$$\int \frac{x-1}{x^2+x} = \left\{ \text{partial fraction: } -\frac{1}{x} + \frac{2}{x+1} \right\}$$

$$= -\int \frac{1}{x}\, dx + \int \frac{2}{x+1}\, dx = -\ln|x| + 2\ln|x+1| + C$$

(b)
$$\int \frac{x}{(x+1)(x-4)} = \left\{ \text{partial fraction: } \frac{1}{5(x+1)} + \frac{4}{5(x-4)} \right\}$$

$$= \int \frac{1}{5(x+1)}\, dx + \int \frac{4}{5(x-4)}\, dx$$

$$= \frac{1}{5}\ln|x+1| + \frac{4}{5}\ln|x-4| + C$$

(c)
$$\int \frac{x+2}{x^3-2x^2} = \left\{ \text{partial fraction: } -\frac{1}{x} - \frac{1}{x^2} + \frac{1}{x-2} \right\}$$

$$= \int -\frac{1}{x}\, dx - \int \frac{1}{x^2}\, dx + \int \frac{1}{x-2}\, dx$$

$$= -\ln|x| + \frac{1}{x} + \ln|x-2| + C$$

(d)
$$\int \frac{2x^2}{x^4-1} = \left\{ \text{partial fraction: } \frac{1}{x^2+1} - \frac{1}{2(x+1)} + \frac{1}{2(x-1)} \right\}$$

$$= \arctan x - \frac{1}{2}\ln|x+1| + \frac{1}{2}\ln|x-1| + C$$

(e)
$$\int \frac{1}{x^3-1} = \left\{ \text{partial fraction: } \frac{1}{3(x-1)} - \frac{x+2}{3(x^2+x+1)} \right\}$$

$$= \frac{1}{3}\int \frac{1}{x-1}\, dx - \frac{1}{3}\int \frac{x+2}{x^2+x+1}\, dx$$

$$= \frac{1}{3}\ln|x-1| - \frac{1}{3}\int \frac{x}{x^2+x+1}\, dx - \frac{2}{3}\int \frac{1}{x^2+x+1}\, dx$$

(e)   *(cont.)* The last two integrals we can solve are as follows:

$$\int \frac{x}{x^2 + x + 1}\, dx$$

$$= \left\{ x^2 + x + 1 = x^2 + 2\frac{1}{2}x + \left(\frac{1}{2}\right)^2 - \left(\frac{1}{2}\right)^2 + 1 = \left(x + \frac{1}{2}\right)^2 + \frac{3}{4} \right\}$$

$$= \int \frac{x}{\left(x + \frac{1}{2}\right)^2 + \frac{3}{4}}\, dx = \left\{ t = x + \frac{1}{2} \quad \therefore \quad dt = dx \text{ and } x = \frac{2t - 1}{2} \right\}$$

$$= \int \frac{2(2t - 1)}{4t^2 + 3}\, dt = 2 \int \frac{2t}{4t^2 + 3}\, dt - 2 \int \frac{1}{4t^2 + 3}\, dt$$

$$\{ z = 4t^2 + 3 \quad \therefore \quad dz = 4 \times 2t\, dt \}$$

$$= \frac{1}{2} \ln(4t^2 + 3) - 2 \int \frac{1}{4t^2 + 3}\, dt$$

$$\left\{ \begin{array}{c} \dfrac{1}{4t^2 + 3} = \dfrac{1}{3} \dfrac{1}{\left(\dfrac{2t}{\sqrt{3}}\right)^2 + 1}, \quad n = \dfrac{2t}{\sqrt{3}} \quad \therefore \quad dn = \dfrac{\sqrt{3}}{2}\, dt \\[4mm] \text{so that} \quad \displaystyle\int \dfrac{1}{n^2 + 1} = \arctan n \end{array} \right\}$$

$$= \frac{1}{2} \ln(4t^2 + 3) - \frac{1}{\sqrt{3}} \arctan\left(\frac{2t}{\sqrt{3}}\right)$$

$$= \frac{1}{2} \ln(4x^2 + 4x + 4) - \frac{1}{\sqrt{3}} \arctan\left(\frac{2x + 1}{\sqrt{3}}\right)$$

And, similarly,

$$\int \frac{1}{x^2 + x + 1}\, dx = \frac{2}{\sqrt{3}} \arctan\left(\frac{2x + 1}{\sqrt{3}}\right)$$

so that we write

$$\int \frac{1}{x^3 - 1} = \frac{1}{3} \ln|x - 1|$$

$$- \frac{1}{6} \ln(4x^2 + 4x + 4) + \frac{1}{3\sqrt{3}} \arctan\left(\frac{2x + 1}{\sqrt{3}}\right)$$

$$- \frac{4}{3\sqrt{3}} \arctan\left(\frac{2x + 1}{\sqrt{3}}\right) + C$$

$$= \frac{1}{3} \ln|x - 1| - \frac{1}{6} \ln(4x^2 + 4x + 4) - \frac{1}{3\sqrt{3}} \arctan\left(\frac{2x + 1}{\sqrt{3}}\right) + C$$

1. Form of integrals that are solved by trigonometric substitutions.

(a) $\displaystyle\int \sqrt{9-x^2}\,dx = \{x = 3\sin\theta \quad \therefore \quad dx = 3\cos\theta\,d\theta \text{ and } x^2 = 9\sin^2\theta\}$

$$= \int \sqrt{9 - 9\sin^2\theta}\,3\cos\theta\,d\theta$$

$$= \int 3\sqrt{1 - \sin^2\theta}\,3\cos\theta\,d\theta$$

$$= 9\int \cos\theta\,\cos\theta\,d\theta = 9\int \cos^2\theta\,d\theta$$

$$\left\{\cos^2\theta = \frac{1 + \cos 2\theta}{2}\right\}$$

$$= \frac{9}{2}\left[\int (1 + \cos 2\theta)\,d\theta\right] = \frac{9}{2}\left[\int d\theta + \int \cos 2\theta\,d\theta\right]$$

$$= \frac{9}{2}\left[\theta + \frac{1}{2}\sin 2\theta\right] + C = \{\sin 2\theta = 2\sin\theta\cos\theta\}$$

$$= \frac{9}{2}[\theta + \sin\theta\cos\theta] + C$$

$$\begin{cases} x = 3\sin\theta & \therefore \quad \sin\theta = \dfrac{x}{3} \quad \therefore \quad \theta = \arcsin\dfrac{x}{3} \\[2mm] 3\,\cos\theta = \sqrt{9 - x^2} & \therefore \quad \cos\theta = \dfrac{\sqrt{9 - x^2}}{3} \end{cases}$$

$$= \frac{9}{2}\arcsin\frac{x}{3} + \frac{\cancel{9}\,x}{2\,\cancel{3}}\frac{\sqrt{9 - x^2}}{\cancel{3}} + C$$

$$= \frac{9}{2}\arcsin\frac{x}{3} + \frac{x}{2}\sqrt{9 - x^2} + C$$

(b) $\displaystyle\int \frac{x^2}{\sqrt{36 - x^2}}\,dx = \{x = 6\sin\theta \quad \therefore \quad dx = 6\cos\theta\,d\theta \text{ and } x^2 = 36\sin^2\theta\}$

$$= \int \frac{36\sin^2\theta}{\sqrt{36 - 36\sin^2\theta}}\,6\cos\theta\,d\theta$$

$$= \int \frac{36\sin^2\theta}{\cancel{6\cos\theta}}\,\cancel{6\cos\theta}\,d\theta = 36\int \sin^2\theta\,d\theta$$

$$\left\{\sin^2\theta = \frac{1 - \cos 2\theta}{2}\right\}$$

$$= 18\int (1 - \cos 2\theta)\,d\theta = 18\int d\theta - 18\int \cos 2\theta\,d\theta$$

$$= 18\theta - 18 \int \cos 2\theta \, d\theta = \{y = 2\theta \quad \therefore \quad dy = 2 \, d\theta\}$$

$$= 18\left(\theta - \frac{1}{2} \int \cos y \, dy\right) = 18\left(\theta - \frac{1}{2} \sin 2\theta\right)$$

$$\{\sin 2\theta = 2 \sin \theta \cos \theta\}$$

$$= 18\left(\theta - \sin \theta \cos \theta\right) + C$$

It is necessary to reintroduce the original variable x. From the trigonometric substitution it follows.

$$x = 6 \sin \theta \quad \therefore \quad \sin \theta = \frac{x}{6} \quad \therefore \quad \theta = \arcsin\left(\frac{x}{6}\right)$$

$$\sqrt{36 - x^2} = 6 \cos \theta \quad \therefore \quad \cos \theta = \frac{\sqrt{36 - x^2}}{6}$$

Therefore,

$$\int \frac{x^2}{\sqrt{36 - x^2}} \, dx = 18\left(\theta - \sin \theta \cos \theta\right) + C$$

$$= 18 \arcsin\left(\frac{x}{6}\right) - \cancel{18} \, 1 \, \frac{x}{\cancel{6}} \frac{\sqrt{36 - x^2}}{\cancel{6} \, 2} + C$$

$$= 18 \arcsin\left(\frac{x}{6}\right) - \frac{x}{2}\sqrt{36 - x^2} + C$$

(c) $\displaystyle \int \frac{\sqrt{25x^2 + 4}}{x^4} \, dx = $

$$\left\{ \begin{array}{l} \sqrt{u^2 + k^2} \text{ substitution is } u = k \tan \theta \\[2mm] \therefore \quad 5x = 2 \tan \theta \quad \therefore \quad dx = \frac{2}{5} \frac{d\theta}{\cos^2 \theta} \\[2mm] \therefore \quad \sqrt{25x^2 + 4} = \sqrt{25 \frac{4}{25} \tan^2 \theta + 4} = \frac{2}{\cos \theta} \end{array} \right\}$$

$$= \int \frac{2}{\cos \theta \left(\frac{2}{5} \tan \theta\right)^4} \frac{2}{5} \frac{d\theta}{\cos^2 \theta}$$

$$= \int \frac{125 \cos^{\cancel{4}} \theta}{4 \cancel{\cos \theta} \sin^4 \theta \, \cancel{\cos^2 \theta}} \, d\theta = \frac{125}{4} \int \frac{\cos \theta}{\sin^4 \theta} \, d\theta$$

$$\{t = \sin \theta \quad \therefore \quad dt = \cos \theta \, d\theta\}$$

$$= -\frac{125}{12} \frac{1}{\sin^3 \theta} + C$$

The original variable is reinserted as follows:

$$5x = 2\tan\theta \quad \therefore \quad \tan\theta = \frac{\sin\theta}{\cos\theta} = \frac{5x}{2} \quad \therefore \quad \sin\theta = \frac{5x}{2}\cos\theta$$

$$\sqrt{25x^2 + 4} = \frac{2}{\cos\theta} \quad \therefore \quad \cos\theta = \frac{2}{\sqrt{25x^2 + 4}}$$

$$\therefore$$

$$\sin\theta = \frac{5x}{\cancel{2}}\frac{\cancel{2}}{\sqrt{25x^2 + 4}}$$

So that the final solution is

$$\int \frac{\sqrt{25x^2 + 4}}{x^4}\,dx = -\frac{125}{12}\frac{1}{\sin^3\theta} + C = -\frac{125}{12}\left(\frac{\sqrt{25x^2 + 4}}{\cancel{5}x}\right)^3 + C$$

$$= -\frac{(\sqrt{25x^2 + 4})^3}{12x^3} + C$$

Exercise 11.8, page 250

1. Area under the curve A is calculated as the definite integral bound of the given interval. Note that when calculating finite integrals, the integration constants are cancelled.

(a) This is tabular integral,

$$A = \int_2^3 x^3\,dx = \left\{\int x^n dx = \frac{x^{n+1}}{n}\right\}$$

$$= \frac{x^4}{4}\Bigg|_2^3 = \frac{1}{4}\left(3^4 - 2^4\right) = \frac{1}{4}(81 - 16) = \frac{65}{4}$$

The sign of area A under the curve is always positive, see Fig. 11.7.

Fig. 11.7 Example 11.8-1(a)

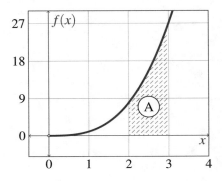

(b)   This is sum of two tabular integrals,

$$A = \int_{-1}^{2} x(1+x^3)\,dx = \int_{-1}^{2} x\,dx + \int_{-1}^{2} x^4 dx = \frac{x^2}{2}\Big|_{-1}^{2} + \frac{x^5}{5}\Big|_{-1}^{2}$$

$$= \frac{1}{2}[2^2 - (-1)^2] + \frac{1}{5}[2^5 - (-1)^5] = \frac{3}{2} + \frac{33}{5} = \frac{81}{10}$$

Note that surface area is "signed area addition", i.e. in the absolute sense the small negative area in interval $[-1, 0]$ is subtracted from the large positive area in interval $[0, 1]$, see Fig. 11.8.

Fig. 11.8 Example 11.8-1(b)

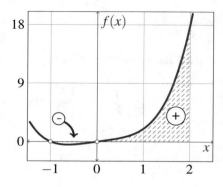

(c)   This is tabular integral (Fig. 11.9),

$$\int_{-\pi/4}^{\pi/4} \cos x\,dx = \sin x\Big|_{-\pi/4}^{\pi/4} = \sin\frac{\pi}{4} - \sin\frac{-\pi}{4} = \left\{\sin(-x) = -\sin(x)\right\}$$

$$= 2\sin\frac{\pi}{4} = 2\frac{\sqrt{2}}{2} = \sqrt{2}$$

Fig. 11.9 Example 11.8-1(c)

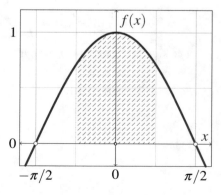

(d) This is tabular integral,

$$\int_0^{2\pi} \cos x \, dx = \sin x \Big|_0^{2\pi} = \sin(2\pi) - \sin(0) = 0$$

Note that the positive and negative areas of sine form function are perfectly matched over one period, and thus the sum is zero, see Fig. 11.10.

Fig. 11.10
Example 11.8-1(d)

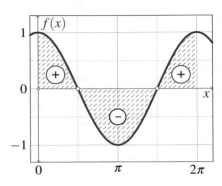

(e) This integral is solved by using the integration by substitution technique (change of variables),

$$\int_0^2 \frac{x}{\sqrt{1+3x^2}} \, dx = \left\{ \begin{array}{l} t = 1+3x^2 \quad \therefore \quad dt = 6x \, dx \quad \therefore \quad x \, dx = \dfrac{dt}{6} \\[2mm] x = 0 \Rightarrow t = 1, \quad \text{and} \quad x = 2 \Rightarrow t = 13 \end{array} \right\}$$

$$= \int_1^{13} \frac{1}{6} \frac{dt}{\sqrt{t}} = \left\{ \sqrt[n]{x^m} = x^{m/n}, \quad \frac{1}{x^n} = x^{-n} \right\}$$

$$= \frac{1}{6} \int_1^{13} t^{-1/2} \, dt = \frac{1}{\cancel{6}3} \frac{t^{1/2}}{\cancel{1/2}} \Big|_1^{13} = \frac{1}{3}[\sqrt{13} - \sqrt{1}]$$

Fig. 11.11
Example 11.8-1(e)

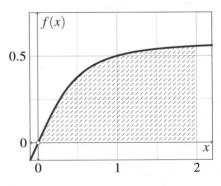

Note that surface area equals an irrational number smaller than one, see Fig. 11.11.

(f)   This integral is solved by using the partial integration technique two times.

$$\int_0^\pi x^2 \cos x \, dx = \left\{ \int u \, dv = u \, v - \int v \, du \right\}$$

$$= \left\{ \begin{array}{l} u = x^2 \Rightarrow \dfrac{du}{dx} = 2x \quad \therefore \quad du = 2x \, dx \\[2mm] dv = \cos x \, dx \Rightarrow \int dv = \int \cos x \, dx \quad \therefore \quad v = \sin x \end{array} \right\}$$

$$= x^2 \sin x - 2 \int x \sin x \, dx$$

$$= \left\{ \begin{array}{l} u = x \Rightarrow du = \, dx \\[2mm] dv = \sin x \, dx \Rightarrow \int dv = \int \sin x \, dx \quad \therefore \quad v = -\cos x \\[2mm] \therefore \\[2mm] \int x \sin x \, dx = -x \cos x + \int \cos x \, dx = -x \cos x + \sin x \end{array} \right\}$$

$$= \left[x^2 \sin x - 2(-x \cos x + \sin x) \right]_0^\pi$$

$$= \left[(x^2 - 2) \sin x + 2x \cos x \right]_0^\pi$$

$$= \left[(\pi^2 - 2) \underbrace{\sin \pi}_{0} - (0 - 2) \underbrace{\sin 0}_{0} + 2\pi \underbrace{\cos \pi}_{-1} - \underbrace{2(0) \cos 0}_{0} \right]$$

$$= -2\pi$$

The negative result is illustrated in Fig. 11.12, where it is clearly visible that the negative signed area under the curve dominates.

Fig. 11.12
Example 11.8-1(f)

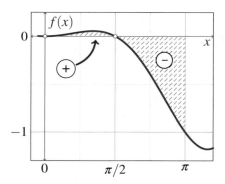

2. In order to calculate the area bound by two functions, e.g. $f(x)$ and $g(x)$, it is necessary to determine the applicable interval. In this case, we can visualize $f(x)$ "cutting out" $g(x)$ and removing the overlapping area, see Fig. 11.13 (left), which leaves area that represents the difference between the two, Fig. 11.13 (right). The interval boundaries are found at the cross-over points $\mathscr{A}$ and $\mathscr{B}$, i.e. when

$$f(x) = g(x) \quad \therefore \quad x^2 = 8 - x^2$$

$$2x^2 = 8$$

$$x^2 = 4 \quad \therefore \quad \underline{x = \pm 2}$$

Therefore, the "cutting out" operation is a simple difference as

$$A = \int_{-2}^{2} [g(x) - f(x)]\, dx = \int_{-2}^{2} [8 - x^2 - x^2]\, dx = 8 \int_{-2}^{2} dx - 2 \int_{-2}^{2} x^2\, dx$$

$$= 8x \Big|_{-2}^{2} - 2\frac{x^3}{3}\Big|_{-2}^{2} = 8(2 - (-2)) - \frac{2}{3}(2^3 - (-2)^3) = \frac{64}{3}$$

Fig. 11.13 Example 11.8-2

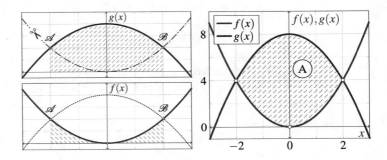

3. Given rational function

$$f(x) = \frac{x}{(x + 1)(x - 4)}$$

and the interval of interest, we use the partial fraction decomposition technique, see Example 11.6-1(b).

$$\frac{x}{(x+1)(x-4)} = \frac{A}{x+1} + \frac{B}{x-4} = \frac{A(x-4) + B(x+1)}{(x+1)(x-4)} = \frac{Ax - 4A + Bx + B}{(x+1)(x-4)}$$

$$= \frac{(A+B)x - 4A + B}{(x+1)(x-4)}$$

$$\therefore$$

$$A + B = 1 \quad \text{and} \quad -4A + B = 0 \quad \therefore \quad B = 1 - A \quad \text{and} \quad B = 4A$$

$$\therefore$$

$$A = \frac{1}{5} \ \text{et} \ B = \frac{4}{5}$$

Introduce two variables: $t = x + 1$ and $r = x - 4$; therefore, their respective intervals are: $t \in (6, 7)$ and $r \in (1, 2)$:

$$\int_5^6 f(x)\,dx = \int_5^6 \frac{x}{x^2 - 3x - 4}\,dx = \frac{1}{5}\int_5^6 \frac{dx}{x+1} + \frac{4}{5}\int_5^6 \frac{dx}{x-4}$$

$$= \frac{1}{5}\int_6^7 \frac{dt}{t} + \frac{4}{5}\int_1^2 \frac{dr}{r} = \frac{1}{5}\ln(t)\Big|_6^7 + \frac{4}{5}\ln(r)\Big|_1^2$$

$$= \frac{1}{5}(\ln 7 - \ln 6) + \frac{4}{5}\left(\ln 2 - \overset{0}{\cancel{\ln 1}}\right)$$

$$\therefore$$

(the answer may be in any of the following forms)

$$= \frac{1}{5}(\ln 7 - \ln 6 + 4\ln 2) = \frac{1}{5}\ln\left(\frac{7 \times 2^4}{6}\right) = \frac{1}{5}\ln\frac{56}{3} = \ln\sqrt[5]{\frac{56}{3}}$$

$$= 0.58534...$$

Graph illustrating the solution is in Fig. 11.14. Note that by visual inspection of the graph, we can confirm that the calculated surface area is indeed approximately 0.6 or so.

Fig. 11.14 Example 11.8-3

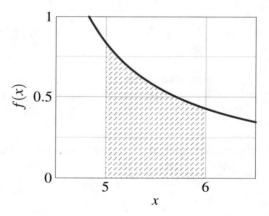

4. The well known high school-level geometry equations are derived by using calculus.

(a) A right-angled triangle is easily mapped into the coordinating system where the two catheti $a = 2$ and $b = 1$ are assumed to be the horizontal and vertical coordinates, Fig. 11.15. In this respect, obviously, hypotenuse is a linear function whose algebraic form is found by knowing that the line must cross two points whose coordinates are $(0, 0)$ and $(2, 1)$. We write two equations as

$$f(x) = ax + b$$

$$(y, x) = (0, 0): \quad 0 = a \times 0 + b \quad \therefore \quad \underline{b = 0}$$

$$(y, x) = (2, 1): \quad 1 = a \times 2 \quad \therefore \quad a = \frac{1}{2}$$

$$\therefore \quad f(x) = \frac{1}{2}x, \quad x \in [0, 2]$$

Therefore, the area of triangular surface is calculated as

$$A = \int_1^2 f(x)\, dx = \int_1^2 \frac{1}{2}x\, dx = \frac{1}{2}\frac{x^2}{2}\bigg|_1^2 = \frac{1}{4}(4 - 0) = 1$$

Of course, the high school formula that gives $A = ab/2 = 2 \times 1/2 = 1$ is a simple summary of the calculus method.

Fig. 11.15
Example 11.8-4(a)

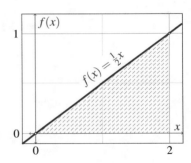

(b) One possible technique to calculate the circle area is as follows. From algebraic equation for circle, given $r = 1$, we write

$$x^2 + y^2 = r^2 \quad \therefore \quad x^2 + y^2 = 1 \quad \therefore \quad y = \sqrt{1 - x^2} = f(x), x \in [0, 1], y \in [0, 1]$$

The plot of $f(x)$, see Fig. 11.16, shows a surface that is equivalent to one-quarter of the full circle area. Thus, at the end, the result of the definite integral of $f(x)$, $x \in [0, 1]$ must be multiplied by four.

Fig. 11.16
Example 11.8-4(b)

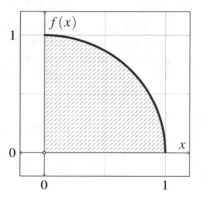

This integral is solved by using the integration by substitution technique (change of variables) two times. The idea is that in a unity circle, the horizontal projection equals $x = r\cos\theta = \cos\theta$, because $r = 1$.

$$A_4 = \int_0^1 f(x)\, dx = \int_0^1 \sqrt{1 - x^2}\, dx$$

$$= \left\{ \begin{aligned} & x = \cos\theta \quad \therefore \quad \frac{dx}{d\theta} = -\sin\theta \quad \therefore \quad dx = -\sin\theta\, d\theta \\ & x = 0 \Rightarrow \theta = \frac{\pi}{2}, \quad x = 1 \Rightarrow \theta = 0 \end{aligned} \right\}$$

$$= \int_{\pi/2}^0 \sqrt{1 - \cos^2\theta}\,(-\sin\theta)\, d\theta$$

$$= \left\{ \sin^2 x + \cos^2 x = 1 \quad \therefore \quad \sin x = \sqrt{1 - \cos^2 x} \right\}$$

$$= \int_{\pi/2}^0 -\sin^2\theta\, d\theta = \int_0^{\pi/2} \sin^2\theta\, d\theta = \left\{ \sin^2 x = \frac{1}{2}(1 - \cos(2x)) \right\}$$

$$= \frac{1}{2}\int_0^{\pi/2}(1 - \cos(2\theta))\, d\theta = \frac{1}{2}\int_0^{\pi/2} d\theta - \frac{1}{2}\int_0^{\pi/2}\cos(2\theta)\, d\theta$$

we introduce second change of variables and continue as

$$A_4 = \left\{ \begin{aligned} & t = 2\theta \quad \therefore \quad dt = 2d\theta \quad \therefore \quad d\theta = \frac{dt}{2} \\ & \theta = 0 \quad \therefore \quad t = 0, \quad \theta = \frac{\pi}{2} \quad \therefore \quad t = \pi \end{aligned} \right\} = \frac{1}{2}\theta \Big|_0^{\pi/2} - \frac{1}{2}\int_0^\pi \cos t\, \frac{dt}{2}$$

$$= \frac{1}{2}\left[\frac{\pi}{2} - 0\right]\frac{1}{4}\sin t\,\Big|_0^\pi = \frac{\pi}{4} - \frac{1}{4}\left(\sin\pi^{\,0} - \sin 0^{\,0}\right) = \frac{\pi}{4}$$

$$\therefore \quad A = 4A_4 = 4\frac{\pi}{4} = \pi$$

The high school-level formula summarizes the above calculus result by stating $A = r^2\pi = (r = 1) = \pi$.

(c) Calculation of volumes is done by "integral of an integral" or multiple integral, in other words, one integration for each variable. Visually, a surface is created by adding infinitely many lines in parallel next to each other, where the respective length of each line equals $f(x)$ (i.e. first dimension) and the lines are added within the given range (i.e. second dimension).

For example, the lowest horizontal surface, see Fig. 11.17, is created by aligning y-directional lines ($f(x)$) within range of $x \in [0, 4]$, where each line has the length equal to three. In other words, the surface area A_1 is found as

$$A_1 = \int_0^4 f(x)\, dx = \int_0^4 3\, dx = 3\, x \Big|_0^4 = 3(4 - 0) = 12$$

which is the result that we know from the high school formula for rectangle surface $A_1 = ab = 3 \times 4 = 12$. Visually, this surface is equivalent to one page of a book whose volume we want to calculate. That is to say, we add pages on top of each other in z-direction. Thickness of surface is infinitely small, that is to say it is equal to dz. We have to simply "pile them up" between zero and two. Mathematically, this addition of surfaces is another integral in z-direction where $z \in [0, 2]$. This integral (i.e. addition) of surfaces is written as

$$V = \int_0^2 A_1\, dz = \{A_1 = 12 = \text{const.}\} = 12\, z \Big|_0^2 = 12(2 - 0) = 24$$

which is a well known result for cuboid volume $V = abc = 4 \times 3 \times 2 = 24$.
This process of double integration is formally written in one line as

$$V = \int_0^2 \int_0^4 f(x)\, dx\, dz = \int_0^2 \int_0^4 3\, dx\, dz = 3 \int_0^2 dz \int_0^4 dx$$

$$= 3 \int_0^2 dz\, x \Big|_0^4 = 3 \int_0^2 4\, dz = 12\, z \Big|_0^2 = 12 \times 2 = 24$$

Note that the order of addition is not relevant, and thus we choose most convenient one. In addition, this trivial example illustrates the principle and way to reason the volume calculations in 3D space. Nevertheless, the same reasoning and techniques apply to higher order spaces and non-trivial objects.

Fig. 11.17
Example 11.8-4(c)

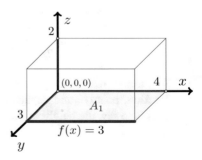

5. One possible geometric interpretation of "average" may be by comparing (signed) area under the function with the area of a rectangle in the same interval. Surface area under function is calculated with the finite integral. Surface area of rectangle A equals its length a (equal to the interval) multiplied by its height h. Therefore, the height of the rectangle is calculated by dividing its area by its length. If the two surface areas happen to be equal, then the height h of the rectangle is called the "average" of the function.

For example, see Fig. 11.18 (left), by inspection we find the area under piecewise linear function $f(x)$ in $x \in [0, 4]$ interval to be equal to $0 + 2 + 0 + 2 = 4$ units. At the same time, rectangular surface in Fig. 11.18 (right) in $x \in [0, 4]$ interval is $A = a \times h = 4 \times 1 = 4$. Thus, in this special case, we write

$$h = \frac{A}{a} = \frac{1}{4} \int_0^4 f(x)\, dx$$

In this case, we say that h is the average value of $f(x)$ and we use syntax $\langle f(x) \rangle$ to convey that information. In general case, i.e. any continuous function $f(x)$ in $x \in [a, b]$, we write the formal definition of average as

$$\underbrace{\langle f(x) \rangle}_{\text{height}} = \underbrace{\frac{1}{b-a}}_{\text{length}} \underbrace{\int_a^b f(x)\, dx}_{\text{surface area}}$$

Fig. 11.18 Example 11.8-5

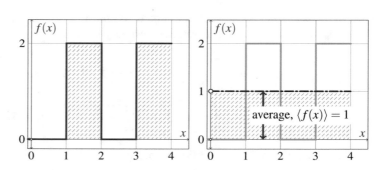

(a)   Average of a periodic function as calculated over one period is

$$\langle \sin x \rangle = \frac{1}{2\pi} \int_0^{2\pi} \sin x\, dx = \frac{1}{2\pi} (-\cos x) \Big|_0^{2\pi} = \frac{1}{2\pi}(-(-1) - 1) = 0$$

To interpret result that $\langle \sin x \rangle = 0$, $x \in [0, 2\pi]$, we look at Fig. 11.19. Within one period of sine function, positive and negative surface areas are perfectly matched and thus cancelled to zero. That is to say, the height of the equivalent rectangle equals zero. The same conclusion is valid for $\langle \cos x \rangle = 0$, $x \in [0, 2\pi]$, Fig. 11.19 (right).

Fig. 11.19
Example 11.8-5(a)

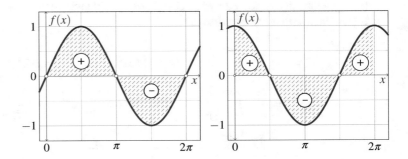

(b)   Average of sine and cos functions over interval, for example from $\pi/2$ to $3p/2$ (i.e. half-period long, but only over negative surface section), is

$$\langle \cos x \rangle = \frac{1}{\pi} \int_{\pi/2}^{3\pi/2} \cos x \; dx = \frac{1}{\pi} \; \sin x \; \Big|_{\pi/2}^{3\pi/2} = \frac{1}{\pi}(-1 - 1) = -\frac{2}{\pi}$$

which is not equal to zero. However, see Fig. 11.19 (right), we find that the average $\langle \cos x \rangle = 0$ if $x \in [0, \pi]$, which is also a half-period long but over both positive and negative surface sections.

(c)   Average over one period of a sine function ("AC") that is added to a constant ("DC") is found as follows:

$$\langle f(x) \rangle = \frac{1}{2\pi} \int_0^{2\pi} (1 + \cos x) \; dx = \frac{1}{2\pi} \int_0^{2\pi} dx + \frac{1}{2\pi} \underset{\nearrow^{0}}{\int_0^{2\pi} \cos x \; dx}$$

$$= \frac{1}{2\pi} \; x \; \Big|_0^{2\pi} = \frac{1}{2\pi}(2\pi - 0) = 1$$

This is an important case to notice, and the graphical interpretation is shown in Fig. 11.20.

Fig. 11.20
Example 11.8-5(c)

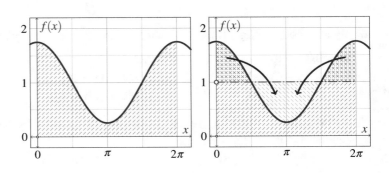

(d)   Integral that is solved by the change of variable technique in addition to using an important trigonometric identity transformation.

$$\langle f(x) \rangle = \frac{1}{2\pi} \int_0^{2\pi} \cos^2 x \; dx = \left\{ \cos^2 x = \frac{1}{2}(1 + \cos(2x)) \right\}$$

$$= \frac{1}{2\pi} \int_0^{2\pi} \frac{1}{2}(1 + \cos(2x)) \; dx = \frac{1}{4\pi} \left. x \right|_0^{2\pi} + \frac{1}{2\pi} \int_0^{2\pi} \frac{1}{2} \cos(2x) \; dx$$

$$= \left\{ \begin{array}{l} t = 2x \quad \therefore \quad dt = 2 \, dx \quad \therefore \quad dx = \dfrac{dt}{2} \\[2mm] x = 0 \quad \therefore \quad t = 0 \;\; \text{and} \;\; x = 2\pi \quad \therefore \quad t = 4\pi \end{array} \right\}$$

$$= \frac{1}{4\pi}(2\pi - 0) + \frac{1}{4\pi} \int_0^{4\pi} \cos t \; dt = \frac{1}{2} + \frac{1}{4\pi}(\underbrace{\sin(4\pi)}_{0} - \underbrace{\sin(0)}_{0})$$

$$= \frac{1}{2}$$

Due to symmetrical regions of this function, similar to the illustration in Fig. 11.20 (right) again, we can see how the area under $f(x)$ is "rearranged" into rectangle whose height equals one-half, Fig. 11.21.

Fig. 11.21
Example 11.8-5(d)

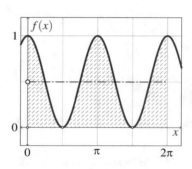

6. By inspection of Fig. 11.2, we calculate surface area divided by the interval as

(a) $0/(1 - 0) = 0$ (b) $2/(2 - 0) = 1$ (c) $2/(3 - 0) = 2/3$
(d) $4/(4 - 0) = 1$ (e) $2/(2 - 1) = 2$ (f) $0/(3 - 2) = 0$
(g) $2/(4 - 2) = 1$ (h) $2/(3 - 0.5) = 4/5$ (i) $2/(3.5 - 1.5) = 1$

7. By inspection of Fig. 11.3, we calculate surface area divided by the interval for the top and bottom functions to find

(a) $\langle f \rangle = 0, \;\; \langle g \rangle = 0$ (b) $\langle f \rangle = 1, \;\; \langle g \rangle = 0$
(c) $\langle f \rangle = 3/2, \;\; \langle g \rangle = 3/2$ (d) $\langle f \rangle = 1, \;\; \langle g \rangle = 1$
(e) $\langle f \rangle = 1, \;\; \langle g \rangle = 4/5$ (f) $\langle f \rangle = 1, \;\; \langle g \rangle = 1$
(g) $\langle f \rangle = 8/7, \;\; \langle g \rangle = 8/7$ (h) $\langle f \rangle = 6/7, \;\; \langle g \rangle = 6/7$
(i) $\langle f \rangle = 1, \;\; \langle g \rangle = 1$

8. By inspection of Fig. 11.4, we calculate surface area divided by the interval for the top and bottom functions to find

(a) $\langle f \rangle = 0$, $\langle g \rangle = 2$

(b) $\langle f \rangle = 1$, $\langle g \rangle = 1$

(c) $\langle f \rangle = 2/3$, $\langle g \rangle = 4/3$

(d) $\langle f \rangle = 1/2$, $\langle g \rangle = 3/2$

(e) $\langle f \rangle = 2/5$, $\langle g \rangle = 8/5$

(f) $\langle f \rangle = 1/2$, $\langle g \rangle = 3/2$

(g) $\langle f \rangle = 4/7$, $\langle g \rangle = 10/7$

(h) $\langle f \rangle = 1/3$, $\langle g \rangle = 5/3$

(i) $\langle f \rangle = 0$, $\langle g \rangle = 2$

Exercise 11.9, page 252

1. In case of unbound functions, it is necessary to find if they converge or diverge. The idea is to see if the function's limit is finite or infinite. Definite integrals give the surface area under function, and thus if the function reaches infinite value, the surface area is also infinite (i.e. diverge). It is not always obvious to determine the function's behavior by simple inspection.

(a) $$\lim_{a \to \infty} \int_1^a \frac{1}{x^2}\, dx = \lim_{a \to \infty} \left(-\frac{1}{x}\Big|_1^a \right) = \lim_{a \to \infty} \left(-\frac{1}{a} - \left(-\frac{1}{1} \right) \right)$$
$$= (0 - (-1)) = 1$$

(b) $$\lim_{a \to \infty} \int_1^a \frac{1}{x}\, dx = \lim_{a \to \infty} \left(\ln|x|\Big|_1^a \right) = \lim_{a \to \infty} (\ln a - \ln 1)$$
$$= \infty - 0 = \infty \quad \therefore \quad \text{diverges}$$

The difference between functions $1/x$ and $1/x^2$ is illustrated in Fig. 11.22. Although, by inspection, both surface areas seem to converge, it is clearly visible that their respective integrals behave very differently: $\ln x$ is divergent and $1/x$ is convergent for $x \to \infty$.

Fig. 11.22
Example 11.9-1(a)(b)

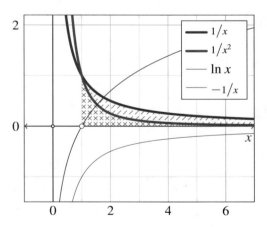

(c)   This function is not defined for $x = 0$, because $f(0) = \infty$. Therefore, we evaluate its integral as two improper integrals, one approaching zero from the left side and one approaching zero from the right side.

$$\lim_{a \to 0_-} \int_{-1}^{a} \frac{1}{\sqrt[3]{x^2}}\, dx + \lim_{a \to 0_+} \int_{a}^{1} \frac{1}{\sqrt[3]{x^2}}\, dx$$

$$= \lim_{a \to 0_-} \int_{-1}^{a} x^{-2/3}\, dx + \lim_{a \to 0_+} \int_{a}^{1} x^{-2/3}\, dx$$

$$= 3 \lim_{a \to 0_-} \sqrt[3]{x}\,\Big|_{-1}^{a} + 3 \lim_{a \to 0_+} \sqrt[3]{x}\,\Big|_{a}^{1}$$

$$= 3 \lim_{a \to 0_-} \left(\sqrt[3]{a} - \sqrt[3]{-1} \right) + 3 \lim_{a \to 0_+} \left(\sqrt[3]{1} - \sqrt[3]{a} \right)$$

$$= 3 \left(\sqrt[3]{0} + 1 \right) + 3 \left(1 - \sqrt[3]{0} \right)$$

$$= 6$$

In this case, even though there is vertical asymptote within the given interval $x \in [-1, 1]$, the surface area converges to 3 on each side, Fig. 11.23.

Fig. 11.23
Example 11.9-1(c)

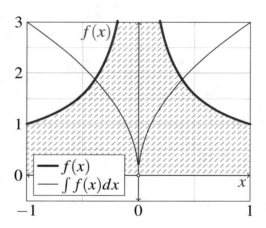

(d) $$\lim_{a \to \infty} \int_{-a}^{a} x e^{-x^2}\, dx = \lim_{a \to \infty} -\frac{1}{2} e^{-x^2}\,\Big|_{-a}^{a} = -\frac{1}{2} \lim_{a \to \infty} \left(e^{-a^2} - e^{-(-a)^2} \right)$$

$$= -\frac{1}{2}(0 - 0) = 0$$

(e) $\displaystyle\lim_{a\to-\infty}\int_a^0 \frac{1}{\sqrt{1-x}}\,dx = \lim_{a\to-\infty} -2\sqrt{1-x}\;\Big|_a^0$

$$= -2\lim_{a\to-\infty}\left(\sqrt{1-0}-\sqrt{1-a}\right)$$

$$= -2(1-\infty) = \infty \quad \therefore \quad \text{diverges}$$

Even though, by a quick look inspection, it may appear that area under $f(x)$ converges, its integral diverges, Fig. 11.24.

Fig. 11.24
Example 11.9-1(e)

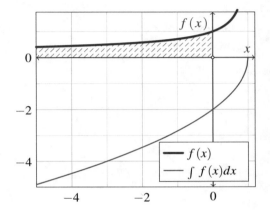

(f) $\displaystyle\lim_{a\to\infty}\int_0^a \frac{1}{1+x^2},\,dx = \lim_{a\to\infty}\arctan x\,|_0^a = \lim_{a\to\infty}(\arctan a - \arctan 0)$

$$= \frac{\pi}{2} - 0 = \frac{\pi}{2}$$

Exercise 11.10, page 252

1. Knowing that definite integrals give the total *signed* surface area within the given interval, sometimes we can intuitively deduce the correct result.

(a) Inverse function has vertical asymptote for $x = 0$; thus, it is logical to calculate limit $a \to 0$ after applying the Newton–Leibniz formula; that is to say,

$$\int_1^{-1} \frac{1}{x}\,dx = \lim_{a\to 0_-}\int_{-1}^a \frac{1}{x}\,dx + \lim_{a\to 0_+}\int_a^1 \frac{1}{x}\,dx = \lim_{a\to 0_-} \ln|a|\,\Big|_{-1}^a + \lim_{a\to 0_+}\ln|a|\,\Big|_a^1$$

$$= \lim_{a\to 0}(\ln|a| - \ln|-1| + \ln|1| - \ln|a|) = -\infty - 0 + 0 - \infty$$

$$\therefore$$

 (undermined, incorrect conclusion)

Mathematical software that blindly implements the Newton–Leibniz formula reports this result. However, the inverse function is odd, that is to say the origin-symetric, Fig. 11.25. One should conclude that the total sum is zero. Indeed, the evaluation of undermined result should be modified as

$$\int_{1}^{-1} \frac{1}{x} \, dx = \lim_{a \to 0} (\ln |a| - \underbrace{\ln |-1|}_{0} + \underbrace{\ln |1|}_{0} - \ln |a|)$$

$$= \lim_{a \to 0} \ln \frac{|a|}{|a|} = \lim_{a \to 0} \ln 1 = 0$$

$$\therefore$$

(correct)

Fig. 11.25
Example 11.10-1(a)

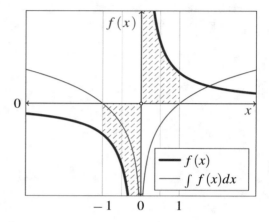

(b) Periodic functions may present a challenge as well. Direct implementation of the Newton–Leibniz formula may lead into wrong conclusion.

$$\int_{0}^{100\pi} \sqrt{1 - \cos 2x} \, dx = \left\{ 1 - \cos x = 2 \sin^2 \frac{x}{2} \right\}$$

$$= \sqrt{2} \int_{0}^{100\pi} |\sin x| \, dx = \sqrt{2} \left| \int_{0}^{100\pi} \sin x \, dx \right|$$

$$= \sqrt{2} \left| \cos x \, \right|_{0}^{100\pi} \right| = \sqrt{2} \, |\cos(50 \times 2\pi) - \cos 0|$$

$$= \sqrt{2} \, |1 - 1| = 0$$

However, by inspection, area under $f(x)$ is obviously non-zero and positive ($|\sin x| \geq 0$), and it is periodic with the period $T = \pi$, see Fig. 11.26. One way to make correct conclusion is to calculate the surface area within one period.

$$\int_0^\pi \sqrt{1 - \cos 2x} \, dx = \sqrt{2} \left| \int_0^\pi \sin x \, dx \right|$$

$$= \sqrt{2} \left| \cos x \Big|_0^\pi \right| = \sqrt{2} \left| \cos(\pi) - \cos 0 \right|$$

$$= \sqrt{2} |-1 - 1| = 2\sqrt{2}$$

Therefore, the surface area within interval $x \in [0, 100\pi]$ equals $200\sqrt{2}$.

Fig. 11.26
Example 11.10-1(b)

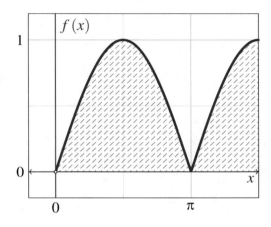

Multivariable Functions

12

Important to Know

Multivariable functions are in the form $f(x, y, z, \ldots)$.

Limits:

$$\lim_{\substack{x \to x_0, \\ y \to y_0, \\ z \to z_0, \\ \ldots}} f(x, y, z, \ldots)$$

exists if all variable limits converge to the same value; otherwise, the limit does not exist. In its basic form, for example, the following limits converge to the same value (note, the limit is calculated along one variable at the time, while all the others are kept constant):

$$\lim_{\substack{x \to x_0, \\ y_0, \\ z_0, \\ \ldots}} f(x, y, z, \ldots) = a$$

$$\lim_{\substack{x_0, \\ y \to y_0, \\ z, \\ \ldots}} f(x, y, z, \ldots) = a$$

$$\lim_{\substack{x_0, \\ y_0, \\ z \to z_0, \\ \ldots}} f(x, y, z, \ldots) = a$$

$$\ldots$$

Note that this condition is necessary but not sufficient to prove the existence of the global limit.

Partial Derivatives: these are calculated for each variable relative to all the others. For example, in the case of a two-variable function $f(x, y)$, we can use the following equivalent syntax form notations to specifically write the order of second partial derivatives as

$$\left(f'_x\right)_x = f''_{xx} = \frac{\partial f}{\partial x}\left(\frac{\partial f}{\partial x}\right) = \frac{\partial^2 f}{\partial x^2}$$

© Springer Nature Switzerland AG 2021
R. Sobot, *Engineering Mathematics by Example*,
https://doi.org/10.1007/978-3-030-79545-0_12

$$\left(f'_x\right)_y = f''_{xy} = \frac{\partial f}{\partial y}\left(\frac{\partial f}{\partial x}\right) = \frac{\partial^2 f}{\partial y \partial x}$$

$$\left(f'_y\right)_x = f''_{yx} = \frac{\partial f}{\partial x}\left(\frac{\partial f}{\partial y}\right) = \frac{\partial^2 f}{\partial x \partial y}$$

$$\left(f'_y\right)_y = f''_{yy} = \frac{\partial f}{\partial y}\left(\frac{\partial f}{\partial y}\right) = \frac{\partial^2 f}{\partial y^2}$$

Integrals: the volume of multidimensional space is calculated as integral along each variable (the integral sum is a commutative operation). For example, in the case of three-dimensional space, we systematically reduce the number of variables as

$$V_{3D} = \iiint_{x,y,z} f(x,y,z)\ dx\ dy\ dz$$

$$= \iint_{y,z} dy\ dz \underbrace{\int_x f(x,y,z)\ dx}_{f(y,z)}$$

$$= \int_z dz \underbrace{\int_y f(y,z)\ dy}_{f(z)}$$

$$= \int_z f(z)\ dz$$

where, in the each integration step along given variable, the reminding variables are treated as constants.

12.1   Exercises

12.1 ***   Domain of Multivariable Functions

1. Sketch plot of the following surfaces:

(a) $f(x,y) = x^2 + y^2$ (b) $f(x,y) = x^2$

(c) $r^2 = x^2 + y^2 + z^2$ (d) $z^2 = x^2 + y^2$

2. Determine domains of the following functions:

(a) $f(x,y) = x + \sqrt{y}$.   (b) $f(x,y) = \ln(-x - y)$.

(c) $g(x,y) = \sqrt{9 - x^2 - y^2}$.   (d) $h(x,y) = \sqrt{1 - x^2} + \sqrt{y^2 - 1}$.

12.2 *** Limits of Multivariable Functions

1. Calculate the following limits:

(a) $\lim\limits_{(x,y)\to(1,2)} (x^2y^3 - x^3y^2 + 3x + 2y)$

(b) $\lim\limits_{(x,y)\to(0,0)} \dfrac{x^2 - y^2}{x^2 + y^2}$

(c) $\lim\limits_{(x,y)\to(0,0)} \dfrac{xy}{x^2 + y^2}$

(d) $\lim\limits_{(x,y)\to(0,0)} \dfrac{3x^2y}{x^2 + y^2}$

12.3 *** Derivatives of Multivariable Functions

1. (a) Derive partial derivatives of $f(x, y) = 4 - x^2 - 2y^2$ and then calculate $f'_x(1, 1)$ and $f'_y(1, 1)$.

(b) Derive partial derivatives of $f(x, y) = \sin \dfrac{x}{1 + y}$.

(c) Given $x^3 + y^3 + z^3 + 6xyz = 1$, derive $\frac{\partial z}{\partial x}$ and $\frac{\partial z}{\partial y}$.

(d) Given function $f(x, y) = x^3 + x^2y^3 - 2y^2$, derive $f''_{xx}, f''_{xy}, f''_{yx}, f''_{yy}$.

2. Derive partial derivatives of the following composite functions:

(a) Derive partial derivatives of $z = e^x \sin y$, where $x = st^2$ and x^2t.

(b) Given $u = \ln \sqrt{(x - a)^2 + (y - b)^2}$, where a and b are constants, prove the following equation:

$$\frac{\partial^2 u}{\partial x^2} + \frac{\partial^2 u}{\partial y^2} = 0$$

12.4 *** Integrals of Multivariable Functions

1. (a) Calculate the following integral: $\displaystyle\int_0^3 \int_1^2 x^2y \, dx \, dy$

(b) Calculate the following integral: $\displaystyle\iint_R \sqrt{1 - x^2} \, dA$ where R is bound of the two variables forming surface A, given that $|y| \leq 2$.

Solutions

Exercise 12.1, page 284

1. Two-variable functions define surfaces whose elevation is measured along z-axis.

(a) Starting with a point at $(x, y) = (0, 0)$, function $f(x, y) = x^2 + y^2$ generates a circle whose radius is larger and larger for each subsequent (x, y) pair and its value is assigned to z-axis, Fig. 12.1.

(b) Function $f(x, y) = x^2$ does not depend on y, and thus when looking into y-axis there is only quadratic function $z = x^2$ visible. However, this quadratic function is found at any position along $y \in [-\infty, \infty]$, Fig. 12.2.

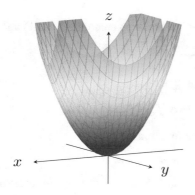

Fig. 12.1 Example 12.1-1(a)

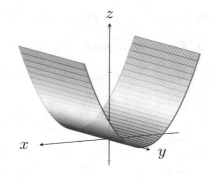

Fig. 12.2 Example 12.1-1(b)

(c) As the next expansion of 2D circle equation, function $r^2 = x^2 + y^2 + z^2$ defines a sphere whose radius equals r, Fig. 12.3.

(d) Two-sided conus function $z^2 = x^2 + y^2$ also starts with a single point at $(x, y) = (0, 0)$ and then progressively moves in both directions of z-axis, while the radius of circles changes linearly, Fig. 12.4.

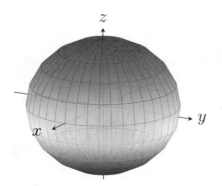

Fig. 12.3 Example 12.1-1(c)

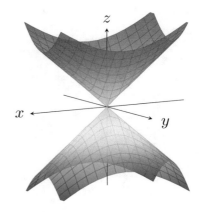

Fig. 12.4 Example 12.1-1(d)

2. Domain of a multivariable function is limited by the limitations of each variable.

(a) Function $f(x, y) = x + \sqrt{y}$ is defined for $x \in \mathbb{R}$ and $y \geq 0$.

(b) Function $f(x, y) = \ln(-x - y)$ is limited by $\ln t$, and thus $-x - y > 0 \quad \therefore \quad x + y < 0$.

(c) Given $g(x, y) = \sqrt{9 - x^2 - y^2}$, we conclude

$$D = \left\{ (x, y) \,\middle|\, 9 - x^2 - y^2 \geq 0 \right\}$$

$$\therefore \ \left\{ (x, y) \,\middle|\, x^2 + y^2 \leq 9 \right\}$$

Therefore, in x, y plane, $g(x, y)$ is contained within a circle whose radius is $r = 3$. At the same time, z is limited only to $z \geq 0$ and $\sqrt{9 - x^2 - y^2} \leq 3$ (i.e. when $(x, y) = (0, 0)$), that is to say, $0 \leq z \leq 3$ or, in other words, the upper half of a sphere.

(d)   Function
$$h(x, y) = \sqrt{1 - x^2} + \sqrt{y^2 - 1}$$
is limited by two square roots, Fig. 12.5, and thus we write

$$1 - x^2 \geq 0 \quad \therefore \quad x^2 \leq 1 \quad \therefore \quad |x| \leq 1$$
$$y^2 - 1 \geq 0 \quad \therefore \quad y^2 \geq 1 \quad \therefore \quad |y| \geq 1$$

Fig. 12.5 Example 12.1-2(d)

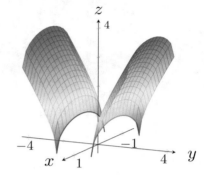

Exercise 12.2, page 285

1. In the first iteration, limits of a multivariable function are calculated along each variable, in the given examples along (x, y). However, in order to be defined (i.e. the surface is continuous in any direction), *all* limits along any arbitrary directions must be equal.

(a) $\lim\limits_{(x,y)\to(1,2)} (x^2 y^3 - x^3 y^2 + 3x + 2y) = 1^2 2^3 - 1^3 2^2 + 3(1) + 2(2) = 11$

(b) $\lim\limits_{(x,y)\to(0,0)} \dfrac{x^2 - y^2}{x^2 + y^2} = \dfrac{0 - 0}{0 + 0} \quad \therefore \quad$ does not exist

We can confirm this conclusion by the two limits along x-axis and y-axis as

$$\lim\limits_{(x,y)\to(x,0)} \frac{x^2 - 0^2}{x^2 + 0^2} = \frac{\cancel{x^2}}{\cancel{x^2}} = 1$$

$$\lim\limits_{(x,y)\to(0,y)} \frac{0^2 - y^2}{0^2 + y^2} = \frac{-\cancel{y^2}}{\cancel{y^2}} = -1$$

These two limits are not equal, and thus we conclude that in this case $\lim_{(x,y)\to(x,0)} f(x, y)$ does not exist.

(c)   However, even if the two limits along x- and y-axes are equal, it is necessary but not sufficient condition. For example,

$$\lim_{(x,y)\to(x,0)} \frac{xy}{x^2+y^2} = \frac{x\,(0)}{x^2+0} = 0$$

$$\lim_{(x,y)\to(0,y)} \frac{xy}{x^2+y^2} = \frac{(0)\,y}{0+y^2} = 0$$

however, limit along $y = x$ is

$$\lim_{(x,y)\to(x,x)} \frac{xy}{x^2+y^2} = \frac{x\,x}{x^2+x^2} = \frac{x^2}{2x^2} = \frac{1}{2}$$

therefore, even though the first the two limits are equal, they are different from the third, and thus the limit does not exist.

(d)   This limit appears to not exist, because

$$\lim_{(x,y)\to(0,0)} \frac{3x^2y}{x^2+y^2} = \left(\frac{0}{0}\right) \quad \text{however,} \quad \lim_{(x,y)\to(x,0)} \frac{3x^2y}{x^2+y^2} = \frac{0}{x^2} = 0$$

$$\lim_{(x,y)\to(0,y)} \frac{3x^2y}{x^2+y^2} = \frac{0}{y^2} = 0$$

$$\lim_{(x,y)\to(x,x)} \frac{3x^3}{2x^2} = \lim_{(x,y)\to(0,0)} \frac{3x}{2} = 0$$

Even though the three limits are equal, the question is how to prove that this limit exists along any arbitrary line? We can use "the squeeze theorem" (sometimes known as "the two police officers and burglar" or "the sandwich theorem"). The idea is to determine two extreme boundaries of a given function and then to bring these two boundaries together; the function itself must always stay in between the two extremes. In the given analogy, if a burglar is squeezed between two officers, and if the two officers are walking towards prison building, the burglar is forced to end up in the prison.

Since we are looking for $\lim_{(x,y)\to(0,0)} f(x)$, we express the distance between the origin point $(0, 0)$ and any point of $f(x)$ as

$$\left| \frac{3x^2y}{x^2+y^2} - 0 \right| = \left| \frac{3x^2y}{x^2+y^2} \right| = \frac{3x^2\,|y|}{x^2+y^2} \tag{12.1}$$

where the last absolute value expression is valid because number "3", x^2, and y^2 are always positive, and thus they are equal to their respective absolute values. Consequently, the lower side limit of (12.1) is zero.

Accordingly, because $x^2 \leq x^2 + y^2$ is always true, the upper side limit is

$$x^2 \leq x^2 + y^2 \quad \therefore \quad \frac{x^2}{x^2 + y^2} \leq 1 \quad \therefore \quad \frac{x^2}{x^2 + y^2} \overset{\leq 1}{} 3\,|y| \leq 3\,|y|$$

In conclusion, two extremes of (12.1) are on the lower side it is limited by "0" and on the upper side by $3|y|$, that is to say,

$$0 \leq \frac{3x^2 y}{x^2 + y^2} \leq 3\,|y|$$

$$\therefore$$

$$\lim_{(x,y)\to(0,0)} 0 \leq \lim_{(x,y)\to(0,0)} \frac{3x^2 y}{x^2 + y^2} \leq \lim_{(x,y)\to(0,0)} 3\,|y| \qquad (12.2)$$

Now, the left and right side inequalities in (12.2) are heading towards the origin, i.e. $(x, y) \to (0, 0)$; then, by the virtue of its position between the two extremes, the distance is "squeezed" between two limits and must also converge to zero, Fig. 12.6.
In summary, the limit exists in any direction as

$$\lim_{(x,y)\to(0,0)} \frac{3x^2 y}{x^2 + y^2} = 0$$

Fig. 12.6 Example 12.1-1(d)

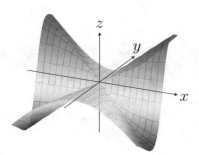

Exercise 12.3, page 285

1. Partial derivatives are calculated as follows:

(a)   Calculate derivates relative to each variable and simply substitute given coordinates.

$$f'_x = \frac{\partial}{\partial x}(4 - x^2 - 2y^2) = -2x \quad \therefore \quad f'_x(1, 1) = -2$$

$$f'_y = \frac{\partial}{\partial y}(4 - x^2 - 2y^2) = -4y \quad \therefore \quad f'_x(1, 1) = -4$$

(b) Derivatives of composite fractions are found as

$$f'_x = \frac{\partial}{\partial x} \sin \frac{x}{1+y} = \cos \frac{x}{1+y} \frac{\partial}{\partial x} \left(\frac{x}{1+y} \right) = \frac{1}{1+y} \cos \frac{x}{1+y}$$

$$f'_y = \frac{\partial}{\partial y} \sin \frac{x}{1+y} = \cos \frac{x}{1+y} \frac{\partial}{\partial y} \left(\frac{x}{1+y} \right) = -\frac{x}{(1+y)^2} \cos \frac{x}{1+y}$$

(c) Implicit partial derivation is done by holding all variables constant except the derivative variable. Note the derivatives of a two functions product.

$$\frac{\partial z}{\partial x} \left(x^3 + y^3 + z^3 + \underbrace{6xyz}_{f(x)f(z)} = 1 \right) \quad \therefore \quad 3x^2 + 3z^2 \frac{\partial z}{\partial x} + 6yz + 6xy \frac{\partial z}{\partial x} = 0$$

$$\therefore \quad \frac{\partial z}{\partial x} \left(3z^2 + 6\,2xy \right) = -3x^2 - 6\,2yz$$

$$\therefore \quad \frac{\partial z}{\partial x} = -\frac{x^2 + 2yz}{z^2 + 2xy}$$

$$\frac{\partial z}{\partial y} \left(x^3 + y^3 + z^3 + \underbrace{6xyz}_{f(y)f(z)} = 1 \right) \quad \therefore \quad 3y^2 + 3z^2 \frac{\partial z}{\partial y} + 6xz + 6xy \frac{\partial z}{\partial y} = 0$$

$$\therefore \quad \frac{\partial z}{\partial y} = -\frac{y^2 + 2xz}{z^2 + 2xy}$$

(d) Second derivatives may be with the same variable as the first derivative or mixed.

$$f'_x \left(x^3 + x^2 y^3 - 2y^2 \right) = 3x^2 + 2xy^3$$

$$f'_y \left(x^3 + x^2 y^3 - 2y^2 \right) = 3x^2 y^2 - 4y$$

$$\therefore$$

$$f''_{xx} = \frac{\partial f'_x}{\partial x} = 6x + 2y^3; \quad f''_{xy} = \frac{\partial f'_x}{\partial y} = 6xy^2$$

$$f''_{yx} = \frac{\partial f'_y}{\partial x} = 6xy^2; \quad f''_{yy} = \frac{\partial f'_y}{\partial y} = 6x^2 y - 4$$

2. Relative to each variable, derivatives of composite and parametric functions follow the same rules and techniques as for single variable functions.

(a)   Given parametric form of function $z = e^x \sin y$, where $x = st^2$ and $x^2 t$, we find its derivatives as follows:

$$\frac{\partial z}{\partial s} = \frac{\partial z}{\partial x}\frac{\partial x}{\partial s} + \frac{\partial z}{\partial y}\frac{\partial y}{\partial s} = e^x(\sin y)\,t^2 + e^x(\cos y)\,2st$$

$$= t^2\,e^{st^2}(\sin s^2 t) + 2st\,e^{st^2}(\cos s^2 t)$$

$$\frac{\partial z}{\partial t} = \frac{\partial z}{\partial x}\frac{\partial x}{\partial t} + \frac{\partial z}{\partial y}\frac{\partial y}{\partial t} = e^x(\sin y)\,2st + e^x(\cos y)\,s^2$$

$$= 2st\,e^{st^2}(\sin s^2 t) + s^2\,e^{st^2}(\cos s^2 t)$$

(b)   Given

$$u = \ln\sqrt{(x-a)^2 + (y-b)^2}$$

we find

$$\frac{\partial u}{\partial x} = \frac{1}{\sqrt{(x-a)^2 + (y-b)^2}}\,\frac{\partial}{\partial x}\sqrt{(x-a)^2 + (y-b)^2}$$

$$= \frac{1}{\sqrt{(x-a)^2 + (y-b)^2}}\,\frac{1}{2\sqrt{(x-a)^2 + (y-b)^2}}$$

$$\times \frac{\partial}{\partial x}\left[(x-a)^2 + (y-b)^2\right]$$

$$= \frac{1}{\sqrt{(x-a)^2 + (y-b)^2}}\,\frac{1}{2\sqrt{(x-a)^2 + (y-b)^2}}\,[2(x-a)]$$

$$= \frac{x-a}{(x-a)^2 + (y-b)^2}$$

$$\therefore$$

$$\frac{\partial^2 u}{\partial x^2} = \frac{-(x-a)^2 + (y-b)^2}{\left[(x-a)^2 + (y-b)^2\right]^2}$$

Similarly,

$$\frac{\partial u}{\partial y} = \frac{y-a}{(x-a)^2 + (y-b)^2} \qquad \therefore \qquad \frac{\partial^2 u}{\partial y^2} = \frac{(x-a)^2 - (y-b)^2}{\left[(x-a)^2 + (y-b)^2\right]^2}$$

Therefore,

$$\frac{\partial^2 u}{\partial x^2} + \frac{\partial^2 u}{\partial x^2} = \frac{-(x-a)^2 + (y-b)^2}{\left[(x-a)^2 + (y-b)^2\right]^2} + \frac{(x-a)^2 - (y-b)^2}{\left[(x-a)^2 + (y-b)^2\right]^2} = 0$$

Exercise 12.4, page 285

1. Fundamentally, integration is an addition, and thus it is commutative operation and the order of integration variables can be changed as necessary.

(a) $\displaystyle \int_0^3 \int_1^2 x^2 y \, dx \, dy = \int_0^3 x^2 \, dx \int_1^2 y \, dy = \int_0^3 x^2 \, dx \, \left. \frac{y^2}{2} \right|_1^2 = \frac{3}{2} \int_0^3 x^2 \, dx$

$$= \frac{3}{2} \left. \frac{x^3}{3} \right|_0^3 = \frac{\cancel{3}}{2} \frac{27}{\cancel{3}} = \frac{27}{2}$$

(b) Double integral delivers surface area bound by the integral limits. Thus, given

$$\iint_R \sqrt{1 - x^2} \, dA \tag{12.3}$$

and interval $|y| \leq 2$, it is necessary to determine the bounds as well as the shape of this surface. Operation of square root imposes limit $1 - x^2 \geq 0$ $\therefore$ $|x| \leq 1$. In addition, $z = \sqrt{1 - x^2}$ defines a positive semicircle whose radius is $r = 1$. Because

$$-2 \leq y \leq 2$$

than (12.3) defines surface of a half-cylinder whose y-direction length is $l = 4$, see Fig. 12.7. The length of semicircle equals $s = r^2 \pi / 2 = \pi / 2$, and therefore rectangular area equals

$$A = s \times l = \frac{\pi}{2} \times 4 = 2\pi$$

$$\therefore$$

$$\iint_R \sqrt{1 - x^2} \, dA = 2\pi$$

Fig. 12.7 Example 12.1-1(b)

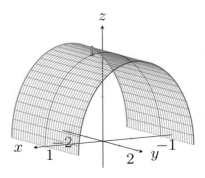

Complex Functions in Engineering and Science

13

Important to Know

Decibel unit: it is used as a *relative* measure G between two quantities A_1 and A_2

$$G_{\mathrm{dB}} \overset{\text{def}}{=} 10 \log \frac{A_2}{A_1}$$

"dBm" unit: it is used as an *absolute* measure G of quantity A_2 normalized to 10^{-3} (i.e. "milli" on the engineering scale)

$$G_{\mathrm{dBm}} \overset{\text{def}}{=} 10 \log \frac{A_2}{10^{-3}}$$

Decibel scale: (note the number of zeros in the ration number x and the value of its respective function $\log x$)

Ratio	Calculation	[dB]	Ratio	Calculation	[dB]
⋮	⋮	⋮	⋮	⋮	⋮
1/1000	$10 \ \log(1/1000) = 10 \times (-3)$	−30	1/8	$10 \ \log(1/8) = 10 \times (-0.9)$	−9
1/100	$10 \ \log(1/100) = 10 \times (-2)$	−20	1/4	$10 \ \log(1/4) = 10 \times (-0.6)$	−6
1/10	$10 \ \log(1/10) = 10 \times (-1)$	−10	1/2	$10 \ \log(1/2) = 10 \times (-0.3)$	−3
1	$10 \ \log(1/1) = 10 \times (0)$	0	1	$10 \ \log(1/1) = 10 \times (0)$	0
10	$10 \ \log(10) = 10 \times (1)$	10	2	$10 \ \log(2) = 10 \times (0.3)$	3
100	$10 \ \log(100) = 10 \times (2)$	20	4	$10 \ \log(4) = 10 \times (0.6)$	6
1000	$10 \ \log(1000) = 10 \times (3)$	30	8	$10 \ \log(8) = 10 \times (0.9)$	9
⋮	⋮	⋮	⋮	⋮	⋮

Basic transfer function forms: in engineering, instead of using letter i as the complex number, letter j is used (the "i" is already used to indicate AC current). In this book, we do not distinguish between the two.

$$H_1(jx) = a_0 \quad (a_0 = \mathrm{const.} \in \mathfrak{R}) \tag{13.1}$$

© Springer Nature Switzerland AG 2021
R. Sobot, *Engineering Mathematics by Example*,
https://doi.org/10.1007/978-3-030-79545-0_13

$$H_2(jx) = j\frac{x}{x_0} \quad (j^2 = -1, \ x_0 = \text{const.} \in \Re) \tag{13.2}$$

$$H_3(jx) = 1 + j\frac{x}{x_0} \tag{13.3}$$

$$H_4(jx) = \frac{1}{1 + j\dfrac{x}{x_0}} \tag{13.4}$$

13.1   Exercises

13.1 **   Basic Forms of Complex Functions $z(x)$

1. For each of $z(x)$ functions, derive $H(x) = 20 \log |z(x)|$ and $P(x) = \angle z(x)$ functions in $\log(x)$ scale. Then, calculate the following limits:

$$\lim_{x \to +\infty} H(x) \quad \lim_{x \to 0} H(x) \quad \lim_{x \to x_0} H(x) \quad \text{and}$$

$$\lim_{x \to +\infty} P(x) \quad \lim_{x \to 0} P(x) \quad \lim_{x \to x_0} P(x)$$

where x_0 is the number in each of the respective x/x_0 fractions. Finally, calculate $H(x = 0.1x_0)$, $H(x = 10x_0)$, $P(x = 0.1x_0)$, and $P(x = 10x_0)$.

(a) $z(x) = j\dfrac{x}{2}$ (b) $z(x) = j\dfrac{x}{10}$ (c) $z(x) = 1 + j\dfrac{x}{2}$

(d) $z(x) = 1 + j\dfrac{x}{10}$ (e) $z(x) = 1 - j\dfrac{x}{2}$ (f) $z(x) = 1 - j\dfrac{x}{10}$

(g) $z(x) = \dfrac{1}{1 + j\dfrac{x}{2}}$ (h) $z(x) = \dfrac{1}{1 + j\dfrac{x}{10}}$ (i) $z(x) = \dfrac{1}{1 - j\dfrac{x}{10}}$

(j) $z(x) = \dfrac{1}{1 - j\dfrac{x}{2}}$

13.2 ***   Piecewise Linear Approximation of $z(x)$

1. For each $z(x)$ function in Example 13.1-1, using $\log(x)$ scale, show the piecewise linear approximation graphs of $H(x) = 20 \log |z(x)|$ and $P(x) = \angle z(x)$.

13.3 ****   Complex Functions: Case Study

1. Show piecewise linear approximation graphs of $H(x)$ and $P(x)$ functions for the following complex function:

$$z(x) = 20{,}000 \, \frac{2 + jx}{220 \, jx + 4000 - x^2} \tag{13.5}$$

Solutions

Exercise 13.1, page 296

1. After using (13.1)–(13.4), we write

   (a)   Given $z(x) = jx/2$, then $x_0 = 2$ and
   $$|z(x)| = \left| j\frac{x}{2} \right| = \frac{x}{2} \quad \therefore$$

   $$H(x) = 20 \log |z(x)| = 20 \log \left(\frac{x}{2} \right) = 20 \log(x) - 20 \log(2)$$

   $$\lim_{x \to 0} H(x) = \lim_{x \to 0} 20 \log(x) - 20 \log(2) = -\infty$$

   $$\lim_{x \to 2} H(x) = \lim_{x \to 2} 20 \log(x) - 20 \log(2) = 0$$

   $$\lim_{x \to +\infty} H(x) = \lim_{x \to +\infty} 20 \log(x) - 20 \log(2) = +\infty$$

   and, phase function $P(x)$ is found by definition as
   $$P(x) = \arctan \left[\frac{\Im(z(x))}{\Re(z(x))} \right] = \arctan \left(\frac{x/2}{0} \right) = \arctan(+\infty) = \frac{\pi}{2} = \text{const.}$$
   therefore, all limits of $P(x) = \pi/2$, i.e. not a function of x.
   We calculate $H(x)$ at points $x = 0.2$, i.e. ten times smaller, and $x = 20$, i.e. ten times greater than $x_0 = 2$ as

   $$H(x = 0.2) = 20 \log \left(\frac{0.2}{2} \right) = 20 \log \left(\frac{1}{10} \right) = -20$$

   $$H(x = 20) = 20 \log \left(\frac{20}{2} \right) = 20 \log (10) = +20$$

   (b)   Given $z(x) = j\frac{x}{10}$, then $x_0 = 10$ and
   $$|z(x)| = \left| j\frac{x}{10} \right| = \frac{x}{10} \quad \therefore$$

   $$H(x) = 20 \log |z(x)| = 20 \log \left(\frac{x}{10} \right) = 20 \log(x) - \underset{20}{\underline{20 \log(10)}}$$

   $$\lim_{x \to +\infty} H(x) = \lim_{x \to +\infty} 20 \log(x) - 20 = +\infty$$

   $$\lim_{x \to 10} H(x) = \lim_{x \to 10} 20 \log(x) - 20 = 0$$

   $$\lim_{x \to 0} H(x) = \lim_{x \to 0} 20 \log(x) - 20 = -\infty$$

   and, phase function $P(x)$ is found by definition as
   $$P(x) = \arctan \left[\frac{\Im(z(x))}{\Re(z(x))} \right] = \arctan \left(\frac{x/10}{0} \right) = \arctan(+\infty) = \frac{\pi}{2} = \text{const.}$$
   therefore, all limits of $P(x) = \pi/2$, i.e. not a function of x.

We calculate $H(x)$ at points $x = 1$, i.e. ten times smaller, and $x = 100$, i.e. ten times greater than $x_0 = 10$ as

$$H(x = 1) = 20 \log \left(\frac{1}{10}\right) = -20$$

$$H(x = 100) = 20 \log \left(\frac{100}{10}\right) = +20$$

(c)　　Given $z(x) = 1 + j\frac{x}{2}$, then $x_0 = 2$ and

$$|z(x)| = \left|1 + j\frac{x}{2}\right| = \sqrt{1^2 + \left(\frac{x}{2}\right)^2} \quad \therefore$$

$$H(x) = 20 \log |z(x)| = 20 \log \sqrt{1 + \left(\frac{x}{2}\right)^2}$$

$$\lim_{x \to 0} H(x) = 20 \log \left[\lim_{x \to 0} \sqrt{1 + \left(\frac{x}{2}\right)^2}\right] = 20 \log(1) = 0$$

$$\lim_{x \to 2} H(x) = 20 \log \left[\lim_{x \to 2} \sqrt{1 + \left(\frac{x}{2}\right)^2}\right] = 20 \log \sqrt{2} = 3$$

$$\lim_{x \gg x_0} H(x) = 20 \log \left[\lim_{x \gg x_0} \sqrt{1 + \left(\frac{x}{2}\right)^2}\right] \approx 20 \log \left[\lim_{x \gg x_0} \sqrt{\left(\frac{x}{2}\right)^2}\right]$$

$$\approx 20 \log \left[\lim_{x \gg x_0} \left(\frac{x}{2}\right)\right] \approx 20 \log |x|$$

and, phase function $P(x)$ is found by definition as

$$P(x) = \arctan \left[\frac{\Im(z(x))}{\Re(z(x))}\right] = \arctan \left(\frac{x/2}{1}\right) = \arctan \left(\frac{x}{2}\right)$$

$$\lim_{x \to 0} P(x) = \lim_{x \to 0} \arctan \left(\frac{x}{2}\right) = \arctan(0) = 0°$$

$$\lim_{x \to 2} P(x) = \lim_{x \to 2} \arctan \left(\frac{x}{2}\right) = \arctan \left(\frac{1}{1}\right) = \frac{\pi}{4} = 45°$$

$$\lim_{x \to +\infty} P(x) = \lim_{x \to +\infty} \arctan \left(\frac{x}{2}\right) = \arctan(\infty) = \frac{\pi}{2} = 90°$$

We calculate $H(x)$ and $P(x)$ at points $x = 0.2$, i.e. ten times smaller, and $x = 20$, i.e. ten times greater than $x_0 = 2$ as

$$H(x = 0.2) = 20 \log \sqrt{1 + \left(\frac{0.2}{2}\right)^2} \approx 20 \log(1) = 0$$

$$H(x = 20) = 20 \log \sqrt{1 + \left(\frac{20}{2}\right)^2} \approx 20 \log(10) = +20$$

$$P(x = 0.2) = \arctan\left(\frac{0.2}{2}\right) = \arctan(0.1) = 5.7° \approx 0$$

$$P(x = 20) = \arctan\left(\frac{20}{2}\right) = \arctan(10) = 84.3° \approx -90° = \frac{\pi}{2}$$

(d)   Given $z(x) = 1 + j\dfrac{x}{10}$, then $x_0 = 10$ and

$$|z(x)| = \left|1 + j\frac{x}{10}\right| = \sqrt{1^2 + \left(\frac{x}{10}\right)^2} \quad \therefore$$

$$H(x) = 20 \log |z(x)| = 20 \log \sqrt{1 + \left(\frac{x}{10}\right)^2}$$

$$\lim_{x \to 0} H(x) = 20 \log \left[\lim_{x \to 0} \sqrt{1 + \left(\frac{x}{10}\right)^2}\right] = 20 \log(1) = 0$$

$$\lim_{x \to 10} H(x) = 20 \log \left[\lim_{x \to 10} \sqrt{1 + \left(\frac{x}{10}\right)^2}\right] = 20 \log \sqrt{2} = 3$$

$$\lim_{x \gg x_0} H(x) = 20 \log \left[\lim_{x \gg x_0} \sqrt{1 + \left(\frac{x}{10}\right)^2}\right] \approx 20 \log \left[\lim_{x \gg x_0} \sqrt{\left(\frac{x}{10}\right)^2}\right]$$

$$\approx 20 \log \left[\lim_{x \gg x_0} \left(\frac{x}{10}\right)\right] \approx 20 \log |x|$$

and, phase function $P(x)$ is found by definition as

$$P(x) = \arctan \left[\frac{\Im(z(x))}{\Re(z(x))}\right] = \arctan\left(\frac{x/10}{1}\right) = \arctan\left(\frac{x}{10}\right)$$

$$\lim_{x \to 0} P(x) = \lim_{x \to 0} \arctan\left(\frac{x}{10}\right) = \arctan(0) = 0°$$

$$\lim_{x \to 10} P(x) = \lim_{x \to 10} \arctan\left(\frac{x}{10}\right) = \arctan\left(\frac{1}{1}\right) = \frac{\pi}{4} = 45°$$

$$\lim_{x \to +\infty} P(x) = \lim_{x \to +\infty} \arctan\left(\frac{x}{10}\right) = \arctan(\infty) = \frac{\pi}{2} = 90°$$

We calculate $H(x)$ and $P(x)$ at points $x = 1$, i.e. ten times smaller, and $x = 100$, i.e. ten times greater than $x_0 = 10$ as

$$H(x = 1) = 20 \log \sqrt{1 + \left(\frac{0.1}{10}\right)^2} \approx 20 \log(1) = 0$$

$$H(x = 100) = 20 \log \sqrt{1 + \left(\frac{100}{10}\right)^2} \approx 20 \log(10) = +20$$

$$P(x = 1) = \arctan\left(\frac{1}{10}\right) = \arctan(0.1) = 5.7° \approx 0$$

$$P(x = 100) = \arctan\left(\frac{100}{10}\right) = \arctan(10) = 84.3° \approx 90° = \frac{\pi}{2}$$

(e)   Given $z(x) = 1 - j\frac{x}{2}$, then $x_0 = 2$ and

$$|z(x)| = \left|1 - j\frac{x}{2}\right| = \sqrt{1^2 + \left(\frac{x}{2}\right)^2} \quad \therefore$$

$$H(x) = 20 \log |z(x)| = 20 \log \sqrt{1 + \left(\frac{x}{2}\right)^2}$$

$$\lim_{x \to 0} H(x) = 20 \log \left[\lim_{x \to 0} \sqrt{1 + \left(\frac{x}{2}\right)^2}\right] = 20 \log(1) = 0$$

$$\lim_{x \to 2} H(x) = 20 \log \left[\lim_{x \to 2} \sqrt{1 + \left(\frac{x}{2}\right)^2}\right] = 20 \log \sqrt{2} = 3$$

$$\lim_{x \gg x_0} H(x) = 20 \log \left[\lim_{x \gg x_0} \sqrt{1 + \left(\frac{x}{2}\right)^2}\right] \approx 20 \log \left[\lim_{x \gg x_0} \sqrt{\left(\frac{x}{2}\right)^2}\right]$$

$$\approx 20 \log \left[\lim_{x \gg x_0} \left(\frac{x}{2}\right)\right] \approx 20 \log |x|$$

and, phase function $P(x)$ is found by definition as

$$P(x) = \arctan\left[\frac{\Im(z(x))}{\Re(z(x))}\right] = \arctan\left(\frac{-x/2}{1}\right) = \arctan\left(\frac{-x}{2}\right)$$

$$\lim_{x \to 0} P(x) = \lim_{x \to 0} \arctan\left(\frac{-x}{2}\right) = \arctan(0) = 0°$$

$$\lim_{x \to 2} P(x) = \lim_{x \to 2} \arctan\left(\frac{-x}{2}\right) = \arctan\left(\frac{-1}{1}\right) = -\frac{\pi}{4} = -45°$$

$$\lim_{x \to +\infty} P(x) = \lim_{x \to +\infty} \arctan\left(\frac{-x}{2}\right) = \arctan(-\infty) = -\frac{\pi}{2} = -90°$$

We calculate $H(x)$ and $P(x)$ at points $x = 0.2$, i.e. ten times smaller, and $x = 20$, i.e. ten times greater than $x_0 = 2$ as

$$H(x = 0.2) = 20 \log \sqrt{1 + \left(\frac{0.2}{2}\right)^2} \approx 20 \log(1) = 0$$

$$H(x = 20) = 20 \log \sqrt{1 + \left(\frac{20}{2}\right)^2} \approx 20 \log(10) = +20$$

$$P(x = 0.2) = \arctan\left(\frac{-0.2}{2}\right) = \arctan(-0.1) = -5.7° \approx 0$$

$$P(x = 20) = \arctan\left(\frac{-20}{2}\right) = \arctan(-10) = -84.3° \approx -90° = -\frac{\pi}{2}$$

(f)   Given $z(x) = 1 - j\dfrac{x}{10}$, then $x_0 = 10$ and

$$|z(x)| = \left|1 - j\frac{x}{10}\right| = \sqrt{1 + \left(\frac{x}{10}\right)^2} \quad \therefore$$

$$H(x) = 20 \log |z(x)| = 20 \log \sqrt{1 + \left(\frac{x}{10}\right)^2}$$

$$\lim_{x \to 0} H(x) = 20 \log \left[\lim_{x \to 0} \sqrt{1 + \left(\frac{x}{10}\right)^2}\right] = 20 \log(1) = 0$$

$$\lim_{x \to 10} H(x) = 20 \log \left[\lim_{x \to 10} \sqrt{1 + \left(\frac{x}{10}\right)^2}\right] = 20 \log \sqrt{2} = 3$$

$$\lim_{x \gg x_0} H(x) = 20 \log \left[\lim_{x \gg x_0} \sqrt{1 + \left(\frac{x}{10}\right)^2}\right] \approx 20 \log \left[\lim_{x \gg x_0} \sqrt{\left(\frac{x}{10}\right)^2}\right]$$

$$\approx 20 \log \left[\lim_{x \gg x_0} \left(\frac{x}{10}\right)\right] \approx 20 \log |x|$$

and, phase function $P(x)$ is found by definition as

$$P(x) = \arctan\left[\frac{\Im(z(x))}{\Re(z(x))}\right] = \arctan\left(\frac{-x/10}{1}\right) = \arctan\left(\frac{-x}{10}\right)$$

$$\lim_{x \to 0} P(x) = \lim_{x \to 0} \arctan\left(\frac{-x}{10}\right) = \arctan(0) = 0°$$

$$\lim_{x \to 10} P(x) = \lim_{x \to 10} \arctan\left(\frac{-x}{10}\right) = \arctan\left(\frac{-1}{1}\right) = -\frac{\pi}{4} = -45°$$

$$\lim_{x \to +\infty} P(x) = \lim_{x \to +\infty} \arctan\left(\frac{-x}{10}\right) = \arctan(-\infty) = -\frac{\pi}{2} = -90°$$

We calculate $H(x)$ and $P(x)$ at points $x = 0.2$, i.e. ten times smaller, and $x = 20$, i.e. ten times greater than $x_0 = 2$ as

$$H(x = 1) = 20 \log \sqrt{1 + \left(\frac{1}{10}\right)^2} \approx 20 \log(1) = 0$$

$$H(x = 100) = 20 \log \sqrt{1 + \left(\frac{100}{10}\right)^2} \approx 20 \log(10) = +20$$

$$P(x = 1) = \arctan\left(\frac{-1}{10}\right) = \arctan(-0.1) = -5.7° \approx 0$$

$$P(x = 100) = \arctan\left(\frac{-100}{10}\right) = \arctan(-10) = -84.3° \approx -90° = -\frac{\pi}{2}$$

(g)   Given $z(x) = \dfrac{1}{1 + j\frac{x}{2}}$, then $x_0 = 2$ and

$$|z(x)| = \left| \frac{1}{1 + j\frac{x}{2}} \right| = \frac{1}{\sqrt{1 + \left(\frac{x}{2}\right)^2}} \quad \therefore$$

$$H(x) = 20 \log |z(x)| = \underset{0}{\underbrace{20 \log(1)}} - 20 \log \sqrt{1 + \left(\frac{x}{2}\right)^2}$$

$$\lim_{x \to 0} H(x) = -20 \log \left[\lim_{x \to 0} \sqrt{1 + \left(\frac{x}{2}\right)^2} \right] = -20 \log(1) = 0$$

$$\lim_{x \to 2} H(x) = -20 \log \left[\lim_{x \to 2} \sqrt{1 + \left(\frac{x}{2}\right)^2} \right] = -20 \log \sqrt{2} = -3$$

$$\lim_{x \to +\infty} H(x) = -20 \log \left[\lim_{x \to +\infty} \sqrt{1 + \left(\frac{x}{2}\right)^2} \right] \approx -20 \log |x|$$

and, phase function $P(x)$ is found by definition, after explicitly deriving the real and imaginary parts of $z(x)$, as

$$z(x) = \frac{1}{1 + jx/2} \frac{1 - jx/2}{1 - jx/2} = \frac{1}{\sqrt{1 + (x/2)^2}} - j \frac{x/2}{\sqrt{1 + (x/2)^2}} \quad \therefore$$

$$\Re(z(x)) = \frac{1}{\sqrt{1 + (x/2)^2}} \quad \text{and} \quad \Im(z(x)) = -\frac{x/2}{\sqrt{1 + (x/2)^2}}$$

which again is reduced to

$$P(x) = \arctan\left[\frac{\Im(z(x))}{\Re(z(x))}\right] = \arctan\left(\frac{-x/2}{1}\right) = \arctan\left(\frac{-x}{2}\right)$$

$$\lim_{x\to 0} P(x) = \lim_{x\to 0}\arctan\left(\frac{-x}{2}\right) = \arctan(0) = 0°$$

$$\lim_{x\to 2} P(x) = \lim_{x\to 2}\arctan\left(\frac{-x}{2}\right) = \arctan(-1) = -\frac{\pi}{4} = -45°$$

$$\lim_{x\to +\infty} P(x) = \lim_{x\to +\infty}\arctan\left(\frac{-x}{2}\right) = \arctan(-\infty) = -\frac{\pi}{2} = -90°$$

We calculate $H(x)$ and $P(x)$ at points $x = 0.2$, i.e. ten times smaller, and $x = 20$, i.e. ten times greater than $x_0 = 2$ as

$$H(x = 0.2) = -20\log\sqrt{1 + \left(\frac{0.2}{2}\right)^2} \approx -20\log(1) = 0$$

$$H(x = 20) = -20\log\sqrt{1 + \left(\frac{20}{2}\right)^2} \approx -20\log(10) = -20$$

$$P(x = 0.2) = \arctan\left(\frac{-0.2}{2}\right) = \arctan(-0.1) = -5.7° \approx 0$$

$$P(x = 20) = \arctan\left(\frac{-20}{2}\right) = \arctan(-10) = -84.3° \approx -90° = -\frac{\pi}{2}$$

(h) Given $z(x) = \dfrac{1}{1 + j\frac{x}{10}}$, then $x_0 = 10$ and

$$|z(x)| = \left|\frac{1}{1 + j\dfrac{x}{10}}\right| = \frac{1}{\sqrt{1^2 + \left(\dfrac{x}{10}\right)^2}} \quad \therefore$$

$$H(x) = 20\log|z(x)| = \underbrace{20\log(1)}_{0} - 20\log\sqrt{1 + \left(\frac{x}{10}\right)^2}$$

$$\lim_{x\to 0} H(x) = -20\log\left[\lim_{x\to 0}\sqrt{1 + \left(\frac{x}{10}\right)^2}\right] = -20\log(1) = 0$$

$$\lim_{x\to 10} H(x) = -20\log\left[\lim_{x\to 10}\sqrt{1 + \left(\frac{x}{10}\right)^2}\right] = -20\log\sqrt{2} = -3$$

$$\lim_{x\gg x_0} H(x) = -20\log\left[\lim_{x\gg x_0}\sqrt{1 + \left(\frac{x}{10}\right)^2}\right] \approx -20\log|x|$$

and, phase function $P(x)$ is found by definition, after explicitly deriving the real and imaginary parts of $z(x)$, as

$$z(x) = \frac{1}{1 + j^x/10} \frac{1 - j^x/10}{1 - j^x/10} = \frac{1}{\sqrt{1 + (^x/10)^2}} - j \frac{^x/10}{\sqrt{1 + (^x/10)^2}} \quad \therefore$$

$$\Re(z(x)) = \frac{1}{\sqrt{1 + (^x/10)^2}} \quad \text{and} \quad \Im(z(x)) = -\frac{^x/10}{\sqrt{1 + (^x/10)^2}}$$

which again is reduced to

$$P(x) = \arctan \left[\frac{\Im(z(x))}{\Re(z(x))} \right] = \arctan \left(\frac{-^x/10}{1} \right) = \arctan \left(\frac{-x}{10} \right) \quad \text{therefore}$$

$$\lim_{x \to 0} P(x) = \lim_{x \to 0} \arctan \left(\frac{-x}{10} \right) = \arctan(0) = 0°$$

$$\lim_{x \to 10} P(x) = \lim_{x \to 10} \arctan \left(\frac{-x}{10} \right) = \arctan(-1) = -\frac{\pi}{4} = -45°$$

$$\lim_{x \to +\infty} P(x) = \lim_{x \to +\infty} \arctan \left(\frac{-x}{10} \right) = \arctan(-\infty) = -\frac{\pi}{2} = -90°$$

We calculate $H(x)$ and $P(x)$ at points $x = 1$, i.e. ten times smaller, and $x = 100$, i.e. ten times greater than $x_0 \leftharpoondown 10$ as

$$H(x = 1) = -20 \log \sqrt{1 + \left(\frac{1}{10} \right)^2} \approx -20 \log(1) = 0$$

$$H(x = 100) = -20 \log \sqrt{1 + \left(\frac{100}{10} \right)^2} \approx -20 \log(10) = -20$$

$$P(x = 1) = \arctan \left(\frac{-1}{10} \right) = \arctan(-0.1) = -5.7° \approx 0$$

$$P(x = 100) = \arctan \left(\frac{-100}{10} \right) = \arctan(-10) = -84.3° \approx -90° = -\frac{\pi}{2}$$

(i) Given $z(x) = \dfrac{1}{1 - j\frac{x}{10}}$, then $x_0 = 10$ and

$$|z(x)| = \left| \frac{1}{1 - j\dfrac{x}{10}} \right| = \frac{1}{\sqrt{1^2 + \left(\dfrac{x}{10}\right)^2}} \quad \therefore$$

$$H(x) = \underbrace{20\log(1)}_{0} - 20\log\sqrt{1 + \left(\frac{x}{10}\right)^2}$$

$$\lim_{x \to 0} H(x) = -20\log\left[\lim_{x \to 0} \sqrt{1 + \left(\frac{x}{10}\right)^2} \right] = -20\log(1) = 0$$

$$\lim_{x \to 10} H(x) = -20\log\left[\lim_{x \to 10} \sqrt{1 + \left(\frac{x}{10}\right)^2} \right] = -20\log\sqrt{2} = -3$$

$$\lim_{x \gg x_0} H(x) = -20\log\left[\lim_{x \gg x_0} \sqrt{1 + \left(\frac{x}{10}\right)^2} \right] \approx -20\log|x|$$

and, phase function $P(x)$ is found by definition, after explicitly deriving the real and imaginary parts of $z(x)$, as

$$z(x) = \frac{1}{1 - j^{x}/10} \frac{1 + j^{x}/10}{1 + j^{x}/10} = \frac{1}{\sqrt{1 + (^{x}/10)^2}} + j\frac{^{x}/10}{\sqrt{1 + (^{x}/10)^2}} \quad \therefore$$

$$\Re(z(x)) = \frac{1}{\sqrt{1 + (^{x}/10)^2}} \quad \text{and} \quad \Im(z(x)) = \frac{^{x}/10}{\sqrt{1 + (^{x}/10)^2}}$$

which again is reduced to

$$P(x) = \arctan\left(\frac{^{x}/10}{1} \right) = \arctan\left(\frac{x}{10} \right) \quad \text{therefore}$$

$$\lim_{x \to 0} P(x) = \lim_{x \to 0} \arctan\left(\frac{x}{10} \right) = \arctan(0) = 0°$$

$$\lim_{x \to 10} P(x) = \lim_{x \to 10} \arctan\left(\frac{x}{10} \right) = \arctan(1) = \frac{\pi}{4} = 45°$$

$$\lim_{x \to +\infty} P(x) = \lim_{x \to +\infty} \arctan\left(\frac{x}{10} \right) = \arctan(+\infty) = \frac{\pi}{2} = 90°$$

We calculate $H(x)$ and $P(x)$ at points $x = 1$, i.e. ten times smaller, and $x = 100$, i.e. ten times greater than $x_0 = 10$ as

$$H(x = 1) = -20\log\sqrt{1 + \left(\frac{1}{10}\right)^2} \approx -20\log(1) = 0$$

$$H(x = 100) = -20\log\sqrt{1 + \left(\frac{100}{10}\right)^2} \approx -20\log(10) = -20$$

$$P(x = 1) = \arctan\left(\frac{1}{10} \right) = \arctan(0.1) = 5.7° \approx 0$$

$$P(x = 100) = \arctan\left(\frac{100}{10} \right) = \arctan(10) = 84.3° \approx 90° = \frac{\pi}{2}$$

(j)   Given $z(x) = \dfrac{1}{1 - j\frac{x}{2}}$, then $x_0 = 2$ and

$$|z(x)| = \left| \frac{1}{1 - j\frac{x}{2}} \right| = \frac{1}{\sqrt{1^2 + \left(\frac{x}{2}\right)^2}} \quad \therefore$$

$$H(x) = \underbrace{20\log(1)}_{0} - 20\log\sqrt{1 + \left(\frac{x}{2}\right)^2}$$

$$\lim_{x \to 0} H(x) = -20\log\left[\lim_{x \to 0}\sqrt{1 + \left(\frac{x}{2}\right)^2}\right] = -20\log(1) = 0$$

$$\lim_{x \to 2} H(x) = -20\log\left[\lim_{x \to 2}\sqrt{1 + \left(\frac{x}{2}\right)^2}\right] = -20\log\sqrt{2} = -3$$

$$\lim_{x \gg x_0} H(x) = -20\log\left[\lim_{x \gg x_0}\sqrt{1 + \left(\frac{x}{2}\right)^2}\right] \approx -20\log|x|$$

and, phase function $P(x)$ is found by definition, after explicitly deriving the real and imaginary parts of $z(x)$, as

$$z(x) = \frac{1}{1 - j^x/2}\frac{1 + j^x/2}{1 + j^x/2} = \frac{1}{\sqrt{1 + (x/2)^2}} + j\frac{x/2}{\sqrt{1 + (x/2)^2}} \quad \therefore$$

$$\Re(z(x)) = \frac{1}{\sqrt{1 + (x/2)^2}} \quad \text{and} \quad \Im(z(x)) = \frac{x/2}{\sqrt{1 + (x/2)^2}}$$

which again is reduced to

$$P(x) = \arctan\left(\frac{x/2}{1}\right) = \arctan\left(\frac{x}{2}\right) \quad \text{therefore}$$

$$\lim_{x \to 0} P(x) = \lim_{x \to 0}\arctan\left(\frac{x}{2}\right) = \arctan(0) = 0°$$

$$\lim_{x \to 2} P(x) = \lim_{x \to 2}\arctan\left(\frac{x}{2}\right) = \arctan(1) = \frac{\pi}{4} = 45°$$

$$\lim_{x \to +\infty} P(x) = \lim_{x \to +\infty}\arctan\left(\frac{x}{2}\right) = \arctan(+\infty) = \frac{\pi}{2} = 90°$$

We calculate $H(x)$ and $P(x)$ at points $x = 0.2$, i.e. ten times smaller, and $x = 20$, i.e. ten times greater than $x_0 = 2$ as

$$H(x = 0.2) = -20\log\sqrt{1 + \left(\frac{0.2}{2}\right)^2} \approx -20\log(1) = 0$$

$$H(x = 20) = -20\log\sqrt{1 + \left(\frac{20}{2}\right)^2} \approx -20\log(10) = -20$$

$$P(x = 0.2) = \arctan\left(\frac{0.2}{2}\right) = \arctan(0.1) = 5.7° \approx 0$$

$$P(x = 20) = \arctan\left(\frac{20}{2}\right) = \arctan(10) = 84.3° \approx 90° = \frac{\pi}{2}$$

Exercise 13.2, page 296

1. The following graphs are commonly known as the Bode plots:

(a) For $z(x) = j^x/2$, Example 13.1-1(a), we found that when $x = 2$, the value of $H(x = 2) = 0$, and at the same time, angle is constant $\angle P(x) = +\pi/2$. In addition, every tenfold increase x results in $+20$ increase of $H(x)$ (see Fig. 13.1).

(b) For $z(x) = j^x/10$, Example 13.1-1(b), we found that when $x = 10$, the value of $H(x = 10) = 0$, and at the same time, angle is constant $\angle P(x) = +\pi/2$. In addition, every tenfold increase x results in $+20$ increase of $H(x)$ (see Fig. 13.2).

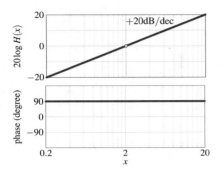

Fig. 13.1 Example 13.2-1(a)

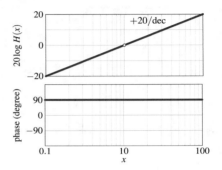

Fig. 13.2 Example 13.2-1(b)

(c) For $z(x) = 1 + j^x/2$, Example 13.1-1(c), we found that when $x = 2$, the value of $H(x = 2) = 3$, and at the same time, angle is $\angle P(x = 2) = +\pi/4$. In addition, angle limits are $0°$ and $+90°$, while every tenfold increase x results in $+20$ increase of $H(x)$. After accounting for all limits of $\lim H(x)$ (see Fig. 13.3).

(d) For $z(x) = 1 + j^x/10$, Example 13.1-1(d), we found that when $x = 10$, the value of $H(x = 10) = 3$, and at the same time, angle is $\angle P(x = 10) = +\pi/4$. In addition, angle limits are $0°$ and $+90°$, while every tenfold increase x results in $+20$ increase of $H(x)$. After accounting for all limits of $\lim H(x)$ (see Fig. 13.4).

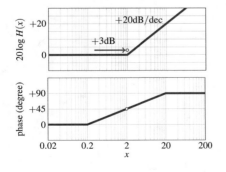

Fig. 13.3 Example 13.2-1(c)

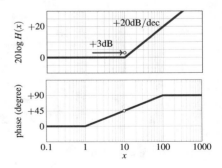

Fig. 13.4 Example 13.2-1(d)

(e) For $z(x) = 1 - j^x/2$, Example 13.1-1(e), we found that when $x = 2$, the value of $H(x = 2) = 3$, and at the same time, angle is $\angle P(x = 2) = -\pi/4$. In addition, angle limits are $0°$ and $-90°$, while every tenfold increase x results in $+20$ increase of $H(x)$. After accounting for all limits of $\lim H(x)$ (see Fig. 13.5).

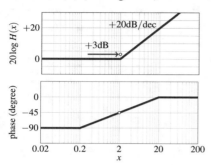

Fig. 13.5 Example 13.2-1(e)

(f) For $z(x) = 1 - j^x/10$, Example 13.1-1(f), we found that when $x = 10$, the value of $H(x = 10) = 3$, and at the same time, angle is $\angle P(x = 10) = -\pi/4$. In addition, angle limits are $0°$ and $-90°$, while every tenfold increase x results in $+20$ increase of $H(x)$. After accounting for all limits of $\lim H(x)$ (see Fig. 13.6).

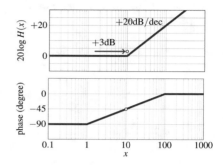

Fig. 13.6 Example 13.2-1(f)

(g) For $z(x) = \dfrac{1}{1 + j\frac{x}{2}}$, Example 13.1-1(g), we found that when $x = 2$, the value of $H(x = 2) = -3$, and at the same time, angle is $\angle P(x = 2) = -\pi/4$. In addition, angle limits are $0°$ and $-90°$, while every tenfold increase x results in -20 increase of $H(x)$. After accounting for all limits of $\lim H(x)$ (see Fig. 13.7).

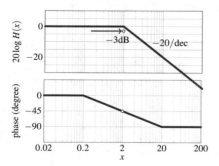

Fig. 13.7 Example 13.2-1(g)

(h) For $z(x) = \dfrac{1}{1 + j\frac{x}{10}}$, Example 13.1-1(h), we found that when $x = 10$, the value of $H(x = 10) = -3$, and at the same time, angle is $\angle P(x = 10) = -\pi/4$. In addition, angle limits are $0°$ and $-90°$, while every tenfold increase x results in -20 increase of $H(x)$. After accounting for all limits of $\lim H(x)$ (see Fig. 13.8).

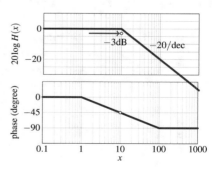

Fig. 13.8 Example 13.2-1(h)

(i)   For $z(x) = \dfrac{1}{1 - j\frac{x}{10}}$, Example 13.1-1(i), we found that when $x = 10$, the value of $H(x = 10) = -3$, and at the same time, angle is $\angle P(x = 10) = \pi/4$. In addition, angle limits are $0°$ and $+90°$, while every tenfold increase x results in -20 increase of $H(x)$. After accounting for all limits of $\lim H(x)$ (see Fig. 13.9).

(j)   For $z(x) = \dfrac{1}{1 - j\frac{x}{2}}$, Example 13.1-1(j), we found that when $x = 2$, the value of $H(x = 2) = -3$, and at the same time, angle is $\angle P(x = 2) = \pi/4$. In addition, angle limits are $0°$ and $+90°$, while every tenfold increase x results in -20 increase of $H(x)$. After accounting for all limits of $\lim H(x)$ (see Fig. 13.10).

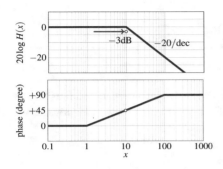

Fig. 13.9 Example 13.2-1(i)

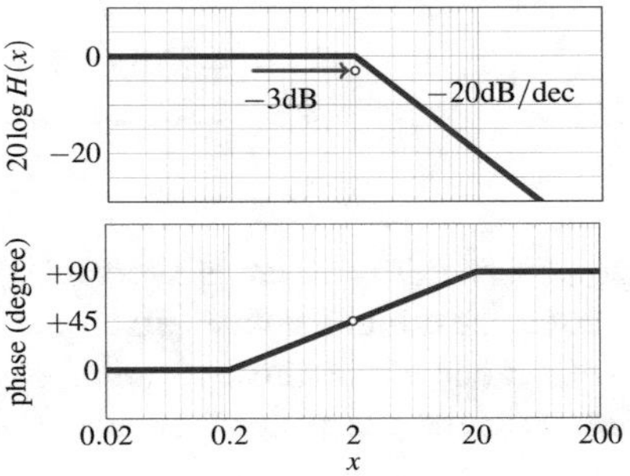

Fig. 13.10 Example 13.2-1(j)

It is important to note that, as long as the form of analytical function $z(x)$ is same where the only change is the value of x_0, then all respective graphs have the same shapes. That is to say, the graphs are only shifted to $x = x_0$. This property greatly simplifies the analysis of more complicated $z(x)$ functions. General idea is to factorize complicated $z(x)$ functions into these basic simpler "building blocks" and then "assemble" the final response by adding these basic functions.

Exercise 13.3, page 296

1. This is the second order function; thus, in order to find out if roots of its denominator are reel or complex, it is necessary to factorize its denominator.

$$H(j\omega) = 2\,000\,\frac{2 + jx}{220jx + 4\,000 - x^2} = 2\,000\,\frac{2 + jx}{j^2x^2 + 220jx + 4\,000}$$

$$= 2\,000\,\frac{2 + jx}{j^2x^2 + 20jx + 200jx + 4\,000} = 2\,000\,\frac{2 + jx}{jx\,(20 + jx) + 200\,(20 + jx)}$$

$$= 2\,000\,\frac{2 + jx}{(20 + jx)(200 + jx)}$$

$$= \cancel{2\,000}\,\frac{\cancel{2}\left(1 + j\dfrac{x}{2}\right)}{\cancel{20}\left(1 + j\dfrac{x}{20}\right)\cancel{200}\left(1 + j\dfrac{x}{200}\right)}$$

$$= \frac{1 + j\,\dfrac{x}{2}}{\left(1 + j\,\dfrac{x}{20}\right)\left(1 + j\,\dfrac{x}{200}\right)}$$

This factorized form is suitable for conversion into the sum of the basic form simply by rewriting it in its logarithmic form,

$$20\log z(x) = 20\log\left[\frac{1 + j\,\dfrac{x}{2}}{\left(1 + j\,\dfrac{x}{20}\right)\left(1 + j\,\dfrac{x}{200}\right)}\right]$$

$$= \underbrace{+20\log\left(1 + j\,\frac{x}{2}\right)}_{\textcircled{1}\ \ x_0=2} \underbrace{-20\log\left(1 + j\,\frac{x}{20}\right)}_{\textcircled{2}\ \ x_0=20} \underbrace{-20\log\left(1 + j\,\frac{x}{200}\right)}_{\textcircled{3}\ \ x_0=200}$$

Obviously, this example is a second order function whose poles are real, and thus it is possible to decompose it into the basic first order functions. Each summation term in the logarithmic form is in effect one of the already studied basic forms, as annotated.

Similarly, the phase plot is created as the sum of the corresponding linear terms,

$$\angle z(x) = \underbrace{+ \arctan\frac{x}{2}}_{\textcircled{1}\ \ x_0=2} \underbrace{- \arctan\frac{x}{20}}_{\textcircled{2}\ \ x_0=20} \underbrace{- \arctan\frac{x}{200}}_{\textcircled{3}\ \ x_0=200}$$

Once the gain and phase logarithmic forms are factorized, first we plot each of the summing terms 1, 2 and 3 separately, see Fig. 13.11(left). Then, the summing operation is done by simply adding the linear sections of all three terms to produce the Body plots for gain and phase, see Fig. 13.11 (right).

Fig. 13.11 Example 13.3-1

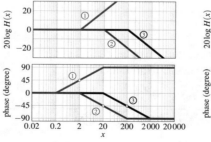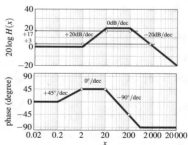

Differential Equations

14

Important to Know

First order differential equations take the form of

$$f(x, y, y') = 0 \text{ where } y' \equiv \frac{dy}{dx}$$

and the common types are as follows:

1. equations with separable variable take the form of

$$y' = \frac{f(x)}{g(x)}$$

2. homogeneous equations take the form of

$$y' = f\left(\frac{y}{x}\right)$$

3. linear equations take the form of

$$y' + f(x)\, y = g(x)$$

Orthogonality between two functions, for example, at a given point (x, y) relation $f(x) \perp g(x)$ is satisfied for

$$F(x, y, y'_f) \text{ and } G\left(x, y, -\frac{1}{y'_f}\right)$$

where y'_f is derivative of $f(x)$ and $(-1/y'_f)$ is derivative of $g(x)$, that is to say, tangent, at a given point (x, y).

© Springer Nature Switzerland AG 2021
R. Sobot, *Engineering Mathematics by Example*,
https://doi.org/10.1007/978-3-030-79545-0_14

Homogeneous linear differential equations with the constant coefficients take the form of

$$a_1 y' + a_0 y = 0 \quad \text{(1st order)}$$
$$a_2 y'' + a_1 y' + a_0 y = 0 \quad \text{(2nd order)}$$
$$a_3 y''' + a_2 y'' + a_1 y' + a_0 y = 0 \quad \text{(2nd order)}$$
$$\cdots$$

Characteristic polynomial $P_n(r)$ is written as

$$a_1 r + a_0 = 0 \quad \text{(1st order)}$$
$$a_2 r^2 + a_1 r + a_0 = 0 \quad \text{(2nd order)}$$
$$a_3 r^3 + a_2 r^2 + a_1 r + a_0 = 0 \quad \text{(2nd order)}$$
$$\cdots$$

Solutions of homogeneous linear differential equations with the constant coefficients are written after calculating the roots of $P_n(r) = 0$, where n indicates nth order polynomial, and thus there are n roots $r_1, r_2, \cdots, r_n$.

$$1: \ (r_1 \neq r_2 \cdots \neq r_n) \ \therefore \ y(x) = C_1 \exp(r_1 x) + C_2 \exp(r_2 x) + \cdots + C_n \exp(r_n x)$$
$$2: \ (r = r_1 = r_2 \neq r_{i>2}) \ \therefore \ y(x) = C_1 \exp(r x) + C_2 x \exp(r x) + \cdots + C_n \exp(r_n x)$$
$$= \exp(r x) [C_1 + C_2 x] + \cdots + C_n \exp(r_n x)$$

where, in order to derive more compact $y(x)$ function, the complex conjugate roots may be further transformed by algebra techniques.

14.1   Exercises

14.1 *   Separable Variables

1. Solve equations:

   (a) $y' + y = a, \ (a \in \mathbb{R})$ (b) $y - 2xy' = 1$

2. Solve equations:

   (a) $\sqrt{x}\, dy = \sqrt{y}\, dx$ (b) $x(1 + y^2) = yy'$

14.2 **   First Order Homogenous Equations

1. Solve equations:

   (a) $y' = \dfrac{y^2 + x^2}{xy}$ (b) $y' = \dfrac{y^2 - x^2}{2xy}$

2. Solve equations:

   (a) $(x + y)\,dx - x\,dy = 0$ (b) $xy^2\,dy = (x^3 + y^3)\,dx$

14.3 *   Linear Equations

1. Solve equations:

   (a) $y' - 4xy = x$ (b) $y' - 2y - 3 = 0$

   (c) $y' + 2xy = 2xe^{-x^2}$ (d) $y' - \dfrac{2y}{x+1} - (x+1)^3 = 0$

14.4 *   Linear Equations with Constant Coefficients

1. Solve equations:

   (a) $y'' - 9y = 0$ (b) $y'' - 2y' + y = 0$

   (c) $y'' + 4y' + 4y = 0$ (d) $y'' + 6y' + 25y = 0$

14.5 ***   Engineering Examples

1. In a series RL circuit, see Fig. 14.1, at moment $t = 0$, current $i(0) = 0$. Derive expression for $i(t)$ $(t \geq 0)$, i.e. when the switch is closed, and sketch its time domain graph.

Fig. 14.1 Example 14.5-1

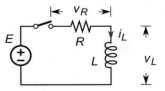

2. At moment $t = 0$, a free falling ball whose mass is m and the initial velocity (a vector quantity) $\vec{v}(0) = 0$ is dropped, Fig. 14.2. Assuming the air resistance to be proportional to velocity as $k\,\vec{v}$, derive expression for the ball's speed (a scalar quantity) $v(t)$ as a function of time.

Fig. 14.2 Example 14.5-2

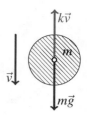

3. Given an RLC circuit, Fig. 14.3, where the switch is closed at moment $t = 0$. Assuming the initial capacitor charge is $q(0) = q_0$, derive the expressions for $i(t)$ and $v(t)$.

Fig. 14.3 Example 14.5-3

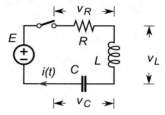

Solutions

Exercise 14.1, page 312

1. Separation of variables technique is used to solve differential equations whose form can be written as

$$y' = \frac{dy}{dx} = \frac{f(x)}{g(y)} \quad \therefore \quad g(y)\,dy = f(x)\,dx$$

Therefore, the solution $y(x)$ is found by integration of both sides of the equation.

(a) Separation of variables technique
$y' - y = a, \quad (a \in \mathbb{R})$

$$\frac{dy}{dx} = a - y \quad \therefore \quad \frac{dy}{a - y} = dx \quad \therefore \quad \frac{dy}{y - a} = -\,dx \quad \therefore \quad \int \frac{dy}{y - a} = -\int dx$$

$$= \left\{ y - a = t \quad \therefore \quad dy = dt \right\}$$

$$\therefore \quad \int \frac{dt}{t} = -x + C_1 \quad \therefore \quad \ln|t| = -x + C_1 \quad \therefore \quad \ln|y - a| = -x + C_1$$

$$\therefore \quad |y - a| = e^{-x + C_1} = e^{-x}\,e^{C_1} = \left\{ C = \pm e^{C_1} \right\}$$

$$\underline{y(x) = a + Ce^{-x}}$$

(b) Separation of variables technique

$y - 2xy' = 1$

$$y = 2xy' + 1 \quad \therefore \quad y - 1 = 2xy' = 2x\frac{dy}{dx} \quad \therefore \quad \frac{dy}{y-1} = \frac{1}{2}\frac{dx}{x}$$

$\therefore$

$$\int \frac{dy}{y-1} = \frac{1}{2}\int \frac{dx}{x} = \left\{y - 1 = t \quad \therefore \quad dy = dt\right\} \quad \therefore \quad \ln|y-1| = \frac{1}{2}\ln|x| + \ln C_1$$

$$\ln|y-1| = \frac{1}{2}\ln(C_1|x|) = \ln(C_1|x|)^{1/2} = \left\{\ln x, \quad (x>0) \quad \therefore \quad \ln|x| = \ln x\right\}$$

$\therefore$

$$y - 1 = \sqrt{C_1|x|} = \sqrt{C_1}\sqrt{|x|} = C\sqrt{|x|}$$

$$\underline{y(x) = 1 + C\sqrt{|x|}}$$

2. Separation of variables technique

(a) $\sqrt{x}\,dy = \sqrt{y}\,dx \quad \therefore \quad \int \frac{dy}{\sqrt{y}} = \int \frac{dx}{\sqrt{x}} \quad \therefore \quad \sqrt{y} - \sqrt{x} = C$

(b) Constant C is arbitrary, so it may as well be $\ln(C)$ or e^C if it helps to write a more compact solution.

$$x(1+y^2) = yy' \quad \therefore \quad x = \frac{y}{1+y^2}\frac{dy}{dx} \quad \therefore \quad \int x\,dx = \int \frac{y}{1+y^2}\,dy$$

$$\left\{1 + y^2 = t \quad \therefore \quad 2y\,dy = dt\right\}$$

$$\therefore \quad \int x\,dx = \frac{1}{2}\int \frac{dt}{t}$$

$$\frac{x^2}{2} = \frac{1}{2}\left(\ln(1+y^2) + \ln C\right) = \ln[C(1+y^2)]$$

or,

$$y = \sqrt{C_1 e^{x^2} - 1}$$

where $C_1 = 1/C$.

Exercise 14.2, page 313

1. Differential equations that can be written in form

$$y' = f\left(\frac{y}{x}\right)$$

are known as "homogeneous" and are solved by the following change of variables technique:

$$\frac{y}{t} = t \quad \therefore \quad \underline{y = tx} = \left\{[f(x)g(x)]' = f'g + fg'\right\}$$

$$\therefore$$

$$dy = t\,dx + x\,dt \quad \Rightarrow \quad \frac{dy}{dx} = t + x\,\frac{dt}{dx}$$

followed by the method of variable separation.

(a) Reorganize the equation into the homogeneous form, then change the variables and integrate.

$$\left.\begin{array}{l} y' = \dfrac{x^2 + y^2}{xy} \quad \therefore \quad \dfrac{dy}{dx} = \dfrac{y}{x} + \dfrac{x}{y} = t + \dfrac{1}{t} \\[2.5ex] \dfrac{dy}{dx} = t + x\,\dfrac{dt}{dx} \end{array}\right\} \quad \cancel{t} + \dfrac{1}{t} = \cancel{t} + x\,\dfrac{dt}{dx}$$

$$\frac{1}{t} = x\,\frac{dt}{dx} \quad \therefore \quad \int \frac{dx}{x} = \int t\,dt \quad \therefore \quad \ln|x| + \ln C_1 = \frac{t^2}{2}; \quad \{C = \ln C_1\}$$

$$t = \sqrt{2\,\ln(C|x|)} \quad \therefore \quad \frac{y}{x} = \sqrt{2\,\ln(C|x|)}$$

$$\therefore \quad \underline{y = x\,\sqrt{2\,\ln(C|x|)}}$$

(b) Homogeneous differential equation with parametric form solution.

$$\left.\begin{array}{l} y' = \dfrac{y^2 - x^2}{2xy} \quad \therefore \quad \dfrac{dy}{dx} = \dfrac{y}{2x} - \dfrac{x}{2y} = \dfrac{1}{2}\left(t - \dfrac{1}{t}\right) \\[2.5ex] \dfrac{dy}{dx} = t + x\,\dfrac{dt}{dx} \end{array}\right\}$$

$$t + x\,\frac{dt}{dx} = \frac{1}{2}\left(t - \frac{1}{t}\right) = \frac{t^2 - 1}{2t}$$

$$x\,\frac{dt}{dx} = \frac{t^2 - 1}{2t} - t = \frac{t^2 - 1 - 2t^2}{2t} = -\frac{t^2 + 1}{2t}$$

$$\therefore \quad \int \frac{dx}{x} = -\int \frac{2t}{t^2 + 1}\,dt$$

$$\therefore \quad \ln|x| = -\int \frac{2t\,dt}{t^2 + 1}$$

$$= \{t^2 + 1 = r \quad \therefore \quad 2t\,dt = dr, \quad (t^2 + 1 \neq 0)\}$$

$$= -\int \frac{dr}{r} = -\ln|r| = -\ln[t^2 + 1] + C_1 \quad (C = \ln C_1)$$

$$\ln|x| = \ln\frac{t^2 + 1}{C} \quad \therefore \quad \underline{x = \frac{C}{t^2 + 1}}, \quad \text{and}$$

$$y = t \quad \therefore \quad \underline{y = \frac{Ct}{t^2 + 1}}$$

At this stage, the solution is parametric, i.e. $y = y(t)$ and $x = x(t)$. The parametric variable t may be eliminated as follows:

$$\left.\begin{array}{l} x^2 = \dfrac{C^2}{(t^2+1)^2} \quad \therefore \quad (t^2+1)^2 = \dfrac{C^2}{x^2} \quad \therefore \quad t^2 = \dfrac{C}{x} - 1 \\[4mm] y^2 = \dfrac{C^2\,t^2}{(t^2+1)^2} \end{array}\right\} \quad \therefore$$

$$y^2 = \frac{\cancel{C^2}\,(C/x-1)}{\cancel{C^2}/x^2} = \frac{(C-x)/\cancel{x}}{1/x^{\cancel{2}}} = x(C-x) = Cx - x^2$$

$$\therefore$$

$$\underline{y^2 + x^2 - Cx = 0 \quad \text{or,} \quad y = \sqrt{Cx - x^2}}$$

2. Homogenous differential equations are solved by using $y/x = t$ change of variables.

   (a) After the initial transformations, we can change the variables.

   $$(x+y)\,dx - x\,dy = 0$$

   $$x\,dx + y\,dx = x\,dy \quad \Big| \quad /x\,dx$$

   $$1 + \frac{y}{x} = \frac{dy}{dx}$$

   $$\left\{ \frac{y}{x} = t \quad \therefore \quad y = xt \quad \therefore \quad \frac{dy}{dx} = y' = t + x\,t' \right\}$$

   $$\therefore$$

   $$1 + \cancel{t} = \cancel{t} + x\,t' \quad \therefore \quad \int dt = \int \frac{dx}{x} \quad \therefore \quad t = \ln C\,x$$

   $$\therefore$$

   $$y = x \ln C\,x$$

(b)   After the initial transformations, we can change the variables.

$$xy^2 \, dy = (x^3 + y^3) \, dx \quad \therefore \quad \frac{dy}{dx} = \frac{x^3 + y^3}{xy^2} = \left(\frac{x}{y}\right)^2 + \frac{y}{x}$$

$$\{y = t\, x\}$$

$$\therefore$$

$$\cancel{t} + x\, t' = \frac{1}{t^2} + \cancel{t}$$

$$\therefore$$

$$\int t^2 \, dt = \int \frac{dx}{x} \quad \therefore \quad \frac{t^3}{3} = \ln C \, |x| \quad \therefore \quad y = x\sqrt[3]{3 \ln C \, |x|}$$

Exercise 14.3, page 313

1. Equations in the form of "linear non-homogeneous differential equation"

$$y' + f(x)\, y = g(x)$$

may be solved with the following steps:

1. Calculate the "integration factor" μ as

$$\mu = e^{\int f(x) \, dx}$$

2. Assume the relationship

$$\frac{d}{dx}(y\, \mu) = \mu g(x)$$

3. Integrate and solve per y.

(a) $y' - 4xy = x$ $\therefore$ $f(x) = -4x$ and $g(x) = x$

$$\int f(x)\,dx = -4 \int x\,dx = -4\frac{x^2}{2} = -2x^2 \quad \therefore \quad \mu = e^{-2x^2}$$

$$\frac{d}{dx}\left(e^{-2x^2} y\right) = x\,e^{-2x^2} \quad \therefore \quad \int d\left(e^{-2x^2} y\right) = \int x\,e^{-2x^2}\,dx$$

$$e^{-2x^2} y = \int x\,e^{-2x^2}\,dx$$

$$\left\{-2x^2 = t \quad \therefore \quad -4x\,dx = dt \quad \therefore \quad x\,dx = -\frac{1}{4}\,dt\right\}$$

$$= -\frac{1}{4} \int e^t\,dt = -\frac{1}{4}e^t = -\frac{1}{4}e^{-2x^2} + C$$

$$\therefore$$

$$\underline{y = -\frac{1}{4} + Ce^{2x^2}}$$

(b) We declare $f(x) = -2$ and $g(x) = 3$ so that

$$y' - 2y - 3 = 0 \quad \therefore \quad y' - 2y = 3 \quad \therefore \quad \int f(x) = -2 \int dx = -2x$$

$$\therefore$$

$$\mu = e^{-2x} \quad \therefore \quad \frac{d}{dx}\left(y\,e^{-2x}\right) = 3\,e^{-2x}$$

$$\therefore$$

$$\int d\left(y\,e^{-2x}\right) = 3 \int e^{-2x}\,dx \quad \therefore \quad y\,e^{-2x} = -\frac{3}{2}e^{-2x} + C \quad \bigg| \ /e^{-2x}$$

$$\therefore$$

$$\underline{y = -\frac{3}{2} + C\,e^{2x}}$$

(c)   Given $y' + 2xy = 2xe^{-x^2}$, declare

$$f(x) = 2x \quad \text{and} \quad g(x) = 2x\,e^{-x^2}$$

so that

$$\int f(x)\,dx = 2\int x\,dx = x \qquad\qquad 2 \quad \therefore \quad \mu = e^{x^2}$$

It follows that

$$ye^{x^2} = 2\int e^{x^2} x\,e^{-x^2}\,dx = x^2 + C \quad \therefore \quad \underline{y = e^{-x^2}\left(x^2 + C\right)}$$

(d)   Declare

$$f(x) = \frac{2x}{(x+1)} \quad \text{and} \quad g(x) = (x+1)^3$$

so that

$$y' - \frac{2y}{x+1} - (x+1)^3 = 0 \quad \therefore \quad y' - \frac{2}{x+1}y = (x+1)^3$$

$$\therefore$$

$$\int f(x)\,dx = -\int \frac{2}{x+1}\,dx = -2\ln|x+1| = \ln\frac{1}{(x+1)^2}$$

$$\therefore$$

$$\mu = \exp\left(\ln\frac{1}{(x+1)^2}\right) = \frac{1}{(x+1)^2}$$

$$\therefore$$

$$\int d\left(y\,\frac{1}{(x+1)^2}\right) = \int \frac{1}{(x+1)^2}(x+1)^3\,dx = \{x+1 = t, dx = dt\}$$

$$= \frac{(x+1)^2}{2} + C_1 \quad \therefore \quad \underline{y = \frac{(x+1)^4}{2} + C_1(x+1)^2}$$

Or, if we calculate the last integral without the change of variables technique, the only difference would be in the integration constant C, i.e.

$$\int (x+1)\, dx = \int x\, dx + \int dx = \frac{x^2}{2} + x + C$$

$$\therefore$$

$$y = (x+1)^2 \left(\frac{x^2}{2} + x + C \right)$$

however, if $C = C_1 + 1/2$, which is also a valid integration constant, then the two forms of result are equivalent.

Exercise 14.4, page 313

1. Equations in the form of "linear differential equation with constant coefficients" as

$$a_n y^{(n)} + a_{n-1} y^{(n-1)} + \cdots + a_2 y'' + a_1 y' + a_0 y = 0$$

are, in general, solved with the following steps:

1. Write the equivalent "characteristic polynomial"

$$P_n(r) = a_n r^n + a_{n-1} r^{n-1} + \cdots + a_2 r^2 + a_1 r + a_0 = 0$$

   where the derivative orders in differential equation are replaced with the same order powers.
2. Calculate the roots of nth order polynomial $P_n(r) = 0$, i.e. $(r_1, r_2, \ldots, r_n)$.
3. Depending upon whether all the roots r_i are unique, or multiple but real, or if there are complex roots, the solution is found in the following form:

$$y(x) = \sum_i f\left(x^n, C_i e^{r_i x}\right) \quad (n = 0, 1, 2 \ldots)$$

where the term x^n depends on multiplicity of the $P(r)$ roots, i.e. if a root is unique, then $n = 0$, if the root is a "double root", then there are x^0, x^1 terms, if it is a triple root, then there are x^0, x^1, x^2 terms, etc.

(a) All roots of $P(r)$ are real and unique

$$y'' - 9y = 0 \quad \therefore \quad P_2(r) = r^2 - 9 = (r-3)(r+3) = 0$$

$$\therefore \quad r_1 = 3, \quad r_2 = -3 \quad (r_1 \neq r_2)$$

Therefore, we write the solution in the form

$$y(x) = C_1 e^{r_1 x} + C_2 e^{r_2 x} = C_1 e^{3x} + C_2 e^{-3x}$$

Verification:

$$y(x) = C_1 e^{3x} + C_2 e^{-3x}$$
$$y'(x) = 3C_1 e^{3x} - 3C_2 e^{-3x}$$
$$y''(x) = 9C_1 e^{3x} + 9C_2 e^{-3x}$$

Therefore,

$$y'' - 9y = 0$$
$$\left(9C_1 e^{3x} + 9C_2 e^{-3x}\right) - 9\left(C_1 e^{3x} + C_2 e^{-3x}\right) = 0$$
$$\cancel{9C_1 e^{3x}} + \cancel{9C_2 e^{-3x}} - \cancel{9C_1 e^{3x}} - \cancel{9C_2 e^{-3x}} = 0$$
$$0 = 0 \ \checkmark$$

(b)   Not all real roots of $P(r)$ are unique, there is one double root

$$y'' - 2y' + y = 0 \quad \therefore \quad \therefore \quad P_2(r) = r^2 - 2r + 1 = (r-1)(r-1) = 0$$
$$\therefore \quad r_1 = 1, \quad r_2 = 1 \quad (r_1 = r_2 = r)$$

Therefore, we write the solution in the form

$$y(x) = C_1 e^{rx} + x\, C_2 e^{rx} = C_1 e^x + x\, C_2 e^x = e^x(C_1 + x\, C_2)$$

Otherwise, the solution terms are not independent.

Verification:

$$y(x) = C_1 e^x + xC_2 e^x = e^x(C_1 + x\, C_2)$$
$$\{[f\,g]' = f'g + fg'\}$$
$$y'(x) = C_1 e^x + C_2 e^x + xC_2 e^x = e^x(C_1 + C_2(x+1))$$
$$y''(x) = C_1 e^x + C_2 e^x + C_2 e^x + xC_2 e^x = e^x(C_1 + C_2(x+2))$$

Therefore,

$$y'' - 2y' + y = 0 = 0$$

$$\left[e^x(C_1 + C_2(x+2))\right] - 2\left[e^x(C_1 + C_2(x+1))\right] + \left[e^x(C_1 + x\,C_2)\right] = 0 \quad \div e^x$$

$$\cancel{C_1} + x\cancel{C_2} + 2C_2 - 2\cancel{C_1} - 2x\cancel{C_2} - 2C_2 + \cancel{C_1} + x\cancel{C_2} = 0$$

$$0 = 0 \;\checkmark$$

(c) Not all real roots of $P(r)$ are unique, there is one double root

$$y'' + 4y' + 4y = 0 \quad \therefore \quad P_2(r) = r^2 + 4r + 4 = (r+2)(r+2) = 0$$

$$\therefore \quad r_1 = -2, \quad r_2 = -2 \quad (r_1 = r_2 = r)$$

Therefore, we write the solution in the form

$$y(x) = C_1 e^{rx} + x\,C_2 e^{rx} = C_1 e^{-2x} + x\,C_2 e^{-2x} = e^{-2x}(C_1 + x\,C_2)$$

which can be confirmed by the verification.

(d) Roots of $P(r)$ are complex and therefore conjugated

$$y'' + 6y' + 25y = 0 \quad \therefore \quad P_2(r) = r^2 + 6r + 25$$

$$r_{1,2} = \frac{-6 \pm \sqrt{36 - 4 \times 25}}{2} = -3 \pm 4i$$

$$\therefore$$

$$r_1 = -3 + 4i$$

$$r_2 = -3 - 4i \quad (r_1 \neq r_2^*)$$

In the case of unique roots, we write the solution of form

$$y(x) = C_1 e^{r_1 x} + C_2 e^{r_2 x} = C_1 e^{(-3+4i)x} + C_2 e^{(-3-4i)x}$$

$$= e^{-3x}\left[C_1 e^{4xi} + C_2 e^{-4xi}\right]$$

$$\{e^{ix} = \cos x + i\sin x\}$$

$$= e^{-3x}\left[C_1(\cos 4x + i\sin 4x) + C_2(\cos(-4x) + i\sin(-4x))\right]$$

$$\{\cos(-x) = \cos x, \quad \sin(-x) = -\sin x\}$$

$$= e^{-3x}\left[C_1(\cos 4x + i\sin 4x) + C_2(\cos 4x - i\sin 4x)\right]$$

$$= e^{-3x}\left[(C_1 + C_2)\cos 4x + i(C_1 - C_2)\sin 4x\right]$$

$$\{C_1 + C_2 = D_1, \quad i(C_1 - C_2) = D_2\}$$

$$= e^{-3x} \left(D_1 \cos 4x + D_2 \sin 4x \right) \quad [D_{1,2} \in \mathbb{C}]$$

As an illustration, assuming a simple case of $D_1 = D_2 = 1$, the plot of $y(x)$ module is shown in Fig. 14.4. For $x < 0$, the amplitude of $y(x)$ shows oscillatory behaviour, while for $x \geq 0$ it is damped by the e^{-3x} term.

Fig. 14.4 Example 14.4-1

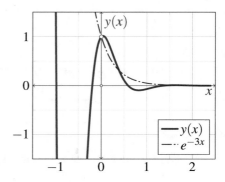

Exercise 14.5, page 313

1. This classic example is analysed by the first order *non-homogeneous* differential equation with constant coefficients. In general, first we solve the homogenous part, then the non-homogeneous part, and then based on the initial conditions, we solve the complete equation as the sum of two partial solutions.

The circuit equation setup: KVL equation of the circuit loop is derived as

$$E = v_R + v_L = i\,R + L\,\frac{di}{dt}$$

$$\therefore$$

$$L\frac{di}{dt} + R\,i = E \quad \therefore \quad \frac{di}{dt} + \frac{R}{L}\,i = \frac{E}{L}$$

Because E, R, L =const., this equation is in the form of first order non-homogeneous differential equation with constant coefficients. The initial condition is $i(0) = 0$.

Homogenous part solution, $i_h(t)$:

$$i' + \frac{R}{L}\,i = 0 \quad \therefore \quad P(r) = r + \frac{R}{L} = 0 \quad \therefore \quad r = -\frac{R}{L} \quad \therefore \quad i_h(t) = C\,e^{-(R/L)t}$$

Non-homogenous part solution, $i_p(t)$:

$$\left. \begin{array}{l} i' + \dfrac{R}{L}\,i = \dfrac{E}{L} \\[2mm] \dfrac{E}{L} = \text{const.} \end{array} \right\} \quad \begin{array}{l} i'_p = 0 \\[2mm] i_p = C_2 \end{array} \quad \therefore \quad 0 + \frac{R}{\cancel{L}}C_2 = \frac{E}{\cancel{L}} \quad \therefore \quad C_2 = \frac{E}{R}$$

Complete solution, $i(t) = i_h(t) + i_p(t)$:

$$\left. \begin{array}{l} i(t) \;\; = C\,e^{-(R/L)t} + \dfrac{E}{R} \\[3mm] i(0) \;\; = 0 \end{array} \right\} \;\; i(0) = C\,\cancel{e^{-(R/L)0}}^{\,1} + \dfrac{E}{R} = 0 \;\; \therefore \;\; C = -\dfrac{E}{R}$$

Therefore, the complete solution is in form

$$i(t) = \frac{E}{R} - \frac{E}{R}e^{-(R/L)t} = \frac{E}{R}\left(1 - e^{-(R/L)t}\right)$$

$$\therefore$$

$$\lim_{t\to\infty} i(t) = \frac{E}{R}$$

It is important to note that at moment $\tau = L/R$, the amplitude of current reached the level

$$i(\tau) = \frac{E}{R}\left(1 - e^{-(R/L)\,(L/R)}\right) = \frac{E}{R}\left(1 - e^{-1}\right) \approx 0.632\,\frac{E}{R}$$

$$\therefore$$

$$i(\tau) = 0.632\,\frac{E}{R} = 0.632\,i(\infty)$$

It is also important to note that at moment $5\tau = 5(L/R)$, the amplitude of current reached the level

$$i(5\tau) = \frac{E}{R}\left(1 - e^{-(R/L)\,5\,(L/R)}\right) = \frac{E}{R}\left(1 - e^{-5}\right) \approx 0.632\,\frac{E}{R}$$

$$\therefore$$

$$i(5\tau) = 0.993\,\frac{E}{R} = 0.993\,i(\infty)$$

Variable τ is commonly known as "time constant" and is a very important "unit of measure" in engineering and science, Fig. 14.5.

Fig. 14.5 Example 14.5-1

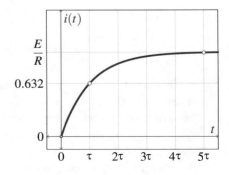

2. This classic example is also analysed by the same type of the first order *non-homogeneous* differential equation with constant coefficients. In general, first we solve the homogenous part, then the non-homogeneous part, and then based on the initial conditions, we solve the complete equation as the sum of two partial solutions.

The force equation setup: in accordance with Newton's laws of motion, we write

$$F = ma = m\frac{dv}{dt} = mg - kv$$

$$\therefore$$

$$mv' + kv = mg \quad \therefore \quad v' + \frac{k}{m}v = g$$

Because k, m, g =const., this equation is in the form of first order non-homogeneous differential equation with constant coefficients. The initial condition is $v(0) = 0$.

Homogenous part solution, $v_h(t)$:

$$v' + \frac{k}{m}v = 0 \quad \therefore \quad P(r) = r + \frac{k}{m} = 0 \quad \therefore \quad r = -\frac{k}{m} \quad \therefore \quad v_h(t) = C\,e^{-(k/m)\,t}$$

Non-homogenous part solution, $v_p(t)$:

$$\left.\begin{array}{l} v' + \dfrac{k}{m}v = g \\[2mm] g = \text{const.} \end{array}\right\} \quad \begin{array}{l} v_p' = 0 \\[2mm] v_p = C_2 \end{array} \quad \therefore \quad 0 + \frac{k}{m}C_2 = g \quad \therefore \quad C_2 = \frac{mg}{k}$$

Complete solution, $v(t) = v_h(t) + v_p(t)$:

$$\left.\begin{array}{l} v(t) \quad = C\,e^{-(k/m)\,t} + \dfrac{mg}{k} \\[2mm] v(0) \quad = 0 \end{array}\right\} \quad v(0) = C\,e^{-(k/m)\,0}{}^{1} + \frac{mg}{k} = 0 \quad \therefore \quad C = -\frac{mg}{k}$$

Therefore, the complete solution is in form

$$i(t) = \frac{mg}{k} - \frac{mg}{k}\,e^{-(k/m)\,t} = \frac{mg}{k}\left(1 - e^{-(k/m)\,t}\right)$$

$$\therefore$$

$$\lim_{t \to \infty} i(t) = \frac{mg}{k}$$

In other words, there is "terminal velocity" that a free falling body can achieve in the resistive media (e.g. air). In addition, exactly the same type of equations describes very different physical processes, for example, the velocity of falling body and current in RL circuit.

3. This example is based on the second order *homogeneous* differential equation with constant coefficients.

The circuit's KVL equation setup:

$$v_L + v_R + v_C = E \quad \therefore \quad L\frac{di}{dt} + Ri + \frac{1}{C}\int i\,dt = E \qquad \frac{d}{dt}$$

$$\therefore$$

$$Li'' + R\,i' + \frac{1}{C}\,i = 0 \quad \therefore \quad i'' + \frac{R}{L}\,i' + \frac{1}{LC}\,i = 0$$

Characteristic equation:

$$P(r) = r^2 + \frac{R}{L} + \frac{1}{LC} = 0$$

$$\therefore$$

$$r_{1,2} = -\frac{R}{2L} \pm \frac{\sqrt{R^2 - \dfrac{4L}{C}}}{2L}$$

$$= -\frac{R}{2L} \pm \sqrt{\left(\frac{R}{2L}\right)^2 - \frac{\cancel{4}\cancel{L}}{\cancel{4}L^{\cancel{2}}C}} \qquad = -\frac{R}{2L} \pm \sqrt{\left(\frac{R}{2L}\right)^2 - \frac{1}{LC}}$$

Case 1: $R^2 > 4L/C$: both roots are real, unique, and negative. They are negative because

$$R^2 - 4L/C < R^2 \quad \therefore \quad \sqrt{R^2 - 4L/C} < R \quad \therefore \quad \begin{cases} -R + \sqrt{R^2 - 4L/C} & < 0 \\ -R - \sqrt{R^2 - 4L/C} & < 0 \end{cases}$$

and thus, the amplitude in this case is exponentially falling in time, see Fig. 14.6.

$$i(t) = C_1 e^{r_1 t} + C_2 e^{r_2 t}$$

Case 2: $R^2 = 4L/C$: both roots are real, negative, and equal; that is to say,

$$r_1 = r_2 = r = \frac{-R}{2L} \quad \therefore \quad i(t) = C_1 e^{rt} + t\,C_2 e^{rt} = e^{rt}(C_1 + t\,C_2)$$

In this case, for ($t > 0$), amplitude quickly reaches maximum, and then it exponentially tends to zero, Fig. 14.6.

Case 3: $R^2 < 4L/C$: both roots are complex (conjugate) and unique; that is to say,

$$r_1 = \frac{-R}{2L} + i\,\sqrt{\frac{1}{LC} - \frac{R^2}{4L^2}} = \alpha + i\omega \qquad r_2 = \frac{-R}{2L} - i\,\sqrt{\frac{1}{LC} - \frac{R^2}{4L^2}} = \alpha - i\omega$$

$$\therefore$$

$$i(t) = C_1 e^{r_1 t} + C_2 e^{r_2 t} = e^{-\alpha t}(D_1 \cos \omega t + D_2 \sin \omega t)$$

In this case, the response is damped oscillatory, Fig. 14.6. However, in the extreme case of no damping, i.e. $\alpha = 0$, the oscillatory behaviour continues indefinitely.

Fig. 14.6 Example 14.5-3

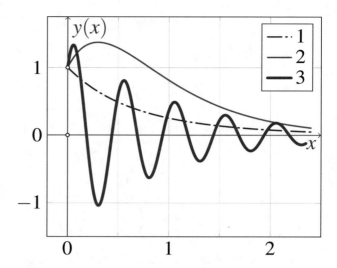

Part II

Mathematics for Signal Processing

Series

15

Important to Know

Arithmetic progression: It is a sequence of numbers where the difference d between any two subsequent terms is constant. That being case, given the first term a_1, a general term a_n is found as

$$a_n = a_1 + (n - 1)\, d$$

The sum of first n terms in arithmetic progression:

$$S_n = \sum_{k=1}^{n} a_k = \frac{n}{2}(a_1 + a_n)$$

Geometric progression: It is a sequence of numbers where the *ratio r* between any two subsequent terms is constant. That being case, given the first term a_1, a general term a_n is found as

$$a_1, \ a_2 = r\, a_1, \ \ a_3 = r\, a_2, \cdots$$

$$\therefore$$

$$a_n = a_1\, r^{n-1}$$

The sum of geometric progression: if $a_1 = a$; $(a \neq 0)$ then the sum of first n terms is found as

$$S_n = \sum_{k=1}^{n} a_k = \sum_{k=1}^{n} a\, r^{k-1} = \underbrace{a + ar + ar^2 + \cdots + ar^{n-1}}_{n \text{ terms}}$$

We consider the following cases:

1. $r = 1$: Assuming $a \neq 0$, the sum is calculated simply as

© Springer Nature Switzerland AG 2021
R. Sobot, *Engineering Mathematics by Example*,
https://doi.org/10.1007/978-3-030-79545-0_15

$$\sum_{n=1}^{n\to\infty} a\,r^{n-1} = \sum_{n=1}^{n\to\infty} a\,1^{n-1} = \sum_{n=1}^{n\to\infty} a = \underbrace{a + a + a + \cdots}_{\text{infinitely terms}} \to \infty$$

2. $r \neq 1$: The sum may be found as follows:

$$S_n = a + ar + ar^2 + \cdots + ar^{n-1} \quad \therefore$$

$$r\,S_n = ar + ar^2 + \cdots + ar^{n-1} + ar^n$$

$$\therefore$$

$$(S_n - r\,S_n) = a - ar^n \quad \Rightarrow \quad S_n = \frac{a\,(1 - r^n)}{1 - r}$$

Convergence of the geometric sum: Depending upon value of $|r| \neq 0$, the infinite sum may be calculated in the following two cases:

(a) $|r| \geq 1$: In this case, each term of the geometric series is greater and greater; therefore,

$$\sum_{n=1}^{\infty} a\,r^{n-1} = \lim_{n\to\infty} \frac{a\,(1 - r^n)}{1 - r} \overset{(=)}{} \left(\frac{\infty}{\infty}\right) \overset{\text{L.H.}}{=} \lim_{n\to\infty} \frac{-anr^{n-1}}{-1} \to \infty$$

that is to say, for $|r| \geq 1$ the sum of geometric series is divergent.

(b) $|r| < 1$: In this case, we recall that powers of numbers inferior to one tend to zero very fast, thus

$$\sum_{n=1}^{\infty} a\,r^{n-1} = \lim_{n\to\infty} \frac{a\,(1 - r^{n}{}^{0})}{1 - r} = \frac{a}{1 - r}$$

that is to say, for $|r| < 1$ the sum of geometric series is convergent.

15.1   Exercises

15.1 *   Basic Sums

1. Write the first few terms of the following series:

(a) $S = \left\{ \sqrt{n - 3} \right\}_{n=3}^{\infty}$

(b) $S = \left\{ \cos \dfrac{n\pi}{6} \right\}_{n \geq 0}^{\infty}$

2. Calculate the following sums:

(a) $\displaystyle\sum_{i=1}^{4} i$

(b) $\displaystyle\sum_{i=1}^{4} 2$

(c) $\displaystyle\sum_{i=1}^{3} i^2$

(d) $\displaystyle\sum_{j=0}^{5} 2^j$

(e) $\displaystyle\sum_{k=1}^{4} \frac{1}{k}$

(f) $\displaystyle\sum_{n=1}^{3} \frac{n-1}{n^2 + 3}$

3. Calculate the following sums:

(a) $\displaystyle\sum_{i=1}^{n} 1$

(b) $\displaystyle\sum_{i=1}^{n} i$

(c) $\displaystyle\sum_{i=1}^{n} i^2$

(d) $\displaystyle\sum_{i=1}^{n} i(4i - 3)$

4. Write the following series as sum of its general term:

(a) $S = 2^3 + 3^3 + \cdots + n^3$

(b) $S = \dfrac{1}{2} + \dfrac{2}{3} + \dfrac{3}{4} + \cdots$

15.2 ***   Limits of Sums

1. Calculate the following limits:

(a) $\displaystyle\lim_{n\to\infty} \sum_{i=1}^{\infty} \dfrac{3}{n}\left[\left(\dfrac{i}{n}\right)^2 + 1\right]$

15.3 ***   Geometric Series

1. Write the following series as sum of its general term:

$$S = a + ar + ar^2 + ar^3 + \cdots + ar^n + \cdots$$

then calculate the sum if

(a) $r = 1$

(b) $r \neq 1$

(c) $|r| < 1$

2. Calculate the following sums:

(a) $S = 5 - \dfrac{10}{3} + \dfrac{20}{9} - \dfrac{40}{27} + \cdots$

(b) $S = \displaystyle\sum_{n=1}^{\infty} 2^{2n}\, 3^{1-n}$

(c) $S = \displaystyle\sum_{n=1}^{\infty} x^n; \quad |x| < 1$

Solutions

Exercise 15.1, page 332

1. Given values for n and the general term, for infinite series we evaluate the general term for several subsequent n.

   (a) $S = \left\{ \sqrt{n-3} \right\}_{n=3}^{\infty} = \sum_{n=3}^{\infty} \sqrt{n-3} = 0 + 1 + \sqrt{2} + \sqrt{3} + \cdots + \sqrt{n-3} + \cdots$

   (b) $S = \left\{ \cos \frac{n\pi}{6} \right\}_{n=0}^{\infty} = \sum_{n\geq 0}^{\infty} \cos \frac{n\pi}{6} = 1 + \frac{\sqrt{3}}{2} + \frac{1}{2} + 0 + \cdots + \cos \frac{n\pi}{6} + \cdots$

2. For limited series (i.e. with less than infinite number of terms), it is possible to add all terms.

   (a) $\sum_{i=1}^{4} i = 1 + 2 + 3 + 4 = 10$

   (b) $\sum_{i=1}^{4} 2 = 2 + 2 + 2 + 2 = 8$

   (c) $\sum_{i=1}^{3} i^2 = 1^2 + 2^2 + 3^2 = 1 + 4 + 9 = 14$

   (d) $\sum_{j=0}^{5} 2^j = 2^0 + 2^1 + 2^2 + 2^3 + 2^4 + 2^5 = 1 + 2 + 3 + 8 + 16 + 32 = 63$

   (e) $\sum_{k=1}^{4} \frac{1}{k} = \frac{1}{1} + \frac{1}{2} + \frac{1}{3} + \frac{1}{4} = \frac{25}{12}$

   (f) $\sum_{n=1}^{3} \frac{n-1}{n^2+3} = \frac{0}{3} + \frac{1}{7} + \frac{2}{12} = \frac{13}{42}$

3. Summing the first n terms results in a general algebraic expression.

   (a) $S = \sum_{i=1}^{n} 1 = \underbrace{1 + 1 + 1 + \cdots + 1}_{n \times 1} = n$

   (b) $S = \sum_{i=1}^{n} i = 1 + 2 + 3 + \cdots + n = \frac{n(n+1)}{2}$

   The legend has it that this classic problem of the sum of an arithmetic sequence was solved by Gauss when he was still in the elementary school. His proof is based on understanding that the addition is a commutative operation (meaning that order does not matter), thus the same sum can be written in both ascending and descending orders, then term by term addition of the two equations gives

$$S = 1 + 2 \quad + 3 \quad + \cdots + (n-1) + n$$
$$S = n + (n-1) + \cdots + 3 \quad + 2 \quad + 1$$
$$\therefore$$
$$2S = \underbrace{(n+1) + (n+1) + (n+1) + \cdots + (n+1) + (n+1)}_{n \times (n+1)}$$

Therefore, we write

$$2S = n(n+1) \quad \therefore \quad S = \frac{n(n+1)}{2}$$

This result is often used and assumed to be well known.

(c) The sum of first n squares is also often used result.

$$S = \sum_{i=1}^{n} i^2 = 1^2 + 2^2 + 3^2 + \cdots + (n-1)^2 + n^2 = \frac{n(n+1)(2n+1)}{6} \tag{15.1}$$

Derivation of sum (15.1) is based on the mathematical induction technique. The main idea is split into two steps: First we prove that equality is valid for a certain number, often for $n = 0$ or $n = 1$ (but, any number $n = N$ could be used as the starting point). In the second step, we make the claim that *if* the equality is valid for a given $n = k$ case, then it must be also valid for the next term, i.e. $n = k + 1$.

Specifically,

1. *Case: $n = 1$* we must show that the following equality is correct:

$$S_1 = \sum_{i=1}^{1} i^2 = 1^2 = 1 = \frac{n(n+1)(2n+1)}{6}$$

Indeed, the last formula gives

$$S_1 = \frac{1 \times (1+1)(2 \times 1 + 1)}{6} = \frac{1 \times 2 \times 3}{6} = \frac{6}{6} = 1$$

which simply proves that (15.1) is correct for $n = 1$.

2. *Induction:* In general case when $n = k$, we say that if

$$S_k = \frac{k(k+1)(2k+1)}{6} \quad \text{then,}$$

$$S_{k+1} = \underbrace{1^2 + 2^2 + \cdots + k^2}_{S_k} + (k+1)^2$$

$$= S_k + (k+1)^2 = \frac{k(k+1)(2k+1)}{6} + (k+1)^2 \; \frac{6}{6}$$

$$= \frac{k(k+1)(2k+1) + 6(k+1)^2}{6} = (k+1)\frac{k(2k+1) + 6(k+1)}{6}$$

$$= (k+1)\frac{2k^2 + 7k + 6}{6} = (k+1)\frac{2k^2 + 4k + 3k + 6}{6}$$

$$= \frac{(k+1)\,(k+2)(2k+3)}{6}$$

$$= \frac{(k+1)\,[(k+1)+1]\,[2(k+1)+1]}{6} \tag{15.2}$$

where the form of (15.1) is identical to (15.2) after each n is replaced by $n + 1$, which is by definition S_{k+1} sum. Thus, after proving the first step, i.e. $n = 1$, then $n = 2$ is also valid, then $n = 3$, etc., similar to the domino effect after the first block falls down.

(d)   Knowing the sums of first n integers and sum of their respective squares, we write

$$\sum_{i=1}^{n} i\,(4i - 3) = \sum_{i=1}^{n} 4i^2 - \sum_{i=1}^{n} 3i = 4\sum_{i=1}^{n} i^2 - 3\sum_{i=1}^{n} i$$

$$= 4\frac{n(n+1)(2n+1)}{6} - 3\frac{n(n+1)}{2}$$

$$= \frac{4n(n+1)(2n+1) - 9n(n+1)}{6} = (n+1)\frac{8n^2 - 5n}{6}$$

$$= \frac{n(n+1)(8n - 5)}{6}$$

4. The series counter can be adjusted to start at any number.

(a) $S = 2^3 + 3^3 + \cdots + n^3 = \displaystyle\sum_{k=2}^{n} k^3$

but also,

$$S = 2^3 + 3^3 + \cdots + n^3 = \sum_{j=1}^{n-1} (j+1)^3$$

or,

$$S = 2^3 + 3^3 + \cdots + n^3 = \sum_{m=0}^{n-2} (m+2)^3$$

(b) $\quad S = \dfrac{1}{2} + \dfrac{2}{3} + \dfrac{3}{4} + \cdots = \displaystyle\sum_{n=1}^{\infty} \dfrac{n}{n+1}$

Exercise 15.2, page 333

1. Note which variable is the sum counter, and which one is the limit counter.

(a) Depending upon which operation is in progress, one or the other variable behaves as a constant.

$$\lim_{n\to\infty} \sum_{i=1}^{\infty} \frac{3}{n}\left[\left(\frac{i}{n}\right)^2 + 1\right] = \lim_{n\to\infty} \sum_{i=1}^{\infty} \left[\frac{3}{n}\frac{i^2}{n^2} + \frac{3}{n}\right] = \lim_{n\to\infty} \sum_{i=1}^{\infty} \left[\frac{3}{n^3}i^2 + \frac{3}{n}\right]$$

$$= \lim_{n\to\infty}\left[\frac{3}{n^3}\sum_{i=1}^{\infty} i^2 + \frac{3}{n}\sum_{i=1}^{\infty} 1\right]$$

$$= \lim_{n\to\infty}\left[\frac{3}{n^3}\frac{n(n+1)(2n+1)}{6} + \frac{3}{n}n\right]$$

$$= \lim_{n\to\infty}\left[\frac{1}{2}\frac{n+1}{n}\frac{2n+1}{n} + 3\right]$$

$$= \lim_{n\to\infty}\left[\frac{1}{2}\frac{n\left(1+\frac{1}{n}\right)}{n}\frac{n\left(2+\frac{1}{n}\right)}{n} + 3\right]$$

$$= \lim_{n\to\infty}\left[\frac{1}{2}\left(1+\frac{1}{n}\right)\left(2+\frac{1}{n}\right) + 3\right]$$

$$= \frac{1}{2}(1+0)(2+0) + 3$$

$$= 4$$

Exercise 15.3, page 333

1. Geometric series

$$S = a + ar + ar^2 + ar^3 + \cdots + ar^n + \cdots = \sum_{n=1}^{\infty} a\,r^{n-1} \quad (a \neq 0)$$

may reach the following sums S_n for the first n terms:

(a) Case: $r = 1$, where $(a \neq 0)$

$$S_n = \sum_{n=1}^{\infty} a\, 1^{n-1} = \sum_{n=1}^{\infty} a \to \pm\infty; \quad (a \neq 0)$$

In this case, therefore, the sum diverges.

(b) Case: $r \neq 1$.

$$S_n = a + ar + ar^2 + ar^3 + \cdots + ar^{n-1} \tag{15.3}$$

$$\therefore$$

$$r\, S_n = ar + ar^2 + ar^3 + \cdots + ar^n \tag{15.4}$$

Consequently, the difference of (15.3) and (15.4)

$$S_n - r\, S_n = a - \cancel{ar} + \cancel{ar} - \cancel{ar^2} + \cancel{ar^2} - \cdots - \cancel{ar^{n-1}} + \cancel{ar^{n-1}} - ar^n$$

$$= a - ar^n$$

$$\therefore$$

$$S_n(1 - r) = a - ar^n \quad \therefore \quad S_n = \frac{a(1 - r^n)}{1 - r}$$

(c) Case: $|r| < 1$ is most important because

$$\lim_{n\to\infty} S_n = \lim_{n\to\infty} \frac{a(1 - \cancel{r^n}^{\,0})}{1 - r}$$

$$\therefore$$

$$\sum_{n=1}^{\infty} a\, r^{n-1} = \frac{a}{1 - r}; \quad |r| < 1$$

2. First, we verify if given series is indeed geometric series, i.e. if the ratio of any two subsequent terms is constant.

(a)   Given,

$$S = 5 - \frac{10}{3} + \frac{20}{9} - \frac{40}{27} + \cdots \quad \therefore \quad \frac{-\frac{10}{3}}{5} = \frac{\frac{20}{9}}{-\frac{10}{3}} = \cdots = -\frac{2}{3}$$

i.e. the ratio of any two subsequent terms equals $-2/3$. Therefore,

$$S = 5 - \frac{10}{3} + \frac{20}{9} - \frac{40}{27} + \cdots = 5\left(-\frac{2}{3}\right)^0 + 5\left(-\frac{2}{3}\right)^1 + 5\left(-\frac{2}{3}\right)^2 + \cdots$$

which is to say that

$$S = \sum_{n=1}^{\infty} a\,r^{n-1} = 5\left(-\frac{2}{3}\right)^{n-1} \quad a = 5, |r| = \left|-\frac{2}{3}\right| = \frac{2}{3} < 1$$

$$\therefore$$

$$S = \frac{a}{1-r} = \frac{5}{1 - \left(-\frac{2}{3}\right)} = \frac{5}{\frac{5}{3}} = 3$$

(b)   We find that this is geometric series as

$$S = \sum_{n=1}^{\infty} 2^{2n}\,3^{1-n} = \sum_{n=1}^{\infty} (2^2)^n\,3^{-(n-1)} = \sum_{n=1}^{\infty} \frac{4^n}{3^{n-1}} = \sum_{n=1}^{\infty} \frac{4 \times 4^{n-1}}{3^{n-1}}$$

$$= \sum_{n=1}^{\infty} 4\left(\frac{4}{3}\right)^{n-1} \quad \therefore \quad |r| = \frac{4}{3} > 1$$

$$\therefore$$

$$S \to \infty$$

(c)   This is geometric series as

$$S = \sum_{n=1}^{\infty} x^n; \quad |x| < 1 \quad \therefore \quad = \sum_{n=1}^{\infty} 1\,x^n = \frac{1}{1-x}$$

Special Functions

16

Important to Know

Delta function: It is defined as

$$\delta(t) = 0; \quad (t \neq 0)$$

$$\int_{-\infty}^{\infty} \delta(t)\, dt = 1$$

Properties of Delta function: As the consequence of $\delta(t)$ non-zero value at only one point, i.e. $t = 0$, at any other point $t \neq 0$ all products of $x(t)\,\delta(t)$ equal zero (which may be either delayed or advanced relative to $t = 0$ by some constant t_0). For that reason, Dirac function is also referred to as the "sampling function."

$$\delta(-t) = \delta(t)$$

$$x(t)\,\delta(t - t_0) = x(t_0)\,\delta(t - t_0)$$

$$\int_{-\infty}^{+\infty} x(t)\,\delta(t)\, dt = x(0)$$

$$\int_{-\infty}^{+\infty} x(t)\,\delta(t - t_0)\, dt = x(t_0)$$

$$\int_{-\infty}^{+\infty} x(\tau)\,\delta(t - \tau)\, d\tau = x(t)$$

$$\int_{-\infty}^{+\infty} x(t + t_0)\,\delta(t)\, dt = x(t_0)$$

$$\int_{t_1}^{t_2} x(t)\,\delta(t - t_0)\, dt = \begin{cases} x(t_0); & t_1 < t_0 < t_2 \\ 0; & \text{otherwise} \end{cases}$$

© Springer Nature Switzerland AG 2021
R. Sobot, *Engineering Mathematics by Example*,
https://doi.org/10.1007/978-3-030-79545-0_16

16.1 Exercises

16.1 *   Special Functions

---

1. Sketch a graph of

$$x(t) = e^{st} \quad \text{where,} \quad s = \sigma + j\omega \quad \text{and} \quad \sigma, \omega \in \mathbb{R}$$

   in the following cases:

   (a) $\sigma = 0, \omega = 0$ (b) $\sigma > 0, \omega = 0$ (c) $\sigma = 0, \omega \neq 0$

   (d) $\sigma > 0, \omega \neq 0$ (e) $\sigma = 0, \omega \neq 0$ (f) $\sigma < 0, \omega \neq 0$

2. Sketch a graph and give definition of Dirac delta (sometimes called "the unit impulse") function $\delta(t)$.

3. Sketch a graph and give definition of Dirac comb (sometimes known as "sampling") function $\text{III}_T(t)$.

4. Sketch a graph and give definition of "step" (sometimes called "switch") function $u(t)$.

5. Sketch a graph and give definition of "sign" function $\text{sign}(t)$.

6. Sketch a graph and give definition of "ramp" function $r(t)$.

7. Sketch a graph and give definition of "rectangular" function $\Pi(t)$.

8. Sketch a graph and give definition of "triangular" (sometimes called "hat" or "tent") function $\Lambda(t)$.

9. Sketch a graph and give definition of "cardinal sine" function $\text{sinc}(t)$.

16.2 **   Special Function, Interrelations

---

1. Problems with $\delta(t)$ function.

   (a) $\displaystyle\int_{-\infty}^{\infty} \sin(2.3\pi t)\, \delta(t-1)\, dt$ (b) $\displaystyle\int_{-\infty}^{\infty} e^{-at^2}\, \delta(t+5)\, dt, \quad (a > 0)$

   (c) $\displaystyle\int_{0}^{\infty} e^{-at^2}\, \delta(t+1)\, dt, \quad (a > 0)$ (d) $\displaystyle\int_{-\infty}^{\infty} x(t)\, \delta(t-a)\, dt, \quad a \in \mathbb{R}$

2. Solve,

   (a) $\displaystyle x(t) = \int_{-\infty}^{t} u(\tau)\, dt, \quad t > 0$ (b) $x(t) = u'(t)$

   (c) $u(t)$ as function of $\text{sign}(t)$ (d) $\Pi(t)$ as function of $\delta(t)$

   (e) $\Pi(t)$ as function of $u(t)$ (f) $\Pi(t)$ as function of $\text{sign}(t)$

Solutions

Exercise 16.1, page 342

1. Exponential function

$$x(t) = e^{st} \quad \text{where,} \quad s = \sigma + j\omega \quad \text{and} \quad \sigma, \omega \in \mathbb{R}$$

may be transformed as follows:

$$x(t) = e^{(\sigma+j\omega)t} = e^{\sigma t}e^{j\omega t} = e^{\sigma t}\left[\underbrace{\cos(\omega t)}_{\Re(x(t))} + j\underbrace{\sin(\omega t)}_{\Im(x(t))}\right]$$

and, its complex conjugate version

$$x^*(t) = e^{(\sigma-j\omega)t} = e^{\sigma t}e^{-j\omega t} = e^{\sigma t}\left[\cos(\omega t) - j\sin(\omega t)\right]$$

That is, by using the Euler's formula, complex exponent function is transformed into product of a real exponential function and trigonometric functions. In other words, the complex exponential function generates both cosine (along real axis) and sine (along perpendicular imaginary axis) functions, Fig. 16.1.

Fig. 16.1 Example 16.1-1

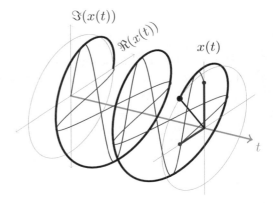

In the following three cases, we find.

(a) $\sigma = 0, \omega = 0$: Therefore, degenerates into a constant.

$$x(t) = e^{(\sigma+j\omega)t} = e^0 = \text{const.}$$

(b) $\sigma > 0, \omega = 0$: Therefore, becomes a real exponential function.

$$x(t) = e^{(\sigma+j\omega)t} = e^{\sigma t}$$

(c) $\sigma = 0, \omega \neq 0$: Therefore, becomes a complex exponential function that can be used to generate real sine/cosine functions.

$$x(t) = e^{(\sigma + j\omega)t} = e^{j\omega t} = [\cos(\omega t) + j\sin(\omega t)]$$

In addition, the sum and difference of complex exponents give

$$x(t) + x^*(t) = e^{j\omega t} + e^{-j\omega t}$$
$$= [\cos(\omega t) + j\sin(\omega t)] + [\cos(\omega t) - j\sin(\omega t)]$$
$$= 2\cos(\omega t)$$

$$\therefore$$

$$\cos(\omega t) = \frac{e^{j\omega t} + e^{-j\omega t}}{2}$$

$$x(t) - x^*(t) = e^{j\omega t} - e^{-j\omega t}$$
$$= [\cos(\omega t) + j\sin(\omega t)] - [\cos(\omega t) - j\sin(\omega t)]$$
$$= 2j\sin(\omega t)$$

$$\therefore$$

$$\sin(\omega t) = \frac{e^{j\omega t} - e^{-j\omega t}}{2j}$$

(d) $\sigma > 0, \omega \neq 0$: Generates cosine function along the real axis, Fig. 16.2, as well as sine function in the complex plane, both with progressively increasing amplitudes.

Fig. 16.2 Example 16.1-1(d)

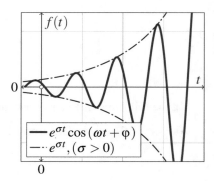

(e) $\sigma = 0, \omega \neq 0$: Generates cosine function along the real axis, Fig. 16.3, as well as sine function in the complex plane, both with constant amplitudes.

Fig. 16.3 Example 16.1-1(e)

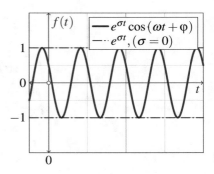

(f) $\sigma < 0, \omega \neq 0$: Generates cosine function along the real axis, Fig. 16.4, as well as sine function in the complex plane, both with progressively decreasing amplitudes.

Fig. 16.4 Example 16.1-1(f)

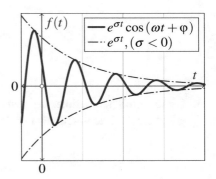

2. Dirac delta function $\delta(t)$, Fig. 16.5, is defined by the means of limited integral as

$$\int_{-\infty}^{\infty} \delta(t)\, dt = 1$$

$$\delta(t) = 0, \quad t \neq 0$$

Main properties of $\delta(t)$ are:

$$\delta(t) = \delta(-t)$$

$$x(t)\delta(t - t_0) = x(t_0)\delta(t - t_0)$$

$$\int_{-\infty}^{\infty} x(t)\delta(t)\, dt = x(0)$$

$$\int_{-\infty}^{\infty} x(t)\delta(t - t_0)\, dt = x(t_0)$$

$$\int_{-\infty}^{\infty} x(t + t_0)\delta(t)\, dt = x(t_0)$$

$$\int_{a}^{b} x(t)\delta(t - t_0)\, dt = \begin{cases} x(t_0), t_0 \in (a, b) \\ 0, \text{otherwise} \end{cases}$$

Fig. 16.5 Example 16.1-2

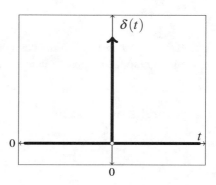

3. Dirac comb (sometimes known as "sampling") function $\text{III}_T(t)$ is created by a series of equidistant Dirac functions, Fig. 16.6.

$$\text{III}_T(t) = \sum_{k=-\infty}^{\infty} \delta(t - kT)$$

where, T is sampling time, thus sampling frequency is $f = 1/T$.

Fig. 16.6 Example 16.1-3

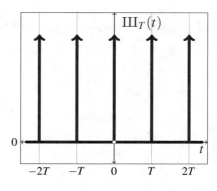

4. Step function $u(t)$, Fig. 16.7, is defined as

$$u(t) = \begin{cases} 0, t < 0 \\ 1, t \geq 0 \end{cases}$$

Fig. 16.7 Example 16.1-4

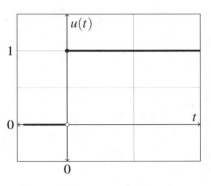

5. Sketch a graph and give definition of "sign" function sign(t), Fig. 16.8.

$$\text{sign}(t) = \begin{cases} -1, t < 0 \\ 0, t = 0 \\ 1, t \geq 0 \end{cases}$$

or, alternatively,

$$\text{sign}(t) = \frac{t}{|t|} = \frac{|t|}{t}, \quad \forall x \neq 0$$

Fig. 16.8 Example 16.1-5

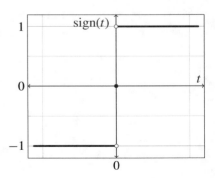

6. Ramp function $r(t)$, Fig. 16.9, is defined as follows:

$$r(x) = \begin{cases} 0, t < 0 \\ t, t \geq 0 \end{cases}$$

Fig. 16.9 Example 16.1-6

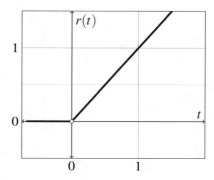

7. Rectangular function $\Pi(t)$, Fig. 16.10, is defined as follows:

$$\Pi\left(\frac{t}{T}\right) = \begin{cases} 1, & -\dfrac{T}{2} \geq t \geq \dfrac{T}{2} \\ \\ 0, \text{ otherwise} \end{cases}$$

Fig. 16.10 Example 16.1-7

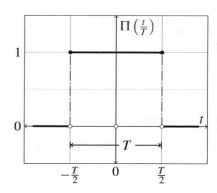

8. Triangle function $\Lambda(t)$, Fig. 16.11, is defined as follows:

$$\Lambda\left(\frac{t}{T}\right) = \begin{cases} \dfrac{t}{T/2} + 1, & -\dfrac{T}{2} \le t \le 0 \\[2mm] 1 - \dfrac{t}{T/2}, & 0 \le t \le \dfrac{T}{2} \\[2mm] 0, \text{ otherwise} \end{cases}$$

Fig. 16.11 Example 16.1-8

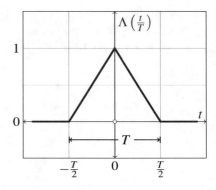

9. Cardinal sine function sinc (t) is defined as follows:

$$\text{sinc}\,(t) = \frac{\sin t}{t} \quad \text{or,} \quad \text{sinc}\,(t) = \frac{\sin(\pi t)}{\pi t}$$

where the last definition is normalized to sine period T, Fig. 16.12 (left). Equally important is $|\text{sinc}\,(t)|$ version, Fig. 16.12 (right).

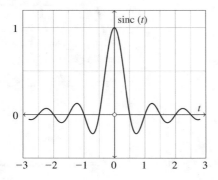 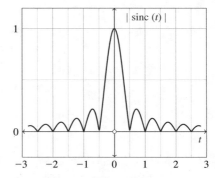

Fig. 16.12 Example 16.1-9

Exercise 16.2, page 342

1. Product of a function $x(t)$ and $\delta(t - t_0)$ equals to zero, except on the point where $\delta(t)$ is located, i.e. $t = t_0$. That is to say, while solving the definite integral, in effect we "search" the location of $\delta(t)$ function. Consequently, only at that point function $x(t_0)$ keeps its value, in any other case $\delta(t) x(t) = 0, \quad t \neq t_0$. Therefore, we write as follows.

(a)   Given,

$$\int_{-\infty}^{\infty} \sin(2.3\pi\, t)\, \delta(t - 1)\, dt$$

we search the location of $\delta(t - 1)$. By definition, it is found at the position when its argument equals zero, i.e. when $t - 1 = 0 \quad \therefore \quad t = 1$. Consequently, the product $\sin(2.3\pi t)\, \delta(t - 1)$ is not equal to zero only for $t = 1$. Thus, we write,

$$\int_{-\infty}^{\infty} \sin(2.3\pi\, t)\, \delta(t - 1)\, dt = \int_{-\infty}^{\infty} \underbrace{\sin(2.3\pi \times 1)}_{\text{const.}}\, \delta(t - 1)\, dt$$

$$= \sin(2.3\pi) \int_{-\infty}^{\infty} \delta(t - 1)\, dt \overset{1}{\nearrow}$$

$$= \sin(2.3\pi) \approx 0.809$$

This integration is illustrated in Fig. 16.13, which shows the position of $\delta(t - 1)$ function relative to $\sin(2.3\pi\, t)$. The multiplication operation is done before the integration and, obviously, only one point i.e. $\sin(2.3\pi)$ is not multiplied by zero. This example illustrates the idea of "sampling" (or, saving) certain points of a given function by using Dirac function.

Fig. 16.13
Example 16.2-1(a)

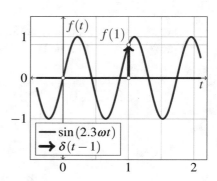

(b)   First, we find position of $\delta(t + 5)$, i.e. when $t + 5 = 0 \quad \therefore \quad t = -5$, which is within the bonds of definite integral, then we write

$$\int_{-\infty}^{\infty} e^{-at^2}\, \delta(t+5)\, dt, \quad (a>0) = \int_{-\infty}^{\infty} e^{-a(-5)^2}\, \delta(t+5)\, dt$$

$$= e^{-25a} \int_{-\infty}^{\infty} \underrightarrow{\delta(t+5)\, dt}^{\,1} = e^{-25a}$$

(c)   However, if $\delta(t)$ is *not* found within the bonds of definite integral, then all relevant products equal zero because $\delta(t) = 0$, see Fig. 16.14.
That is to say,

$$\int_{0}^{\infty} e^{-at^2}\, \delta(t+1)\, dt, \quad (a>0)$$

$$= \left\{ (t=-1) \notin [0,\infty] \right\}$$

$$= \int_{0}^{\infty} e^{-at^2}\, \underrightarrow{\delta(t+1)}^{\,0}\, dt$$

$$= \int_{0}^{\infty} e^{-at^2} \times 0\, dt = 0$$

Fig. 16.14
Example 16.2-1(c)

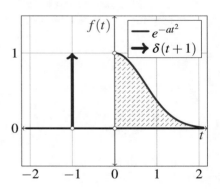

(d)   Considering that $\delta(t-a)$ is certainly found within interval $t \in [-\infty, \infty]$, then in general case we write

$$\int_{-\infty}^{\infty} x(t)\, \delta(t-a)\, dt = \int_{-\infty}^{\infty} x(a)\, \underrightarrow{\delta(t-a)}^{\,1}\, dt = x(a)$$

2. (a) Step function equals to zero for negative t, and it is constant (i.e. equals one) for $t \geq 0$, thus we write (using temporarily variable τ)

$$\int_{-\infty}^{t} u(\tau)\, d\tau = \int_{-\infty}^{0} u(\tau)\, d\tau + \int_{0}^{t} u(\tau)\, d\tau = \int_{-\infty}^{0} \cancelto{0}{0}\, d\tau \quad + \int_{0}^{t} 1\, d\tau$$

$$= \int_{0}^{t} d\tau = \tau\, \Big|_{0}^{t} = t$$

In conclusion,

$$\int_{-\infty}^{t} u(\tau)\, d\tau = \begin{cases} t; & t \geq 0 \\ 0; & t < 0 \end{cases} = r(t)$$

which is logical, considering that by definite integral we calculate surface of the rectangular area (whose height equals to one) and its length linearly increases as t increases.

(b) We can deduce derivative $u'(t)$ by observing graphical interpretation, Fig. 16.15.
While keeping in mind that derivative of a constant equals to zero, and that in general case it shows the function's rapidity of change over given interval, we conclude that if the interval equals to zero, the change must be infinite (due to division by zero). By definition, that infinite change in amplitude over zero interval describes Dirac function.

$$\frac{d\, u(t)}{dt} = \delta(t)$$

Fig. 16.15
Example 16.2-2(b)

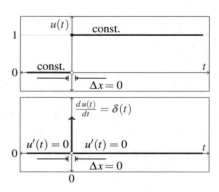

(c) Sign(t) function can be derived from $u(t)$ by simple transformations as illustrated in Fig. 16.16.

In the first step $u(t)$ is multiplied by factor two, which doubles its amplitude for $t > 0$. At this stage we note that sign (t) function has the same form, but appears shifted down by step of one. Thus,

$$\text{sign}(t) = 2u(t) - 1$$

Fig. 16.16
Example 16.2-2(c)

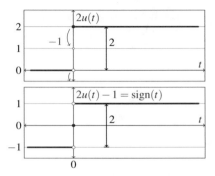

(d) Following the same reasoning as in Example 16.2-2(b), we deduce.

Relation between $\Pi(t)$ and $\delta(t)$ is illustrated in Fig. 16.17. There are two rapid changes at the edges of $\Pi(t)$ that generate Dirac functions at $t = -T/2$ and $t = T/2$ moments. We note that the leading edge of $\Pi(t)$ jumps from zero to one, thus the derivative is positive. However, the trailing edge jumps from one to zero, thus its associated derivative is negative. In summary,

$$\frac{d\Pi}{dt} = \delta\left(t + \frac{T}{2}\right) - \delta\left(t - \frac{T}{2}\right)$$

Fig. 16.17
Example 16.2-2(d)

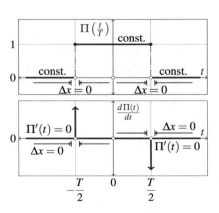

(e) Simple combinations of basic functions can be used to generate another function. One possible relation between $\Pi(t)$ and $u(t)$ is illustrated in Fig. 16.18. For example, one step function can be advanced and one delayed in time. The difference between these two $u(t)$ functions produces $\Pi(t)$ as

$$\Pi\left(\frac{t}{T}\right) = u\left(t + \frac{T}{2}\right) - u\left(t - \frac{T}{2}\right)$$

which is simply deduced by calculating the difference within each piecewise interval. For example, for $t \in [-T/2, T/2]$,

$$u(t + T/2) = 1; \quad u(t - T/2) = 0$$

$$\therefore$$

$$u(t + T/2) - u(t - T/2) = 1 - 0 = 1$$

Fig. 16.18
Example 16.2-2(e)

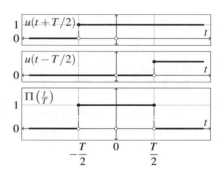

(f) One possible relation between $\Pi(t)$ and sign(t) is illustrated in Fig. 16.19. On the left side, it is shown how to create two sign(t) functions: one advanced by $\tau/2$, and one first delayed by $\tau/2$ then inverted sign. In the next step illustrated on top of Fig. 16.19 (right), the two sign(t) functions are scaled by factor of $1/2$ so that their sum adds up to one. The addition of these two functions produces $\Pi(t/\tau)$, as summarized by equation

$$\Pi\left(\frac{t}{\tau}\right) = \frac{1}{2}\text{sign}\left(t + \frac{\tau}{2}\right) - \frac{1}{2}\text{sign}\left(t - \frac{\tau}{2}\right)$$

Fig. 16.19
Example 16.2-2(f)

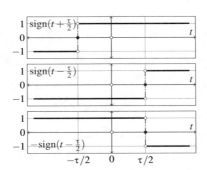

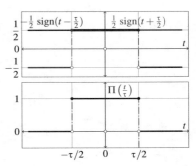

Convolution Integral

17

Important to Know

Convolution integral: It is defined as

$$z(t) = x(t) * y(t) = \int_{-\infty}^{\infty} x(\tau)\, y(t - \tau)\, d\tau \tag{17.1}$$

where, τ is the provisional integration variable. That is to say, at any given instance of time t, the *products* of the two functions $x(\tau)$ and $y(t - \tau)$ are integrated relative to the parametric variable τ.

17.1 Exercises

17.1 * Transformations of Continuous Functions

1. Given $x(t)$ function in Fig. 17.1, show the following graphs:

   (a) Delayed: $x(t - 2)$

   (b) Compressed: $x(2t)$

   (c) Expanded: $x(0.2t)$

   (d) Time inverted: $x(-t)$

Fig. 17.1 Example 17.1-1

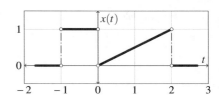

2. Given $x(t)$ function in Fig. 17.2, show graph of $y(t) = x(2t + 3)$.

Fig. 17.2 Example 17.1-2

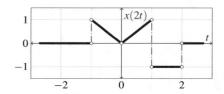

3. Given $x(t)$ function in Fig. 17.3, show graph of $y(t) = x(-\frac{t}{2} - 4)$.

Fig. 17.3 Example 17.1-3

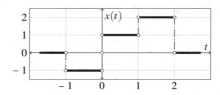

4. Given $x(t)$ function in Fig. 17.4, decompose $x(t)$ into a sum of one even $x_p(t)$ and one odd $x_i(t)$ function so that $x(t) = x_p(t) + x_i(t)$, then show their respective graphs.

Fig. 17.4 Example 17.1-4

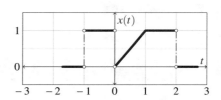

5. Show graphical representation of the following function:

$$x(t) = 3u(t) + r(t) - r(t - 1) - 5u(t - 2)$$

6. Show the analytical form of function $x(t)$ given in Fig. 17.5:

Fig. 17.5 Example 17.1-6

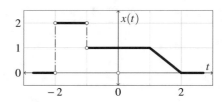

7. Show the analytical form of function $x(t)$ given in Fig. 17.6:

Fig. 17.6 Example 17.1-7

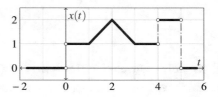

17.2 **   Convolution, Piecewise Linear Functions

1. Calculate $y(t) = x(t) * h(t)$, where $x(t)$ and $h(t)$ are given in Fig. 17.7.

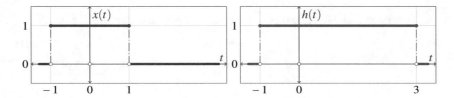

Fig. 17.7 Example 17.2-1

2. Calculate $z(t) = x(t) * y(t)$, where $x(t)$ and $y(t)$ are given in Fig. 17.8.

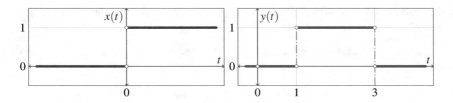

Fig. 17.8 Example 17.2-2

3. Calculate $y(t) = u(t) * u(t)$.

4. Solve convolution $y(t) = x(t) * h(t)$, given

$$x(t) = [u(t + 1) - u(t - 1)] \, t$$
$$h(t) = [u(t + 2) - u(t)]$$

5. Calculate $x(t) = \Lambda\left(\frac{t}{2}\right) * \Lambda\left(\frac{t}{2}\right)$.

17.3 *   Convolution, Continuous Functions

1. Solve convolution $z(t) = x(t) * y(t)$, where $x(t) = t^3 u(t)$, and $y(t) = t^2 u(t)$.
2. Solve convolution $w(t) = x(t) * v(t)$, given

$$x(t) = \begin{cases} 1 - t; & 0 \le t \le 1 \\ 0; & \text{otherwise} \end{cases} \quad \text{and,} \quad v(t) = \begin{cases} e^{-t}; & t \ge 0 \\ 0; & \text{otherwise} \end{cases}$$

17.4 **   Energy and Power of Continuous Functions

1. Given $x(t)$ functions and calculate their respective energies and powers.

(a) $x(t) = A$ (b) $x(t) = \sin(t)$ (c) $x(t) = e^{-t} u(t)$

(d) $x(t) = A e^{j \omega_0 t}$ (e) $x(t) = u(t)$ (f) $x(t) = 2\Pi(t - 1)$

Solutions

Exercise 17.1, page 357

1. In order to carry on the required transformations, visually, it helps to pick some of the typical points in the given function, for example $x(-1), x(0), x(2)$, and follow their respective movements.

(a)   A function is delayed or advanced when a constant is added to the initial argument, for example if

$$x(0) \rightarrow x(t - 2)$$

where "−2" is the shifting constant. New coordinate positions are found by equalizing the two arguments, e.g.

$$0 = t - 2 \quad \therefore \quad t = 2$$

Similarly, we find transformed positions of the other typical points, and conclude that this transformation shifts the original function in the positive direction by two steps, Fig. 17.9.

Fig. 17.9 Example 17.1-1(a)

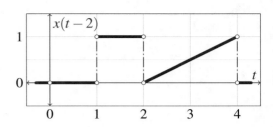

(b)   Similarly, for compressed function $x(2t)$ we write,

$$x(0): \ 2t = 0 \ \therefore \ t = 0$$

$$x(2): \ 2t = 2 \ \therefore \ t = 1$$

$$x(-1): \ 2t = -1 \ \therefore \ t = -\frac{2}{2}$$

$$\cdots$$

here, the origin point does not move. It is evident that when the function's argument is multiplied by a constant greater that one, visually, the overall effect is "compression" of the original form, Fig. 17.10.

Fig. 17.10
Example 17.1-1(b)

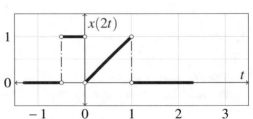

(c)   However, the function's argument is multiplied by a constant lesser than one; visually the overall effect is "expansion" of the original form. In the case of $x(0.2t)$, Fig. 17.11, we find that

$$x(0): \ 0.2t = 0 \ \therefore \ t = 0$$

$$x(2): \ 0.2t = 2 \ \therefore \ t = 10$$

$$x(-1): \ 0.2t = -1 \ \therefore \ t = -5$$

Fig. 17.11
Example 17.1-1(c)

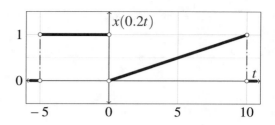

(d)   Time inverted function $x(-t)$ is simply revolved around the vertical axes, Fig. 17.12, where

$$x(0): \quad -t = 0 \quad \therefore \quad t = 0$$

$$x(1): \quad -t = 1 \quad \therefore \quad t = -1$$

. . .

Fig. 17.12
Example 17.1-1(d)

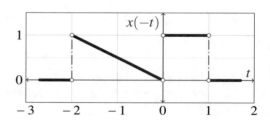

2. When multiple transformations are required, there are multiple ways to execute them.

(a)   *Method 1:*
Transformation $y(t) = x(2t + 3)$ consists of one compression (i.e. $2t$) and one delay (i.e. $+3$). If the delay is applied before the compression, we find

$$x(t) \xrightarrow{\text{delay}} x(t + 3) \xrightarrow{\text{compression}} x(2t + 3)$$

In the first step, the delay $x(t + 3)$, we write

$$x(0): \quad t + 3 = 0 \quad \therefore \quad t = -3$$

which is shown in Fig. 17.13.

Fig. 17.13
Example 17.1-2(a)

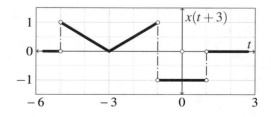

In the second transformation, the compression, we keep the delay and while the time is compressed i.e. $x(2t)$ so that we write

$$x(0): \ 2t = 0 \ \therefore \ t = 0$$
$$x(1): \ 2t = 1 \ \therefore \ t = 1/2$$
$$x(-3): \ 2t = -3 \ \therefore \ t = -3/2$$
$$\cdots$$

as shown in Fig. 17.14.

Fig. 17.14
Example 17.1-2(a)

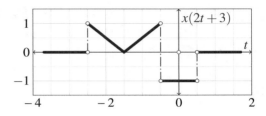

(b) *Method 2:*
Alternatively, the two transformations in $y(t) = x(2t + 3)$ may be executed so that the compression is applied before the delay, thus we find

$$x(t) \xrightarrow{\text{compression}} x(2t) \xrightarrow{\text{delay}} x(2t + 3) = x\left(2\left(t + \frac{3}{2}\right)\right)$$

That is to say, application of compression $x(2t)$ results in the function form as in Fig. 17.15

Fig. 17.15
Example 17.1-2(b)

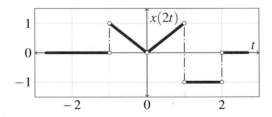

Second transformation, the delay of $(t + 3/2)$ shifts the function horizontally, resulting in the same result as in Fig. 17.14.

3. By focusing, for example, to point $x(0)$ we then compare when the new argument also equals zero. Given $x(t)$ function, one way to derive the total transformation may be to apply delay $x(t - 4)$ first, followed by time inversion $x(-t)$ along with the compression factor of $0.5t$: Delayed function is shown in Fig. 17.16,

Fig. 17.16 Example 17.1-3

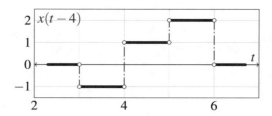

while the time inversion with the scaling factor we find as, for example for point

$$x(4): \quad -\frac{t}{2} = 4 \quad \therefore \quad t = -8$$

so that the total transformation is as in Fig. 17.17,

Fig. 17.17 Example 17.1-3

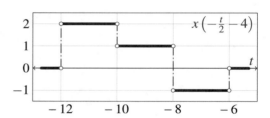

Using some other order of transformations does not change the final result.

4. Given $x(t)$ function, its accompanying even and odd functions are found as

$$x_p(t) = \frac{x(t) + x(-t)}{2} \tag{17.2}$$

$$x_i(t) = \frac{x(t) - x(-t)}{2} \tag{17.3}$$

We verify that

$$\begin{aligned}
x_p(t) + x_i(t) &= \frac{x(t) + x(-t)}{2} + \frac{x(t) - x(-t)}{2} \\
&= \frac{x(t)}{2} + \frac{x(t)}{2} + \frac{x(-t)}{2} - \frac{x(-t)}{2} \\
&= x(t)
\end{aligned}$$

We calculate even and odd functions (17.2) and (17.3) by using graphical technique, i.e. by carrying on calculations within each individual interval.

In order to derive even function $x_p(t)$, as defined by (17.2) first, we derive $x(-t)$ by mirroring $x(t)$ around the vertical axis, Fig. 17.18 (middle). Then, additions are within each piecewise linear interval. For example, we add $x(t) + x(-t)$ as

$$\underline{t = 0_-:} \quad \frac{x(t) + x(-t)}{2} = \frac{1 + 0}{2} = \frac{1}{2}$$

$$t = 0_+ : \quad \frac{x(t) + x(-t)}{2} = \frac{0+1}{2} = \frac{1}{2}$$

that is to say, on both sides of $t = 0$ we find $x_p(0) = 1/2$. Similarly, we calculate $x_p(t)$ at other typical points of $t \in [-3, 3]$, for example $x_p(\pm 2), x_p(\pm 1)$, and create graph in Fig. 17.18 (bottom), which is the literal implementation of (17.2).

Fig. 17.18 Example 17.1-4

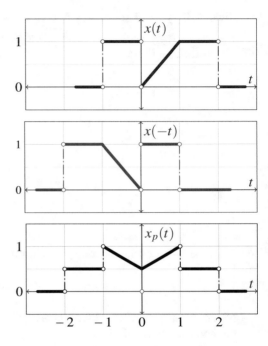

By using the same technique, we derive odd function $x_i(t)$, as defined by (17.3) by calculating the difference at the key points. For example, at $t = 0$ we calculate one minus one and divide by two, thus $x_i(0) = 0$, Fig. 17.19.

Fig. 17.19 Example 17.1-4

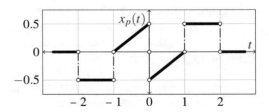

5. Function $x(t)$ is synthesized by adding the basic functions $u(t)$ and $r(t)$ as

$$x(t) = 3u(t) + r(t) - r(t-1) - 5u(t-2)$$

Graphical representation of each term is shown in Fig. 17.20, while the sum is highlighted in the background. The addition is done by simply adding values of the four functions at typical coordinates. For example, for $t = 0_+$, by definition of basic functions we write

$$3u(t) = 3u(0) = 3(1) = 3$$

$$r(t) = r(0) = 0$$

$$-r(t-1) = -r(0-1) = -(0) = 0$$

$$-5u(t-2) = -5u(0-2) = -5(0) = 0$$

thus, $x(0) = 3 + 0 + 0 + 0 = 3$. Similar additions are repeated for $t \in [1, 2, 3]$.

Fig. 17.20 Example 17.1-5

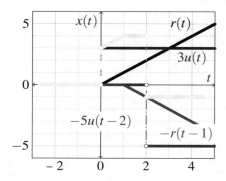

6. By inspection of function $x(t)$ given in Fig. 17.5, we write the following sum:

$$x(t) = 2u(t+2) - u(t+1) - r(t-1) + r(t-2)$$

which is easily verified either by calculating the sums at various points in time, or by graphical method as in Example 17.1-5.

7. By inspection of function $x(t)$ given in Fig. 17.6, we write the following sum:

$$x(t) = u(t) + r(t-1) - 2r(t-2) + r(t-3) + u(t-4) - 2u(t-5)$$

which is easily verified either by calculating the sums at various points in time, or by graphical method as in Example 17.1-5. Note how $x(t)$ is formed in $t \in [1, 3]$ interval by correctly adding "$r(t-1) - 2r(t-2) + r(t-3)$" ramp functions.

Exercise 17.2, page 359

1. Convolution integral (17.1) contains the product of $x(\tau)$ and $h(t-\tau)$ functions. We derive these two functions by using graphical method. In sequence of transformations, we convert $x(t) \rightarrow x(\tau)$, as well as $h(t) \rightarrow h(\tau) \rightarrow y(-\tau) \rightarrow h(t-\tau)$ function, as in Fig. 17.21.

(a)   In the first step, Fig. 17.21a, b, the change of variable t into provisional integration variable τ does not change the original forms of neither $x(t)$ nor $h(t)$ functions, thus only transformations of $h(t)$ function are illustrated.

(b) In the second step, $h(-\tau)$ is the time inverted version of function $h(\tau)$, i.e. is mirrored around the vertical axis, Fig. 17.21c. Note that the origin $\tau = 0$ reference is still known and, visually, the form of $h(t)$ stays unchanged.

(c) In the third step, $h(-\tau)$ is shifted to an arbitrary position t thus $h(t - \tau)$, Fig. 17.21d. Consequently, depending upon relative position of t in reference to the origin point $\tau = 0$ the following cases are possible:

Fig. 17.21 Example 17.2-1

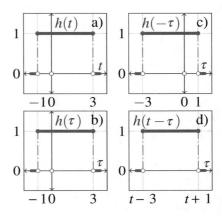

Case 1 : $(t + 1) \leq -1$:
The leading edge of $h(t - \tau)$ still does not surpass $(\tau = -1)$ point, that is to say $h(t - \tau)$ does not overlap $x(\tau)$ as long as

$$t + 1 \leq -1 \quad \therefore \quad t \leq -2$$

Fig. 17.22
Example 17.2-1 (case 1)

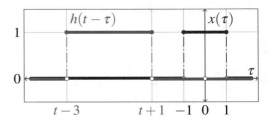

That being the case, it is irrelevant where exactly $(t + 1)$ is relative to the trailing edge of $x(\tau)$, see Fig. 17.22. What is important is that under this condition, due to either $h(t - \tau)$ or $x(\tau)$ equals zero, their product is therefore

$$x(\tau)\, h(t - \tau) = 0, \quad \forall \tau$$

$$\therefore$$

$$y(t) = \int_{-\infty}^{\infty} x(\tau) \, h(t - \tau) \, d\tau = \int_{-\infty}^{\infty} 0 \, d\tau = 0 \quad (t \leq -2)$$

Case 2 : $(t + 1) > -1$ and $(t + 1) \leq 1$:
The leading edge of $h(t - \tau)$ is found within $\tau \in [-1, 1]$ interval determined by $x(\tau)$; however, the trailing edge of $h(t - \tau)$ is still outside of this interval, i.e.

$$t + 1 > -1 \quad \therefore \quad t > -2 \quad \text{and}$$

$$t + 1 \leq 1 \quad \therefore \quad t \leq 0$$

Fig. 17.23
Example 17.2-1 (case 2)

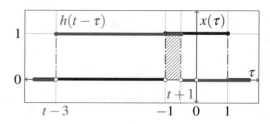

Under this condition, see Fig. 17.23, the overlapping surface is found in the interval $\tau \in [-1, t + 1]$ where $h(t - \tau) = 1$ and $x(\tau) = 1$; therefore, the convolution integral (i.e. the overlapping surface area) is found to be

$$y(t) = \int_{-\infty}^{\infty} x(\tau) \, h(t - \tau) \, d\tau = \int_{-1}^{t+1} (1)(1) \, d\tau = \tau \, \Big|_{-1}^{t+1} = t + 1 - (-1)$$

$$= t + 2 \quad (-2 \leq t \leq 0)$$

Case 3 : $(t + 1) > 1$ and $(t - 3) \leq -1$:
The leading edge of $h(t - \tau)$ is found at $(t + 1 > 1)$ while its trailing edge is still $(t - 3 \leq -1)$, i.e.

$$t + 1 > 1 \quad \therefore \quad t > 0 \quad \text{and}$$

$$t - 3 \leq -1 \quad \therefore \quad t \leq 2$$

Fig. 17.24
Example 17.2-1 (case 3)

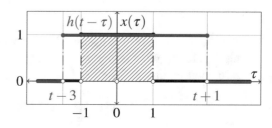

Within these boundaries of t, see Fig. 17.24, the overlapping surface is found in the interval $\tau \in [-1, +1]$ where $h(t - \tau) = 1$ and $x(\tau) = 1$, in other words it is maximal overlapping surface possible whose rectangular area equals two,

$$y(t) = \int_{-\infty}^{\infty} x(\tau)\, h(t - \tau)\, d\tau = \int_{-1}^{1} (1)\,(1)\, d\tau = \tau \Big|_{-1}^{1}$$

$$= 2 \quad (0 \leq t \leq 2)$$

Case 4 : $(t - 3) \geq -1$ and $(t - 3) \leq 1$:
The trailing edge of $h(t - \tau)$ is found within $\tau \in [-1, 1]$ interval determined by $x(\tau)$, i.e.

$$t - 3 \geq -1 \quad \therefore \quad t \leq 2 \ \text{ and}$$

$$t - 3 < 1 \quad \therefore \quad t < 4$$

Fig. 17.25
Example 17.2-1 (case 4)

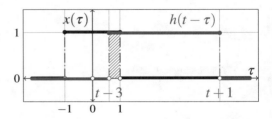

Within these boundaries of t, see Fig. 17.25, the overlapping surface is found in the interval $\tau \in [t - 3, +1]$ where $h(t - \tau) = 1$ and $x(\tau) = 1$, i.e.

$$y(t) = \int_{-\infty}^{\infty} x(\tau)\, h(t - \tau)\, d\tau = \int_{t-3}^{1} (1)\,(1)\, d\tau = \tau \Big|_{t-3}^{1}$$

$$= 4 - t \quad (2 \leq t \leq 4)$$

Case 5 : $(t - 3) \geq 1$:
The trailing edge of $h(t - \tau)$ is found at $(t - 3 \geq 1)$, see Fig. 17.26, i.e.

$$t - 3 \geq 1 \quad \therefore \quad t \geq 4$$

Fig. 17.26
Example 17.2-1 (case 5)

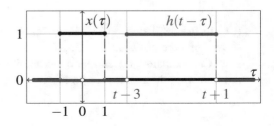

That being the case, it is irrelevant where exactly $(t - 3)$ is relative to the leading edge of $x(\tau)$, see Fig. 17.26. What is important is that within these boundaries of t, due to either $h(t - \tau)$ or $x(\tau)$ equals zero, their product is therefore

$$x(\tau)\,h(t - \tau) = 0, \quad \forall \tau$$

$$\therefore$$

$$y(t) = \int_{-\infty}^{\infty} x(\tau)\,h(t - \tau)\,d\tau = \int_{-\infty}^{\infty} 0\,d\tau = 0 \quad (t \geq 4)$$

Summary: The total convolution integral $y(t)$ shows the progression of the overlapping surface area between $x(\tau)$ and $h(t - \tau)$, Fig. 17.27, as

$$y(t) = \begin{cases} 0 & t \leq -2 \\ 2 + t & -2 \leq t \leq 0 \\ 2 & 0 \leq t \leq 2 \\ 4 - t & 2 \leq t \leq 4 \\ 0 & t \geq 4 \end{cases}$$

Fig. 17.27 Example 17.2-1

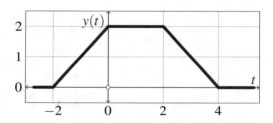

2. Convolution integral (17.1) contains the product of $x(\tau)$ and $y(t - \tau)$ functions. We derive these two functions by using graphical method. In sequence of transformations, we convert $u(t) \rightarrow u(\tau)$, as well as $y(t) \rightarrow y(\tau) \rightarrow y(-\tau) \rightarrow y(t - \tau)$ function, as in Fig. 17.28.

(a)   After the first step, Fig. 17.28a, b, the change of variable t into temporarily integration variable τ does not change the initial form of neither $x(t)$ nor $y(t)$ functions, thus only $y(\tau)$ function is illustrated.

(b)   In the second step, in order to create $y(-\tau)$, function $y(\tau)$ is simply mirrored around the vertical axis, Fig. 17.28c. Note that the origin $\tau = 0$ point is still known, and the form of $y(t)$ visually stays unchanged.

(c)   In the third step, $y(-\tau)$ is shifted to an arbitrary position t, thus $y(t - \tau)$ in Fig. 17.28d shows only the positions of rising (found at $-3 + t$) and falling (found at $-1 + t$) edges relative to τ. Consequently, depending upon relative position of t in reference to the origin point $\tau = 0$ the following three cases are possible:

Fig. 17.28 Example 17.2-2

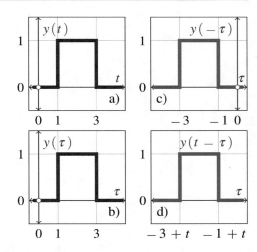

Case 1 : $(t - 1) \leq 0$:
The leading edge of $y(t - \tau)$ is still in the negative side, i.e.

$$t - 1 \leq 0 \quad \therefore \quad t \leq 1$$

Fig. 17.29
Example 17.2-2(case 1)

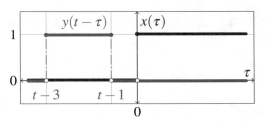

That being the case, it is irrelevant where exactly t is relative to the origin, see Fig. 17.29. What is important is that within these boundaries of t, because always either of the two function equals zero, their product is therefore

$$x(\tau)\, y(t - \tau) = 0, \quad \forall \tau$$

$$\therefore$$

$$z(t) = \int_{-\infty}^{\infty} x(\tau)\, y(t - \tau)\, d\tau = \int_{-\infty}^{\infty} 0\, d\tau$$

$$= 0 \ (t \leq 1)$$

Case 2 : $(t - 1) > 0$ *and* $(t - 3) \leq 0$:
The leading edge of $y(t - \tau)$ is shifted into the positive side; however, the trailing edge is still in the negative side, i.e.

$$t - 1 > 0 \quad \therefore \quad t > 1 \quad \text{and}$$

$$t - 3 \leq 0 \quad \therefore \quad t \leq 3$$

Fig. 17.30
Example 17.2-2(case 2)

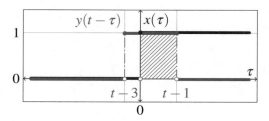

Within these boundaries of t, see Fig. 17.30, the overlapping surface is found in the interval $\tau \in [0, t-1]$ where $y(t-\tau) = 1$ and $x(\tau) = 1$; therefore, the convolution integral (i.e. the overlapping surface area) results in

$$z(t) = \int_{-\infty}^{\infty} x(\tau)\, y(t - \tau)\, d\tau = \int_{0}^{t-1} (1)\,(1)\, d\tau = \tau \Big|_{0}^{t-1}$$

$$= t - 1 \quad (1 \leq t \leq 3)$$

Case 3, $(t - 3) > 0$:
The trailing edge of $y(t - \tau)$ is shifted into the positive side, see Fig. 17.31, i.e.

$$t - 3 > 0 \quad \therefore \quad t > 3$$

Fig. 17.31
Example 17.2-2(case 3)

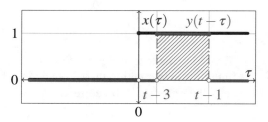

Therefore, the overlapping surface area is at its maximum (i.e. the full area of $\Pi\,(t)$ function, and henceforth stays constant)

$$z(t) = \int_{-\infty}^{\infty} x(\tau)\, y(t - \tau)\, d\tau = \int_{t-3}^{t-1} (1)\,(1)\, d\tau = \tau \Big|_{t-3}^{t-1} = (t - 1) - (t - 3)$$

$$= 2 \quad (t > 3)$$

Summary:
The total convolution integral shows progression of the overlapping, Fig. 17.32, as

$$z(t) = \begin{cases} 0 & t \le 1 \\ t-1 & 1 \le t \le 3 \\ 2 & t > 3 \end{cases}$$

Fig. 17.32 Example 17.2-2

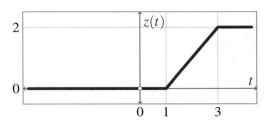

3. By applying the same technique as in the previous examples, we derive $u(\tau)$ and $u(t-\tau)$ functions by using graphical method, see Fig. 17.33.

After the first step, Fig. 17.33a, b, the change of variable t into provisional integration variable τ does not affect the initial form of $u(t)$. Then, $u(\tau)$ is simply mirrored around the vertical axis then shifted horizontally by t, Fig. 17.33c, d.

Fig. 17.33 Example 17.2-3

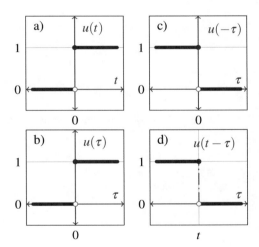

There are two distinct cases to solve.

Case 1 : $t \le 0$:
The leading edge of $u(t-\tau)$ is still in the negative side, i.e.

$$t \le 1$$

It is irrelevant where exactly t is relative to the origin, see Fig. 17.34. What is important is that under this condition, because always either of the two function equals zero, their product is therefore

$$u(\tau)\,u(t-\tau) = 0, \quad \forall \tau$$

$$\therefore$$

$$z(t) = \int_{-\infty}^{\infty} u(\tau)\, u(t-\tau)\, d\tau = \int_{-\infty}^{\infty} 0\, d\tau$$

$$= 0 \quad (t \le 0)$$

Fig. 17.34
Example 17.2-3 (case 1)

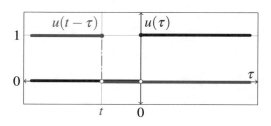

Case 2 : $t > 0$:

The leading edge of $u(t - \tau)$ is shifted into the positive side, Fig. 17.35, i.e.

$$t > 1$$

thus,

$$z(t) = \int_{-\infty}^{\infty} u(\tau)\, u(t-\tau)\, d\tau = \int_{0}^{t} (1)\,(1)\, d\tau = \tau \,\Big|_{0}^{t}$$

$$= t \quad (t > 0)$$

(Surface of the overlapping rectangular area is calculated as height (equals one) times its length (equals t).)

Fig. 17.35
Example 17.2-3 (case 2)

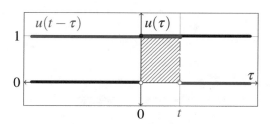

Summary:

The total convolution integral is therefore $r(t)$ function, see Fig. 17.36, or written in equally elegant product form of linear and step functions as

$$z(t) = \begin{cases} 0 & t \le 0 \\ t & t > 0 \end{cases} = r(t) \text{ or, } z(t) = t\, u(t)$$

Fig. 17.36
Example 17.2-3 (summary)

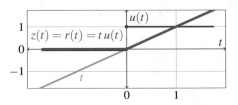

4. We synthesize $h(t) = [u(t + 2) - u(t)]$ as the sum of $u(t)$ functions as illustrated in Fig. 17.37.

Fig. 17.37 Example 17.2-4

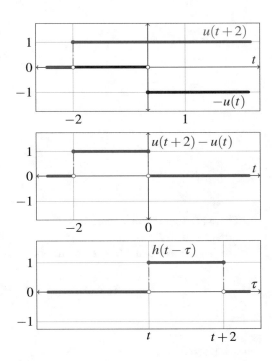

While the syntheses of $x(t) = [u(t + 1) - u(t - 1)]\, t$ as in the following steps, see Fig. 17.38.

Fig. 17.38 Example 17.2-4

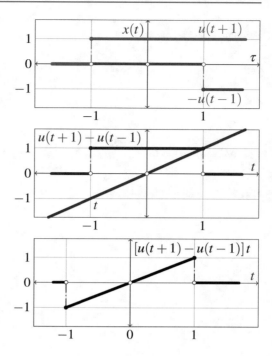

There are four distinct cases to solve.

Case 1 : $t + 2 \leq -1$:

The leading edge of $h(t - \tau)$ is still far away from $\tau = -1$, that is to say

$$t + 2 \leq -1 \quad \therefore \quad t \leq -3$$

Fig. 17.39
Example 17.2-4 (case 1)

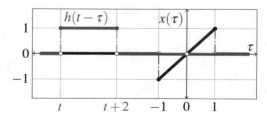

That being the case, it is irrelevant where exactly t is relative to the origin, see Fig. 17.39. What is important is that under this condition, because always either of the two function equals zero, their product is therefore

$$x(\tau)\, h(t - \tau) = 0, \quad \forall \tau$$

$$\therefore$$

$$y(t) = \int_{-\infty}^{\infty} x(\tau)\, h(t - \tau)\, d\tau = \int_{-\infty}^{\infty} 0\, d\tau$$

$$= 0 \quad (t \le -3)$$

Case 2 : $t + 2 > -1$ *and* $t + 2 \le 1$

The leading edge of $h(t - \tau)$ is found within the interval $\tau \in [-1, 1]$, that is to say

$$t + 2 > -1 \quad \therefore \quad t > -3 \ \text{ and}$$

$$t + 2 \le 1 \quad \therefore \quad t \le -1$$

Fig. 17.40
Example 17.2-4 (case 2)

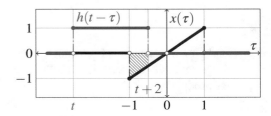

Within these boundaries of t, see Fig. 17.40, the overlapping surface is found in the interval $\tau \in [-1, t + 2]$ where $h(t - \tau) = 1$ and $x(\tau) = t$, consequently

$$y(t) = \int_{-\infty}^{\infty} x(\tau) \, h(t - \tau) \, d\tau = \int_{-1}^{t+2} (\tau) \, (1) \, d\tau = \frac{1}{2} \, \tau^2 \, \Big|_{-1}^{t+2}$$

$$= \frac{1}{2} \left[(t + 2)^2 - (-1)^2 \right] = \frac{1}{2} \left(t^2 + 4t + 3 \right)$$

$$= \frac{1}{2} (t + 1)(t + 3), \quad (-3 \le t \le -1)$$

Case 3 : $t > -1$ *and* $t \le 1$

The trailing edge of $h(t - \tau)$ is found within the interval $\tau \in [-1, 1]$, that is to say

$$t > -1 \ \text{ and } \ t \le 1$$

Fig. 17.41
Example 17.2-4 (case 3)

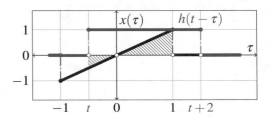

Within these boundaries of t, see Fig. 17.41, the overlapping surface is found in the interval $\tau \in [t, 1]$ where $h(t - \tau) = 1$ and $x(\tau) = t$, consequently

$$y(t) = \int_{-\infty}^{\infty} x(\tau)\, h(t-\tau)\, d\tau = \int_{t}^{1} (\tau)\,(1)\, d\tau = \frac{1}{2}\, \tau^2 \Big|_{t}^{1}$$

$$= \frac{1}{2}(1 - t^2) = \left\{ (a^2 - b^2) = (a-b)(a+b) \right\}$$

$$= -\frac{1}{2}(t-1)(t+1), \quad (-1 \le t \le 1)$$

Case 4 : $t > 1$

The trailing edge of $h(t-\tau)$ is found after $\tau = 1$, that is to say

$$t > 1$$

That being the case, it is irrelevant where exactly t is relative to the origin, see Fig. 17.42. What is important is that under this condition, because always either of the two function equals zero, their product is therefore

$$x(\tau)\, h(t-\tau) = 0, \quad \forall \tau$$

$$\therefore$$

$$y(t) = \int_{-\infty}^{\infty} x(\tau)\, h(t-\tau)\, d\tau = \int_{-\infty}^{\infty} 0\, d\tau$$

$$= 0 \quad (t > 1)$$

Fig. 17.42
Example 17.2-4 (case 4)

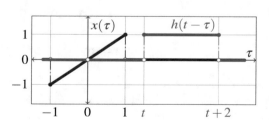

Summary:
The total convolution integral is therefore $y(t)$ function, see Fig. 17.43, as

$$y(t) = \begin{cases} 0 & t < -3 \\ \frac{1}{2}(t+1)(t+3) & (-3 \le t \le -1) \\ -\frac{1}{2}(t-1)(t+1) & (-1 \le t \le 1) \\ 0 & t > 1 \end{cases}$$

Fig. 17.43
Example 17.2-3 (summary)

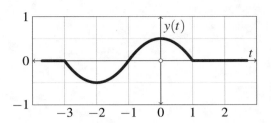

5. We derive $x(t) = \Lambda(\tau)$ and $x(t-\tau) = \Lambda(t-\tau)$ functions by using graphical method, see Fig. 17.44. After the first step, Fig. 17.44a, b, the change of variable t into provisional integration variable τ does not affect the initial form of $x(t)$. Then, $x(\tau)$ is simply mirrored around the vertical axis then shifted horizontally by t, Fig. 17.44c, d.

Fig. 17.44 Example 17.2-5

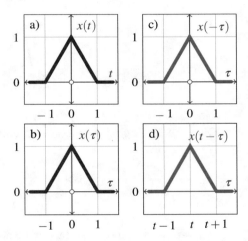

It is important to specifically write linear equations that correspond to each linear section of the given $x(t)$ and $x(t-\tau)$ as:

$$\left.\begin{array}{ll} x(t) = & 1+t, \quad -1 \geq t \geq 0 \\ x(t) = & 1-t, \quad 0 \leq t \leq 1 \\ x(t) = & 0, \quad |t| \geq 1 \end{array}\right\} \quad \therefore \quad \left\{\begin{array}{ll} x(t-\tau) = & 1+\tau-t, \quad t-1 \leq \tau \leq t \\ x(t-\tau) = & 1-\tau+t, \quad t \leq \tau \leq t+1 \\ x(t-\tau) = & 0, \quad \text{sinon} \end{array}\right.$$

We note that, due to $\Lambda(t)$ being a pair function, $x(\tau)$ and $x(-\tau)$ functions have identical linear equations even though the argument τ is transformed into $-\tau$, Fig. 17.44. In order to avoid the calculation errors, we must pay attention which two equations are found in any given interval, as illustrated while resolving the following cases:
There are six distinct cases to solve.
Case 1 : $t+1 < -1$
The leading non-zero point at $(t+1)$ of $x(t-\tau)$ is still in the left side of $\tau = -1$, i.e.

$$t+1 < -1 \quad \therefore \quad t < -2$$

Fig. 17.45
Example 17.2-5 (case 1)

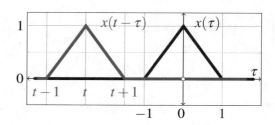

That being the case, it is irrelevant where exactly t is relative to the origin, see Fig. 17.45. What is important is that within these boundaries of t, because always either of the two function equals zero, therefore their product is

$$x(\tau)\,x(t-\tau) = 0, \quad \forall \tau$$

$$\therefore$$

$$z(t) = \int_{-\infty}^{\infty} x(\tau)\,x(t-\tau)\,d\tau = \int_{-\infty}^{\infty} 0\,d\tau$$

$$= 0 \;\; (t < -2)$$

Case 2 : $\; t+1 \geq -1, \; t+1 \leq 0$

Once the leading edge of $x(t-\tau)$ found at $(t+1)$ crosses $\tau = -1$ while at the same time $(t < -1)$, i.e.

$$\left.\begin{array}{ll} t+1 \geq -1 & \therefore \;\; t \;\; \geq -2 \text{ and} \\ & \qquad\qquad t \;\; < -1 \end{array}\right\} \quad \therefore \;\; -2 \leq t \leq -1$$

a non-zero overlapping surface is created by $(1-\tau+t)$ and $(1+\tau)$ linear segments, which defines interval $(a) : [-1, t+1]$ in Fig. 17.46.

Fig. 17.46
Example 17.2-5 (case 2)

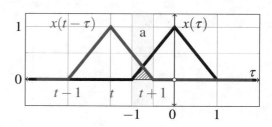

Only in this interval the convolution product is defined with these two linear segments:

$$z(t) = \int_{-\infty}^{\infty} x(\tau)\,x(t-\tau)\,d\tau = \int_{-1}^{t+1} (1+\tau)\,(1-\tau+t)\,d\tau$$

$$= \int_{-1}^{t+1} (1 - \not\tau + t + \not\tau + t\tau - \tau^2)\,d\tau$$

$$= (1+t)\tau \bigg|_{-1}^{t+1} + \frac{t}{2}\tau^2 \bigg|_{-1}^{t+1} - \frac{1}{3}\tau^3 \bigg|_{-1}^{t+1}$$

$$= \cdots = \frac{(t+2)^3}{6}$$

after factorizing third order polynomial (hint: use Pascal triangle for binomial power development), and while keeping in mind that t is considered a constant while integrating relative to τ variable.

Case 3 : $t+1 \geq 0$, $t+1 \leq 1$

The leading edge of $x(t-\tau)$ is found in $\tau \in [0, 1]$, that is to say

$$\left.\begin{aligned} t+1 > 0 \;\; \therefore \;\; t > \;\; -1 \;\; \text{and} \\ t+1 \leq 1 \;\; \therefore \;\; t < \;\; 0 \end{aligned}\right\} \;\; \therefore \;\; -1 < t \leq 0$$

where, non-zero surface is found in $\tau \in [-1, t+1]$ interval, Fig. 17.47. However, actually there are three distinct sub-intervals (a), (b), (c),

$$(a) \; [-1, t],$$

$$(b) \; [t, 0],$$

$$(c) \; [0, t+1]$$

where in each sub-interval the convolution product is created by distinct pairs of linear segments, see Fig. 17.47.

Fig. 17.47
Example 17.2-5 (case 3)

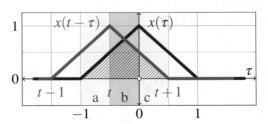

Therefore, we split the convolution integral into the sum of three integrals so that the overlapping areas are correctly calculated as

$$z(t) = \int_{-\infty}^{\infty} x(\tau)\, x(t-\tau)\, d\tau = \int_{-1}^{t} (1+\tau)\,(1+\tau-t)\, d\tau$$

$$+ \int_{t}^{0} (1+\tau)\,(1-\tau+t)\, d\tau$$

$$+ \int_{0}^{t+1} (1-\tau)\,(1-\tau+t)\, d\tau$$

$$= \cdots = \frac{1}{6}(-t^3 + 3t + 2)$$

$$-\frac{1}{6}t\,(t^2 + 6t + 6)$$

$$+\frac{1}{6}(-t^3 + 3t + 2)$$

$$=\frac{1}{3}(-t^3 + 3t + 2) - \frac{1}{6}t\,(t^2 + 6t + 6), \quad (-1 < t \le 0)$$

Case 4 : $t - 1 \ge -1, \; t - 1 \le 0$

The trailing edge of $x(t - \tau)$ is found in $\tau \in [-1, 0]$, that is to say

$$\left.\begin{array}{ll} t - 1 \ge -1 & \therefore \;\; t > 0 \;\; \text{and} \\[4pt] t - 1 < 0 & \therefore \;\; t < 1 \end{array}\right\} \;\; \therefore \;\; 0 < t \le 1$$

where, non-zero surface is found in $\tau \in [t - 1, 1]$ interval, Fig. 17.48. Again, there are three distinct sub-intervals $(a), (b), (c),$

$$(a)\;[t - 1, 0],$$

$$(b)\;[0, t],$$

$$(c)\;[t, 1]$$

say Fig. 17.48, where in each sub-interval the convolution product is created by distinct pairs of linear segments.

Fig. 17.48
Example 17.2-5 (case 4)

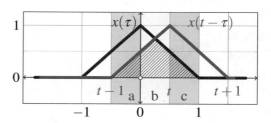

Therefore, we split the convolution integral into the sum of three integrals so that the overlapping areas are correctly calculated as

$$z(t) = \int_{-\infty}^{\infty} x(\tau)\,x(t - \tau)\,d\tau = \int_{t-1}^{0} (1 + \tau)\,(1 + \tau - t)\,d\tau$$

$$+ \int_{0}^{t} (1 - \tau)\,(1 + \tau - t)\,d\tau$$

$$+ \int_{t}^{1} (1 - \tau)\,(1 - \tau + t)\,d\tau$$

$$= \cdots$$

$$= \frac{1}{3}(t^3 - 3t + 2) + \frac{1}{6}t\,(t^2 - 6t + 6), \quad (0 < t \le 1)$$

Case 5 : $t - 1 \geq 0, \quad t - 1 \leq 1$

The trailing edge of $x(t - \tau)$ is found in $\tau \in [0, 1]$, that is to say

$$
\left.
\begin{array}{ll}
t - 1 \geq 0 & \therefore \ t > 1 \ \text{and} \\[4pt]
t - 1 < 1 & \therefore \ t < 2
\end{array}
\right\} \quad \therefore \ 1 \leq t \leq 2
$$

where non-zero surface is found in one distinct interval, see Fig. 17.49:

$$(a) \ [t - 1, 1]$$

Fig. 17.49
Example 17.2-5 (case 5)

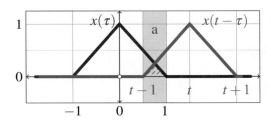

Therefore, we calculate the convolution integral as

$$
z(t) = \int_{t-1}^{1} (1 - \tau) \ (1 + \tau - t) \, d\tau
$$

$$
= \cdots
$$

$$
= -\frac{1}{6}(t - 2)^3
$$

Case 6 : $t - 1 > 1$

The trailing edge of $x(t - \tau)$ is outside of $x(\tau)$ non-zero region, i.e.

$$
t - 1 > 1 \ \therefore \ t > 2
$$

Fig. 17.50
Example 17.2-5 (case 6)

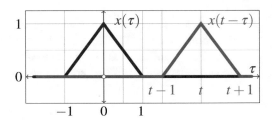

It is irrelevant where exactly t is relative to the origin, see Fig. 17.50. What is important is that under this condition, because always either one of the two function equals zero, the product of

$$x(\tau)\,x(t-\tau) = 0, \quad \forall \tau$$

$$\therefore$$

$$z(t) = \int_{-\infty}^{\infty} x(\tau)\,x(t-\tau)\,d\tau = \int_{-\infty}^{\infty} 0\,d\tau$$

$$= 0 \quad (t \geq 2)$$

Summary:

The total convolution integral is therefore the sum of these six cases, see also Fig. 17.51.

$$z(t) = \begin{cases} 0 & t < -2 \\ \frac{1}{6}(t+2)^3 & -2 \leq t \leq -1 \\ \frac{1}{3}(-t^3 + 3t + 2) - \frac{1}{6}t(t^2 + 6t + 6) & -1 \leq t \leq 0 \\ \frac{1}{3}(t^3 - 3t + 2) + \frac{1}{6}t(t^2 - 6t + 6) & 0 \leq t \leq 1 \\ -\frac{1}{6}(t-2)^3 & 1 \leq t \leq 2 \\ 0 & t > 2 \end{cases}$$

Fig. 17.51 Example 17.2-5

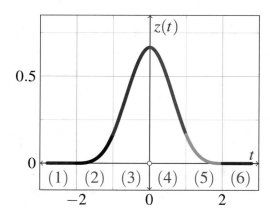

Exercise 17.3, page 360

1. We derive $x(\tau)$ and $y(t-\tau)$ functions by using graphical method, see Fig. 17.52. After the first transformation, Fig. 17.52a, b, the change of variable t into provisional integration variable τ does not affect the initial form of $x(t)$. In the following transformation, $y(\tau)$ is simply mirrored around the vertical axis then shifted horizontally by t, Fig. 17.52c, d.

Fig. 17.52 Example 17.2-1

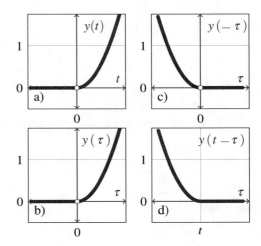

We note that, given

$$y(t) = t^2 \, u(t)$$

$$\therefore$$

$$y(t - \tau) = (t - \tau)^2 \, u(t - \tau)$$

There are two distinct cases to solve.

Case 1 : $t \le 0$:

The last non-zero point t of $y(t - \tau)$ is still in the negative side, i.e.

$$t \le 0$$

Fig. 17.53
Example 17.2-1 (case 1)

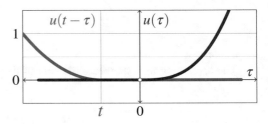

It is irrelevant where exactly t is relative to the origin, see Fig. 17.53. What is important is that under this condition, because always either one of the two function equals zero, the product of

$$u(\tau) \, u(t - \tau) = 0, \quad \forall \tau$$

$$\therefore$$

$$z(t) = \int_{-\infty}^{\infty} u(\tau) \, u(t - \tau) \, d\tau$$

$$= \int_{-\infty}^{\infty} 0 \, d\tau = 0$$

Case 2 :   t > 0:
The last non-zero point t of $y(t - \tau)$ is shifted into the positive side, Fig. 17.54 i.e.

$$t > 1$$

Fig. 17.54
Example 17.2-1 (case 2)

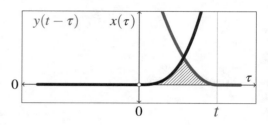

Therefore,

$$z(t) = \int_{-\infty}^{\infty} u(\tau) \, u(t - \tau) \, d\tau = \int_{-\infty}^{\infty} \tau^3 u(\tau) \, (t - \tau)^2 u(t - \tau) \, d\tau$$

$$= \int_{0}^{t} \tau^3 \, (t - \tau)^2 \, d\tau = \int_{0}^{t} (\tau^5 - 2t\tau^4 + t^2\tau^3) \, d\tau$$

$$= \frac{1}{6}\tau^6 \Big|_0^t - \frac{2t}{5}\tau^5 \Big|_0^t + \frac{t^2}{4}\tau^4 \Big|_0^t = \frac{t^6}{6} - \frac{2t^6}{5} + \frac{t^6}{4}$$

$$= \frac{t^6}{60}$$

Summary:
The total integral is therefore the sum of these two cases, see Fig. 17.55,

$$z(t) = \begin{cases} 0 & t \le 0 \\ \dfrac{t^6}{60} & t > 0 \end{cases} = \frac{t^6}{60} \, u(t)$$

Fig. 17.55
Example 17.2-1 (summary)

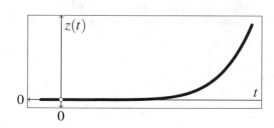

2. Given,

$$x(t) = \begin{cases} 1-t; & 0 \le t \le 1 \\ 0; & \text{otherwise} \end{cases} \quad \text{and,} \quad v(t) = \begin{cases} e^{-t}; & t \ge 0 \\ 0; & \text{otherwise} \end{cases}$$

the two functions $x(t)$ and $v(t)$ are also defined as, see Fig. 17.56.

$$x(t) = (1-t)\,\Pi\,(t) \quad \text{and} \quad v(t) = e^{-t}\,u(t)$$

Fig. 17.56 Example 17.3-2

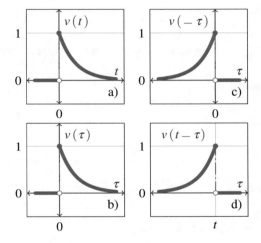

Numerically and geometrically, solution of a convolution integral equals to the area the overlapping surface between $x(\tau)$ and $v(t-\tau)$ functions. There are three distinct cases to be solved.

Case 1 : $t \le 0$

Within this interval, either one of the two function equals zero, see Fig. 17.57, thus the product is

$$x(\tau)\,v(t-\tau) = 0, \quad \forall \tau$$

$$\therefore$$

$$w(t) = \int_{-\infty}^{\infty} x(\tau)\,v(t-\tau)\,d\tau = \int_{0}^{t-1} (0)\,d\tau$$

$$= 0 \ (\ t \le 0)$$

Fig. 17.57
Example 17.3-2 (case 1)

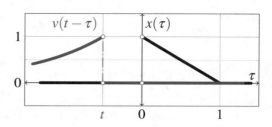

Case 2 : $0 \le t \le 1$:
The leading edge of $v(t - \tau)$ is found in interval $\tau \in [0, 1]$, that is to say

$$t > 0 \ge 0 \ \text{ and } \ t \le 1$$

where non-zero overlapping surface is found in one distinct interval $\tau \in [0, t]$, see Fig. 17.58.

Fig. 17.58
Example 17.3-2 (case 2)

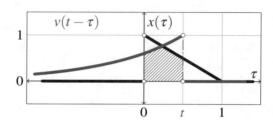

By definition we calculate,

$$w(t) = \int_{-\infty}^{\infty} x(\tau)\, v(t - \tau)\, d\tau = \int_{0}^{t} (1 - \tau)\, e^{-(t-\tau)}\, d\tau = \int_{0}^{t} \left(e^{-t} e^{\tau} - \tau\, e^{-t} e^{\tau} \right) d\tau$$

$$= e^{-t} \left[\int_{0}^{t} e^{\tau}\, d\tau - \int_{0}^{t} \tau\, e^{\tau}\, d\tau \right] = e^{-t} \left[I_1 - I_2 \right]$$

Where,

$$I_1 = \int_{0}^{t} e^{\tau}\, d\tau = e^{\tau} \Big|_{0}^{t} = e^{t} - 1$$

and,

$$I_2 = \int_{0}^{t} \tau\, e^{\tau}\, d\tau = \{\text{partial integration}\} = t\, e^{t} - e^{t} + 1$$

Therefore,

$$w(t) = e^{-t} \left[e^{t} - 1 - t\, e^{t} + e^{t} - 1 \right]$$

$$= 2 - t - 2\, e^{-t} \quad (0 \le t \le 1)$$

Case 3 : $t \geq 1$

The leading edge of $v(t - \tau)$ is found at $t > 1$], that is to say that non-zero overlapping surface is found in one distinct interval $\tau \in [0, 1]$, see Fig. 17.59.

Fig. 17.59
Example 17.3-2 (case 3)

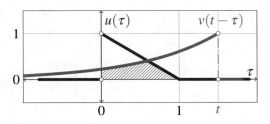

Therefore, by definition we find

$$w(t) = \int_{-\infty}^{\infty} x(\tau)\, v(t - \tau)\, d\tau = \int_{0}^{1} (1 - \tau)\, e^{-(t-\tau)}\, d\tau$$

$$= e^{-t} \int_{0}^{1} (e^{\tau} - \tau\, e^{\tau})\ d\tau = \{\text{etc. as in Case 2.}\,\}$$

$$= e^{-t}(e - 2) \quad (t \geq 1)$$

Summary:

The total convolution integral is therefore, see Fig. 17.60,

$$w(t) = \begin{cases} 0; & t \leq 0 \\ 2 - t - 2\,e^{-t}; & 0 \leq\ t \leq 1 \\ e^{-t}(e - 2); & t \geq 1 \end{cases}$$

Fig. 17.60 Example 17.3-2

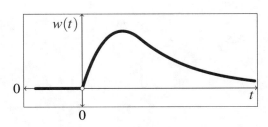

Exercise 17.4, page 360

1. (a) Given a continuous non-periodic infinitely long signal, we calculate its average power as

$$\langle P(t) \rangle \overset{\text{def}}{=} \lim_{T \to \infty} \frac{1}{T} \int_{-T/2}^{T/2} |x(t)|^2 \, dt = \lim_{T \to \infty} \frac{1}{T} \int_{-T/2}^{T/2} |A|^2 \, dt = A^2 \lim_{T \to \infty} \frac{1}{T} \int_{-T/2}^{T/2} dt$$

$$= A^2 \lim_{T \to \infty} \frac{1}{T} \, t \, \Big|_{-T/2}^{T/2} = A^2 \lim_{T \to \infty} \frac{1}{T} \left(\frac{T}{2} - \frac{-T}{2} \right) = A^2 \lim_{T \to \infty} \frac{\cancel{T}}{\cancel{T}}$$

$$= A^2$$

However,

$$E(t) \overset{\text{def}}{=} \int_{-\infty}^{\infty} |x(t)|^2 \, dt = \int_{-\infty}^{\infty} |A|^2 \, dt = A^2 \int_{-\infty}^{\infty} dt$$

$$= A^2 \, t \, \Big|_{-\infty}^{\infty} = A^2 \left(\infty + \infty \right)$$

$$= \infty$$

That is to say, in order to sustain the average non-zero power over infinitely long period of time, evidently, it is necessary to have infinite source of energy.

(b) Given a continuous periodic infinitely long signal, we calculate its average power over one period only as

$$\langle P(t) \rangle \overset{\text{def}}{=} \frac{1}{T} \int_{-T/2}^{T/2} |\sin(t)|^2 \, dt = \frac{1}{T} \int_{-T/2}^{T/2} \sin^2(t) \, dt$$

$$= \left\{ \sin^2(t) = \frac{1 - \cos 2x}{2} \right\}$$

$$= \frac{1}{T} \int_{-T/2}^{T/2} \left(\frac{1 - \cos 2x}{2} \right) dt$$

$$= \frac{1}{T} \left[\frac{1}{2} t - \frac{1}{4} \sin(2t) \right]_{-T/2}^{T/2}$$

$$= \frac{1}{T} \left[\frac{1}{2} \left(\frac{T}{2} + \frac{T}{2} \right) - \frac{1}{4} \left(\sin \left(2\frac{T}{\cancel{2}} \right) - \sin \left(-2\frac{T}{\cancel{2}} \right) \right) \right]$$

$$= \frac{1}{\cancel{T}} \left[\frac{1}{2} \cancel{T} - \frac{1}{4} \left(\cancel{\sin(T)}^{\,0} + \cancel{\sin(T)}^{\,0} \right) \right]$$

$$= \frac{1}{2}$$

Again, in order to sustain the non-zero average power over infinitely long period of time, there must be an infinite source of energy

$$E(t) \overset{\text{def}}{=} \lim_{t \to \infty} \langle P(t) \rangle \, t = \lim_{t \to \infty} \frac{1}{2} \, t = \infty$$

(c)   Given a converging infinitely long signal, we calculate its total energy as

$$E(t) \overset{\text{def}}{=} \int_{-\infty}^{\infty} |x(t)|^2 \, dt = \int_{-\infty}^{\infty} \left| e^{-t} \, u(t) \right|^2 \, dt = \int_{0}^{\infty} e^{-2t} \, (1)^2 \, dt = -\frac{1}{2} \, e^{-2t} \, \Big|_{0}^{\infty}$$

$$= -\frac{1}{2} \, [0 - 1] = \frac{1}{2}$$

Therefore,

$$\langle P(t) \rangle \overset{\text{def}}{=} \lim_{T \to \infty} \frac{1}{T} E(t) = 0$$

That is to say, having a limited source of energy that must be distributed over infinitely long period of time, the average power reduces to zero.

(d)   Given a complex signal, we use Euler's formula to write its period forms as

$$x(t) = A \, e^{j \omega_0 t} = A \left[\cos(\omega_0 t) - j \sin(\omega_0 t) \right] \quad \therefore \quad \omega_0 T_0 = 2\pi \quad \therefore \quad T_0 = \frac{2\pi}{\omega_0}$$

so that average power is calculated as

$$\langle P(t) \rangle \overset{\text{def}}{=} \frac{1}{T_0} \int_{-T_0/2}^{T_0/2} |A \, e^{j \omega_0 t}|^2 \, dt = \frac{1}{T_0} \int_{-T_0/2}^{T_0/2} A^2 \, \underbrace{|e^{j \omega_0 t}|^2}_{1} \, dt$$

$$= \frac{A^2}{T_0} \, t \, \Big|_{-T_0/2}^{T_0/2} = \frac{A^2}{T_0} \left(\frac{T_0}{2} + \frac{T_0}{2} \right) = A^2$$

This is infinitely long signal, thus in order to sustain non-zero average power over infinitely long time it must be that $E(t) = \infty$.

(e) Given an infinitely long signal, its energy is

$$E(t) \stackrel{\text{def}}{=} \int_{-\infty}^{\infty} |x(t)|^2 \, dt = \int_{-\infty}^{\infty} |u(t)|^2 \, dt = \int_{0}^{\infty} (1)^2 \, dt = t \Big|_{0}^{\infty} = \infty$$

Therefore, its average power is

$$\langle P(t) \rangle \stackrel{\text{def}}{=} \lim_{T \to \infty} \frac{1}{T} \int_{-T/2}^{T/2} |x(t)|^2 \, dt = \lim_{T \to \infty} \frac{1}{T} \int_{-T/2}^{T/2} |u(t)|^2 \, dt$$

$$= \lim_{T \to \infty} \frac{1}{T} \int_{0}^{T/2} (1)^2 \, dt = \lim_{T \to \infty} \frac{1}{T} \, t \Big|_{0}^{T/2} = \lim_{T \to \infty} \frac{1}{\cancel{T}} \frac{\cancel{T}}{2}$$

$$= \frac{1}{2}$$

which makes sense because half of the time $u(t)$ signal amplitude equals zero (for $t \leq 0$) and the other half of time (i..e. $t \geq 0$) it equals to one.

(f) Amplitude of delayed port signal $x(t) = 2\Pi(t - 1)$ equals two for $t \in [0, 1]$, otherwise it equals zero. Thus we write,

$$E(t) \stackrel{\text{def}}{=} \int_{-\infty}^{\infty} |x(t)|^2 \, dt = \int_{-\infty}^{\infty} |2\Pi(t - 1)|^2 \, dt = \int_{0}^{1} (2)^2 \, dt = 4 < \infty$$

which is to say that its average power over the infinitely long time interval must be $\langle P(t) \rangle = 0$.

Discrete Convolution Sum

18

Important to Know

Convolution sum: It is defined as

$$y[n] = x[k] * y[n] = \sum_{k=-\infty}^{\infty} x[k]\, y[n-k] \tag{18.1}$$

where, k is the provisional summation variable.

Energy and power definitions: By definition, $E(t) = \int_0^t P(\tau)d\tau$ for continuous and discrete functions. Recall that $|x(t)|^2 = x(t)\, x^*(t)$.

1. Energy within an interval

$$E_x = \int_{t_1}^{t_2} |x(t)|^2 \, dt, \qquad E_n = \sum_{k=n}^{m} |x[k]|^2$$

2. Total energy

$$E_x = \int_{-\infty}^{\infty} |x(t)|^2 \, dt, \qquad E_n = \sum_{k=-\infty}^{\infty} |x[k]|^2$$

3. Average power within an interval

$$\langle P_x \rangle = \frac{1}{t_2 - t_1} \int_{t_1}^{t_2} |x(t)|^2 \, dt, \qquad \langle P_n \rangle = \frac{1}{2n+1} \sum_{k=n}^{n} |x[k]|^2$$

4. Total power

$$P_x = \lim_{\tau \to \infty} \frac{1}{\tau} \int_{-\tau/2}^{\tau/2} |x(t)|^2 \, dt, \qquad P_n = \lim_{n \to \infty} \frac{1}{2n+1} \sum_{k=n}^{n} |x[k]|^2$$

© Springer Nature Switzerland AG 2021
R. Sobot, *Engineering Mathematics by Example*,
https://doi.org/10.1007/978-3-030-79545-0_18

18.1   Exercises

18.1 **   Convolution, Basic Properties

1. Given an arbitrary function $h[k]$ and $\delta[k]$ Dirac function, calculate the convolution sum $y[k] = h[k] * \delta[k - m]$.
2. Derive proof to show that convolution sum is commutative.
3. Show the distributive property of convolution sum.
4. Show the associative property of convolution sum.
5. Derive proof to show the time-shifting property of convolution sum, i.e.

$$h_1[k - m] * h_2[k - n] = f[k - m - n]$$

18.2 ***   Convolution, Discrete Functions

1. Given two short sequences $f[k]$ and $g[k]$, each with only two samples as in Fig. 18.1, calculate

$$h[k] = f[k] * g[k]$$

Fig. 18.1   Example 18.2-1

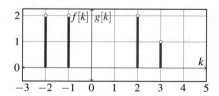

2. Given two long sequences $x[k]$ and $h[k]$,

$$x[k] = \alpha^k u[k]; \quad (0 < \alpha < 1)$$

$$h[k] = u[k]$$

calculate $y[k] = x[k] * h[k]$

3. Given two sequences $h[n] = [1, \underline{2}, 0, -3]$, and $x[n]$ is given as

(a) $x[n] = \delta[n]$

(b) $x[n] = \delta[n+1] + \delta[n-2]$

(c) $x[n] = [\underline{1}, 1, 1]$

(d) $x[n] = [2, 1, \underline{-1}, -2, -3]$

Note: number whose index $n = 0$ is underlined.

4. Given two sequences $x[k]$ and $h[k]$, as

$$x[k] = \begin{cases} 1; & (0 \le k \le 4) \\ 0; & \text{otherwise} \end{cases} \quad \text{and} \quad h[k] = \begin{cases} \alpha^k; & (0 \le k \le 6) \\ 0; & \text{otherwise} \end{cases}$$

calculate $y[k] = x[k] * h[k]$

18.3 ***   Energy and Power of Discrete Functions

1. Given sequences $x[k]$, calculate their respective energies and powers.

(a) $x[k] = \delta[k]$

(b) $x[k] = \left(\dfrac{1}{4}\right)^k u[k]$

(c) $x[k] = e^{j\,10k} u[k]$

(d) $x[k] = \begin{cases} \alpha^k; & k \ge 0 \\ 0; & k < 0 \end{cases}$

(e) $x[k] = u[k]$

(f) $x[k] = \cos\left[\dfrac{\pi}{6} k\right]$

Solutions

Exercise 18.1, page 394

1. The product of an arbitrary function $h[k]$ and Dirac pulse at time m effectively "samples" only one point of the function while products at all other points equal zero. That is to say, using temporary summing variable n, we write

$$y[k] = h[k] * \delta[k-m] = \sum_{n=-\infty}^{\infty} h[n]\,\delta[(k-m)-n]$$

$$= \cdots + 0 + 0 + \cdots + h[n]\,\delta[(k-m)-n]\Big|_{n=k-m} + 0 + 0 + \cdots$$

$$= h[k-m]\,\delta[0]$$

$$= h[k-m]$$

2. The sum operation is commutative, that is to say

$$y[k] = h_1[k] * h_2[k] = h_2[k] * h_1[k]$$

and we can formally prove it as follows:

$$y[k] = h_1[k] * h_2[k] \stackrel{\text{def}}{=} \sum_{m=-\infty}^{\infty} h_1[m]\, h_2[k-m]$$

change of variables:

$$\begin{cases} n &= k - m \quad \therefore \quad m = k - n \\ &\therefore \\ m &= -\infty \quad \therefore \quad n = k - (-\infty) = +\infty \\ m &= \infty \quad \therefore \quad n = k - (+\infty) = -\infty \end{cases}$$

$$= \sum_{n=\infty}^{-\infty} h_1[k-n]\, h_2[n]$$

$$\left\{ \text{the order of addition is not relevant, thus} \right\}$$

$$= \sum_{n=-\infty}^{\infty} h_1[k-n]\, h_2[n]$$

$$= \sum_{n=-\infty}^{\infty} h_2[n]\, h_1[k-n]$$

$$\stackrel{\text{def}}{=} h_2[k] * h_1[k]$$

3. The sum operation is distributive, and we can formally prove it as follows:

$$y[k] = x[k] * \Big(h_1[k] + h_2[k] \Big)$$

$$= \sum_{m=-\infty}^{\infty} x[m] * \Big(h_1[k-m] + h_2[k-m] \Big)$$

$$= \sum_{m=-\infty}^{\infty} \Big(x[m] * h_1[k-m] + x[m] * h_2[k-m] \Big)$$

$$= \sum_{m=-\infty}^{\infty} x[m] * h_1[k-m] + \sum_{m=-\infty}^{\infty} x[m] * h_2[k-m]$$

$$\stackrel{\text{def}}{=} x[k] * h_1[k] + x[k] * h_2[k]$$

4. The sum operation is associative, that is to say

$$\Big(x[k] * h_1[k] \Big) * h_2[k] = x[k] * \Big(h_1[k] * h_2[k] \Big)$$

and we can formally prove it as follows:

$$y[k] = \left(x[k] * h_1[k]\right) * h_2[k] = \left(\sum_{m=-\infty}^{\infty} x[m]h_1[k-m]\right) * h_2[k]$$

$$= \sum_{n=-\infty}^{\infty} \left(\sum_{m=-\infty}^{\infty} x[m]h_1[n-m]\right) h_2[k-n]$$

$$= \sum_{n=-\infty}^{\infty} \sum_{m=-\infty}^{\infty} x[m]\, h_1[n-m]\, h_2[k-n]$$

$$= \sum_{m=-\infty}^{\infty} x[m] \sum_{n=-\infty}^{\infty} h_1[n-m]\, h_2[k-n]$$

Change of variable

$$\begin{cases} r & = k - n \quad \therefore \quad n = k - r \\ & \therefore \\ n & = -\infty \quad \therefore \quad r = k - (-\infty) = +\infty \\ n & = \infty \quad \therefore \quad r = k - (+\infty) = -\infty \end{cases}$$

therefore,

$$y[k] = \sum_{m=-\infty}^{\infty} x[m] \sum_{r=\infty}^{-\infty} h_1[k-r-m]\, h_2[r]$$

$$= \sum_{m=-\infty}^{\infty} x[m] \sum_{r=-\infty}^{\infty} h_2[r]\, h_1[(k-m)-r]$$

$$= \sum_{m=-\infty}^{\infty} x[m] \left(h_2[m] * h_1[k-m]\right)$$

$$= \sum_{m=-\infty}^{\infty} x[m] \left(\underbrace{h_1[k-m] * h_2[m]}_{z[k-m]}\right)$$

$$= \sum_{m=-\infty}^{\infty} x[m]\, z[k-m]$$

$$\overset{\text{def}}{=} x[k] * z[k]$$

$$= x[k] * \left(h_1[k] * h_2[k]\right)$$

5. The time-shifting property of convolution sum, i.e.

$$\underbrace{h_1[k-m]}_{\text{delayed by } m} * \underbrace{h_2[k-n]}_{\text{delayed by } n} = \underbrace{f[k-(m+n)]}_{\text{delayed by } m+n}$$

may be formally proven as follows:

$$f[k] = h_1[k] * h_2[k] = \sum_{r=-\infty}^{\infty} h_1[r]\, h_2[k-r]$$

However, delayed versions of h functions are

$$h_1[k-m] * h_2[k-n] = \sum_{r=-\infty}^{\infty} h_1[r-m]\, h_2[(k-n)-r]$$

$$= \left\{ \begin{array}{l} t \quad = r - m \quad \therefore \quad r = t + m \\[4pt] \quad \therefore \\[4pt] r \quad = -\infty \quad \therefore \quad t = (-\infty) - m = -\infty \\[4pt] r \quad = \infty \quad \therefore \quad t = (+\infty) - m = +\infty \end{array} \right\}$$

$$= \sum_{t=-\infty}^{\infty} h_1[t]\, h_2[(k-n)-(t-m)]$$

$$= \sum_{t=-\infty}^{\infty} h_1[t]\, h_2[(k-m-n)-t]$$

$$= \sum_{r=-\infty}^{\infty} h_1[r]\, h_2[(k-m-n)-r]$$

$$\stackrel{\text{def}}{=} f[k-m-n]$$

Exercise 18.2, page 394

1. Given two short sequences $f[k]$ and $g[k]$, the interval boundaries of convolution $h[k]$ series are determined as follows. Intervals of the two sequences are

$$f[-2] = 2,\ f[-1] = 2 \quad \therefore \quad \text{start point:}\ \ k = -2\ \text{and, end point:}\ \ k = -1$$

$$g[1] = 2,\ f[3] = 1 \quad \therefore \quad \text{start point:}\ \ k = 2\ \text{and, end point:}\ \ k = 3$$

$$\therefore$$

$$h[k] \quad \therefore \quad \text{start point:}\ \ k = -2 + 2 = 0$$

$$\therefore \quad \text{end point:}\ \ k = -1 + 3 = 2$$

that is to say, $h[k] = 0,\ \ 0 > k > 2$; in other words it is necessary to evaluate $h[k]$ for $k = [0, 1, 2]$. We write,

$$h[k] = f[k] * g[k] \stackrel{\text{def}}{=} \sum_{m=-\infty}^{\infty} f[m]\, g[k-m] = \sum_{m=-2}^{-1} f[m]\, g[k-m]$$

where the last sum is limited only to the interval $m = [-2, -1]$ where $f[m] \neq 0$.

$$\underline{k = 0}: \; h[0] = \sum_{m=-2}^{-1} f[m] \, g[0 - m] = f[-2] \, g[0 - (-2)] + f[-1] \, g[0 - (-1)]$$

$$= f[-2] \, g[2] + f[-1] \, g[1] = (2)(2) + (2)(0) = 4$$

$$\underline{k = 1}: \; h[1] = \sum_{m=-2}^{-1} f[m] \, g[1 - m] = f[-2] \, g[1 - (-2)] + f[-1] \, g[1 - (-1)]$$

$$= f[-2] \, g[3] + f[-1] \, g[2] = (2)(1) + (2)(2) = 6$$

$$\underline{k = 2}: \; h[2] = \sum_{m=-2}^{-1} f[m] \, g[2 - m] = f[-2] \, g[2 - (-2)] + f[-1] \, g[2 - (-1)]$$

$$= f[-2] \, g[4] + f[-1] \, g[3] = (2)(0) + (2)(1) = 2$$

In conclusion, $h[k] = [4, 6, 2]$, see Fig. 18.2. Note that value of $h[0]$ is underlined.

Fig. 18.2 Example 18.2-1

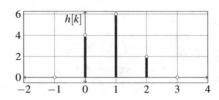

2. Given two long sequences $x[k]$ and $h[k]$, we use graphical technique to determine $x[m]$ and $h[k - m]$, see Fig. 18.3.

$$x[k] = \alpha^k \, u[k]; \quad (0 < \alpha < 1)$$
$$h[k] = u[k]$$

and similarly to Example 17.2-2 we decompose the problem into calculation of convolution integral within multiple intervals.

Fig. 18.3 Example 18.2-2

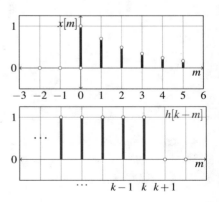

Case 1 : $k < 0$

There is no overlapping interval, consequently all products inside the convolution integrals equal zero, Fig. 18.4.

$$x[m]\, h[k - m] = 0; \quad \forall m$$

$$\therefore$$

$$y[k] = x[k] * h[k] = 0; \quad (k < 0)$$

Fig. 18.4 Example 18.2-2 (case 1)

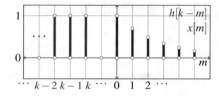

Case 2 : $k > 0$

There is overlapping interval, consequently there are products inside the convolution integrals non-equal zero, Fig. 18.5.

$$x[m]\, h[k - m] \neq 0; \quad m = [0, 1, \ldots, k]$$

Thus,

$$y[k] = x[k] * h[k] \stackrel{\text{def}}{=} \sum_{0}^{k} x[m]\, h[k - m]$$

$$= \underbrace{\sum_{0}^{k} \alpha^m}_{\text{geometric sum}} (1)$$

$$= \left\{ \sum_{n=0}^{N-1} \alpha^n = \begin{cases} N; & \alpha = 1 \\ \frac{1 - \alpha^N}{1 - \alpha}; & \alpha \neq 1 \end{cases} \right\}$$

$$= \frac{1 - \alpha^{k+1}}{1 - \alpha} \quad (k \geq 0)$$

Fig. 18.5 Example 18.2-2 (case 2)

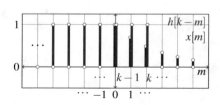

Fig. 18.6 Example 18.2-2

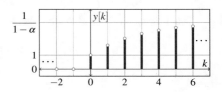

In summary, after adding the two cases together we write the solution in the compact form

$$y[k] = \frac{1 - \alpha^{k+1}}{1 - \alpha} \, u[k]$$

We note that (see Fig. 18.6), because $\alpha < 1$ then

$$\lim_{k \to \infty} \alpha^{k+1} \to 0 \quad \therefore \quad \lim_{k \to \infty} \frac{1 - \alpha^{k+1}}{1 - \alpha} \to \frac{1}{1 - \alpha}$$

3. Given two sequences $h[n] = [1, \underline{2}, 0, -3]$, and $x[n]$ is given as

(a) $y[n] = \delta[n] * h[n] = h[n] * \delta[n] = h[n]$

(b)   Given,

$$x[n] = \delta[n + 1] + \delta[n - 2]$$

First we calculate

$$y[n] = x[n] * h[n] = (\delta[n + 1] + \delta[n - 2]) * h[n]$$
$$= \delta[n + 1] * h[n] + \delta[n - 2] * h[n]$$
$$= h[n + 1] + h[n - 2]$$

Therefore, two shifted sequences $h[n + 1]$ and $h[n - 2]$ are

$$h[n] = [1, \underline{2}, 0, -3]$$

$$h[n + 1] = [1, 2, \underline{0}, -3] \text{ shifted one places to right}$$

$$h[n - 2] = [\,\_\,, 1, 2, 0, -3] \text{ shifted two places to left}$$

We tabulate the convolution sums as

n		-2	-1	0	1	2	3	4
$h[n + 1]$			1	2	0	-3		
$h[n - 2]$	$+$				1	2	0	-3
$y[n]$			1	2	0	-2	2	0	-3

That is to say, (see also Fig. 18.7)

$$y[n] = [1, 2, \underline{0}, -2, 2, 0, -3]$$

Fig. 18.7 Example 18.2-3

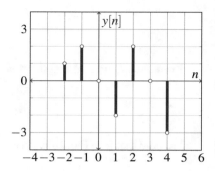

(c)   Given

$$x[n] = [\underline{1}, 1, 1] = \delta[n] + \delta[n-1] + \delta[n-2]$$

we calculate,

$$y[n] = x[n] * h[n] = (\delta[n] + \delta[n-1] + \delta[n-2]) * h[n]$$
$$= h[n] + h[n-1] + h[n-2]$$

We multiply as

n		-1	0	1	2
$h[n]$		1	2	0	-3
$x[n]$	$\times$		1	1	1
$y[n]$		1	2	0	-3

Therefore, knowing $y[n]$ we write

$$h[n-1] = [\underline{1}, 2, 0, -3]$$
$$h[n-2] = [\_, 1, 2, 0, -3]$$

and the total sum $h[n] + h[n-1] + h[n-2]$ is

n		−1	0	1	2	3	4
$h[n]$			1	2	0	−3	
$h[n-1]$				1	2	0	−3
$h[n-2]$	+			1	2	0	−3
$y[n]$			1	3	3	−1	−3	−3

That is to say, (see also Fig. 18.8)

$$y[n] = [1, \underline{3}, 3, -1, -3, -3]$$

Fig. 18.8 Example 18.2-3

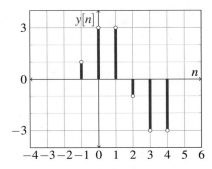

(d)   Given

$$x[n] = [2, 1, -1, -2, -3]$$

$$h[n] = [1, \underline{2}, 0, -3]$$

using the tabular method we calculate,

$$y[n] = x[n] * h[n]$$

for $n = [-2, -1, 0, 1, 2]$. We calculate the interval of the convolution sum

$$x[-2] = 2, \ldots, f[2] = -3 \quad \therefore \quad \text{start point:} \quad n = -2 \text{ and, end point:} \quad n = 2$$

$$h[-1] = 1, \ldots, f[2] = -3 \quad \therefore \quad \text{start point:} \quad n = -1 \text{ and, end point:} \quad n = 2$$

$$\therefore$$

$$y[n] \quad \therefore \quad \text{start point:} \quad n = (-2) + (-1) = -3$$

$$\therefore \quad \text{end point:} \quad k = 2 + 2 = 4$$

that is to say, it is necessary to evaluate $h[n]$ for $n = [-3, -2, -1, 0, 1, 2, 3, 4]$.

We "sweep" index of $x[n]$ and each time perform the multiplication $x[n] \times h[n]$ for each term in $h[n]$ and write the result in the corresponding column. For example, in $n = 2$ column we write

$$x[0]\, h[2] = (-1) \times (-3) = 3$$

$n = 0$

n			-2	-1	0	1	2
$h[n]$				1	2	0	-3
$x[0]$		$\times$			-1		
			2	1		-2	-3
				-1	-2	0	3

And the rest of products are as follows:

$n = -2$

n		-3	-2	-1	0	1	2
$h[n+2]$		1	2	0	-3		
$x[-2]$	$\times$		2				
				1	-1	-2	-3
		2	4	0	-6		

$n = -1$

n			-2	-1	0	1	2
$h[n+1]$			1	2	0	-3	
$x[-1]$		$\times$		1			
			2		-1	-2	-3
			1	2	0	-3	

$n = 1$

n			-2	-1	0	1	2	3
$h[n-1]$					1	2	0	-3
$x[1]$		$\times$				-2		
			2	1	-1		-3	
					-2	-4	0	6

$n = 2$

n	-2	-1	0	1	2	3	4
$h[n-2]$				1	2	0	-3
$x[2]$	$\times$				-3		
	2	1	-1	-2			
				-3	-6	0	9

All intermediate results are tabulated, we execute the convolution sum

$$y[n] = \sum_{k=-2}^{2} x[k]\, h[n-k]$$

in tabular form as,

n	-3	-2	-1	0	1	2	3	4
$x[-2]\,h[n+2]$	2	4	0	-6				
$x[-1]\,h[n+1]$		1	2	0	-3			
$x[0]\,h[n]$			-1	-2	0	3		
$x[1]\,h[n-1]$				-2	-4	0	6	
$x[2]\,h[n-2]$	+				-3	-6	0	9
$y[n]$	2	5	1	-10	-10	-3	6	9

Visually, tabular method illustrates the same "sweep" of $h[k-n]$ as we already saw in the graphics method. In conclusion,

$$y[n] = x[n] * h[n] \stackrel{\text{def}}{=} \sum_{k=-2}^{2} x[k]\, h[n-k] = [2, 5, 1, \underline{-10}, -10, -3, 6, 9]$$

as well as in Fig. 18.9.

Fig. 18.9 Example 18.2-3

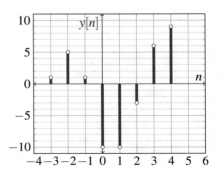

4. Given,

$$x[k] = \begin{cases} 1; & (0 \le k \le 4) \\ 0; & \text{otherwise} \end{cases} \qquad \text{and} \qquad h[k] = \begin{cases} \alpha^k; & (0 \le k \le 6) \\ 0; & \text{otherwise} \end{cases}$$

we calculate convolution sum as $y[k] = x[k] * h[k] \stackrel{\text{def}}{=} \displaystyle\sum_{m=-\infty}^{\infty} x[m]\, h[k-m]$. Using graphical method, we find $x[m]$ and $h[k-m]$ as in Fig. 18.10

Fig. 18.10 Example 18.2-4

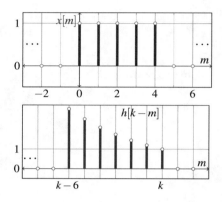

Case 1 : $k < 0$

There is no overlapping interval, Fig. 18.11, consequently all products inside the convolution integrals equal zero.

$$x[m]\, h[k-m] = 0; \quad \forall m$$

$$\therefore$$

$$y[k] = x[k] * h[k] = 0; \quad (k < 0)$$

Fig. 18.11 Example 18.2-4

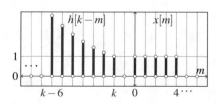

Case 2 : $0 \le k \le 4$

There is overlapping interval $[0, k]$, Fig. 18.12, thus the convolution integrals are calculated as follows:

$$y[k] = x[k] * h[k] = \sum_{m=0}^{k} x[m] \, h[k - m]$$

$$= \sum_{m=0}^{k} (1) \, \alpha^{k-m} = \sum_{m=0}^{k} \alpha^{k-m}$$

$$= \left\{ \begin{array}{ll} r = k - m & m = 0 \; \therefore \; r = k \\ & m = k \; \therefore \; r = 0 \end{array} \right\} \quad = \sum_{r=k}^{0} \alpha^{r} = \sum_{r=0}^{k} \alpha^{r} = \frac{1 - \alpha^{k+1}}{1 - \alpha}$$

Fig. 18.12 Example 18.2-4

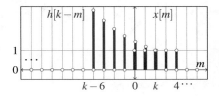

Case 3 : $5 \le k \le 6$

There is overlapping interval $[0, 4]$, Fig. 18.13, thus the convolution integrals are calculated as follows:

$$y[k] = x[k] * h[k] = \sum_{m=0}^{4} x[m] \, h[k - m]$$

$$= \sum_{m=0}^{4} (1) \, \alpha^{k-m} = \sum_{m=0}^{4} \alpha^{k} \, \alpha^{-m}$$

$$= \alpha^{k} \sum_{m=0}^{4} \left(\alpha^{-1} \right)^{m} = \alpha^{k} \frac{1 - \left(\alpha^{-1} \right)^{5}}{1 - \alpha^{-1}} = \frac{\alpha^{k} - \alpha^{k-5}}{1 - \alpha^{-1}} \; \frac{\alpha}{\alpha} = \frac{\alpha^{k+1} - \alpha^{k-4}}{\alpha - 1}$$

Fig. 18.13 Example 18.2-4

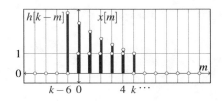

Case 4 : $7 \leq k \leq 10$

There is overlapping interval $[k-6, 4]$, Fig. 18.14, thus the convolution integrals are calculated as follows:

$$y[k] = x[k] * h[k] = \sum_{m=k-6}^{4} x[m]\, h[k-m]$$

$$= \sum_{m=k-6}^{4} (1)\, \alpha^{k-m}$$

$$= \left\{ \begin{array}{l} r = m - k + 6; \quad k - m = 6 - r \\ \qquad\qquad m = k - 6 \ \therefore \ r = 0 \\ \qquad\qquad m = 4 \ \therefore \ r = 10 - k \end{array} \right\}$$

$$= \sum_{r=0}^{10-k} \alpha^{6-r} = \alpha^6 \sum_{r=0}^{10-k} \left(\alpha^{-1}\right)^r = \alpha^6 \frac{1 - \alpha^{k-11}}{1 - \alpha^{-1}} = \frac{\alpha^7 - \alpha^{k-4}}{\alpha - 1}$$

Fig. 18.14 Example 18.2-4

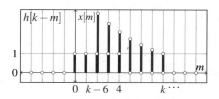

Case 5 : $k > 10$

There is no overlapping interval, Fig. 18.15, consequently all products inside the convolution integrals equal zero.

$$x[m]\, h[k-m] = 0; \quad \forall m$$

$$\therefore$$

$$y[k] = x[k] * h[k] = 0; \quad (k > 10)$$

Fig. 18.15 Example 18.2-4

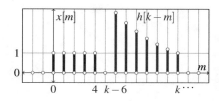

In summary,

$$y[k] = \begin{cases} 0 \; ; & k < 0 \\ \dfrac{1 - \alpha^{k+1}}{1 - \alpha} \; ; & 0 \leq k \leq 4 \\ \dfrac{\alpha^{k+1} - \alpha^{k-4}}{\alpha - 1} \; ; & 5 \leq k \leq 6 \\ \dfrac{\alpha^7 - \alpha^{k-4}}{\alpha - 1} \; ; & 7 \leq k \leq 10 \\ 0 \; ; & k > 10 \end{cases}$$

Exercise 18.3, page 395

1. The energy and power sums of discrete functions are done as follows:

(a) Given Dirac function, by definition we write

$$E[k] \stackrel{\text{def}}{=} \sum_{k=-\infty}^{\infty} |x[k]|^2 = \sum_{k=-\infty}^{\infty} |\delta[k]|^2 = \cdots + 0 + 0 + \cdots + 1^2 + 0 + 0 + \cdots$$

$$= 1 < \infty$$

which is to say that since the energy is distributed over infinitely long time interval $k = \infty$; therefore, at any given moment, the available average power must be

$$\langle P[k] \rangle = \lim_{k \to \infty} \frac{E[k]}{k} = \lim_{k \to \infty} \frac{1}{k} = 0$$

as same as for continuous functions where $t \to \infty$.

(b) Given infinitely long series, we write the expression for its energy as

$$E[k] \stackrel{\text{def}}{=} \sum_{k=-\infty}^{\infty} |x[k]|^2 = \sum_{k=-\infty}^{\infty} \left| \left(\frac{1}{4}\right)^k u[k] \right|^2 = \sum_{k=0}^{\infty} \left(\frac{1}{4}\right)^{2k} = \sum_{k=0}^{\infty} \left(\frac{1}{16}\right)^k$$

$$\left\{ \begin{array}{ll} \text{geometric series,} & \displaystyle\sum_{k=0}^{\infty} r^k = \frac{1}{1 - r} \quad \text{where, } |r| < 1 \\[4mm] \text{as well as,} & \displaystyle\sum_{k=0}^{\infty} ar^{2k} = \frac{a}{1 - r^2} \\[4mm] \text{or,} & \displaystyle\sum_{k=0}^{\infty} ar^{2k+1} = \frac{ar}{1 - r^2} \end{array} \right\}$$

$$\left\{ \therefore \quad a = 1 \text{ and } r = \frac{1}{16} \right\}$$

$$= \frac{1}{1 - \dfrac{1}{16}} = \frac{16}{15} \quad < \infty$$

which is to say that since the energy is distributed over infinitely long time interval $k = \infty$; therefore,

$$\langle P[k] \rangle = \lim_{k \to \infty} \frac{E[k]}{k} = \lim_{k \to \infty} \frac{16/15}{k} = 0$$

as same as for continuous functions where $t \to \infty$.

(c)   Complex exponential series are added for $k \in [-n, n]$; consequently, there are in total $2n+1$ points to add. That is $k = 1, 2, 3, \ldots, n$ (therefore n points in total) on the positive side, plus $k = -1, -2, -3, \ldots, -n$ (i.e. n points in total) on the negative side, plus $k = 0$ (i.e. one additional point) in the middle, thus $2n + 1$ points in total. Accordingly, by definition for power we find

$$\langle P[k] \rangle \overset{\text{def}}{=} \lim_{n \to \infty} \left[\frac{1}{2n + 1} \sum_{k=-n}^{n} |x[k]|^2 \right] = \lim_{n \to \infty} \left[\frac{1}{2n + 1} \sum_{k=-n}^{n} \left| e^{j\,10k}\, u[k] \right|^2 \right]$$

$$= \lim_{n \to \infty} \left[\frac{1}{2n + 1} \sum_{k=0}^{n} \left| e^{j\,10k} \right|^2 \right] = \lim_{n \to \infty} \left[\frac{1}{2n + 1} \underbrace{\sum_{k=0}^{n} |1_k|^2}_{n+1} \right]$$

$$= \lim_{n \to \infty} \left[\frac{n + 1}{2n + 1} \right] = \left(\frac{\infty}{\infty} \right) \overset{\text{L.H.}}{=} \lim_{n \to \infty} \frac{1}{2} = \frac{1}{2} < \infty$$

which is to say that in order to support non-zero average power over infinite long interval $k = \infty$, there must be $E[k] = \langle P[k] \rangle\, k = \infty$.

(d)   Infinite sum of power series where $|\alpha| < 1$ converges, thus by definition we write

$$E[k] \overset{\text{def}}{=} \sum_{k=-\infty}^{\infty} |x[k]|^2 = \sum_{k=-\infty}^{\infty} \left| \alpha^k u[k] \right|^2 = \sum_{k=0}^{\infty} |\alpha^k|^2 = \underbrace{\sum_{k=0}^{\infty} \left(\alpha^2 \right)^k}_{\text{geometric series}}$$

$$= \frac{1}{1 - \alpha^2} < \infty$$

Finite amount of energy must be distributed over infinitely long interval, thus the average power equals

$$\langle P[k] \rangle = \lim_{k \to \infty} \frac{E[k]}{k} = \lim_{k \to \infty} \frac{\frac{1}{1 - \alpha^2}}{k} = 0$$

(e) Infinitely long step function has average power calculated as

$$\langle P[k] \rangle \stackrel{\text{def}}{=} \lim_{n \to \infty} \left[\frac{1}{2n + 1} \sum_{k=-n}^{n} |x[k]|^2 \right] = \lim_{n \to \infty} \left[\frac{1}{2n + 1} \sum_{k=-n}^{n} |u[k]|^2 \right]$$

$$= \lim_{n \to \infty} \left[\frac{1}{2n + 1} \underbrace{\sum_{k=0}^{n} |1_k|^2}_{n+1} \right] = \lim_{n \to \infty} \left[\frac{n + 1}{2n + 1} \right] = \left(\frac{\infty}{\infty} \right) \stackrel{\text{L.H.}}{=} \lim_{n \to \infty} \frac{1}{2}$$

$$= \frac{1}{2} < \infty$$

which is to say that in order to support non-zero average power over infinite long interval $k = \infty$, there must be $E[k] = \langle P[k] \rangle \, k = \infty$.

(f) Average power of an infinitely long periodic series is calculated over one period as

$$\langle P[k] \rangle \stackrel{\text{def}}{=} \lim_{n \to \infty} \left[\frac{1}{2n + 1} \sum_{k=-n}^{n} |x[k]|^2 \right] = \lim_{n \to \infty} \left[\frac{1}{2n + 1} \sum_{k=-n}^{n} \left| \cos \left[\frac{\pi}{6} k \right] \right|^2 \right]$$

$$= \left\{ \cos^2(x) = \left(\frac{1 + \cos(2x)}{2} \right) \right\}$$

$$= \lim_{n \to \infty} \left[\frac{1}{2n + 1} \sum_{k=-n}^{n} \left[\frac{1}{2} + \frac{1}{2} \cos \left[2 \frac{\pi}{6 \, 3} k \right] \right] \right]$$

$$= \lim_{n \to \infty} \left[\frac{1}{2n + 1} \frac{1}{2} \underbrace{\sum_{k=-n}^{n} 1_k}_{2n+1} + \frac{1}{2n + 1} \frac{1}{2} \underbrace{\sum_{k=-n}^{n} \cos \left[\frac{\pi}{3} k \right]}_{=0} \right]$$

$$= \left\{ \begin{array}{l} \text{in total, there are } (2n+1) \text{ terms of 1/2 each;} \\ \text{the sum of sin or cos over one period equals zero ;} \end{array} \right\}$$

$$= \lim_{n \to \infty} \left[\frac{1}{2n + 1} \frac{1}{2} (2n + 1) \right] = \frac{1}{2} < \infty$$

which is to say that in order to support non-zero average power over infinite long interval $k = \infty$, there must be $E[k] = \langle P[k] \rangle \, k = \infty$.

Fourier Transformation Integral

19

Important to Know

Classification of Fourier transformations, sums, and series:

Continuous time (CT) periodic, Fig. 19.1(left), and non-periodic, Fig. 19.1(right), $x(t)$ functions. Periodic functions are transformed by continuous time Fourier series (CT FS), while continuous time non-periodic functions are transformed by continuous time Fourier transformation (CT FT).

Fig. 19.1 Continuous time Fourier series vs. Fourier transformation

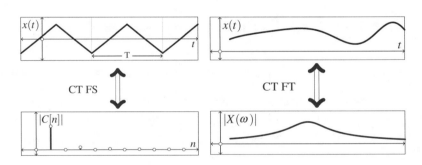

Similarly, discrete time (DT) periodic functions are mapped by discrete time Fourier series (DT FS), while discrete time non-periodic functions are transformed by discrete time Fourier transformation (DT FS).

In addition, in order to carry out FT of continuous signals on numerical processors, CT signals must be sampled before applying discrete Fourier transformation (DFT) algorithm. Version of DFT algorithm that is optimized for fast execution is referred to as fast Fourier transformation (FFT).

Fourier and inverse Fourier integrals: are defined as

$$\mathscr{F}(x(t)) \stackrel{\text{def}}{=} X(\omega) = \int_{-\infty}^{\infty} x(t)\, e^{-j\,\omega\,t}\, dt \qquad (19.1)$$

$$\mathscr{F}^{-1}(X(\omega)) \stackrel{\text{def}}{=} x(t) = \frac{1}{2\pi} \int_{-\infty}^{\infty} X(\omega)\, e^{j\,\omega\,t}\, d\omega \qquad (19.2)$$

© Springer Nature Switzerland AG 2021
R. Sobot, *Engineering Mathematics by Example*,
https://doi.org/10.1007/978-3-030-79545-0_19

19.1   Exercises

19.1 **   Fourier Integral

1. Derive Fourier transformations $X(\omega)$ of the following functions $x(t)$:

   (a) $x(t) = \delta(t)$ (b) $x(t) = a_0$; (const.)

   (c) $x(t) = \delta(t - \tau)$ (d) $x(t) = e^{j\omega_0 t}$

2. Derive Fourier transformations $X(\omega)$ of the following functions $x(t)$:

   (a) $x(t) = \cos(\omega_0 t)$ (b) $x(t) = \sin(\omega_0 t)$

3. Derive Fourier transformations $X(\omega)$ and show its graphical representation of the following functions $x(t)$, where $(a > 0)$:

   (a) $x(t) = e^{-at} u(t)$ (b) $x(t) = e^{at} u(-t)$ (c) $x(t) = e^{-a|t|}$

4. Given $x(t) = a_0 \Pi \left(\dfrac{t}{\tau} \right)$ derive Fourier transformation $X(\omega)$ and show its graphical representation.

5. Given $x(t) = \Lambda \left(\dfrac{t}{\tau} \right)$ derive Fourier transformation $X(\omega)$ and show its graphical representation.

6. Given $x(t) = \text{sign}(t)$ derive Fourier transformation $X(\omega)$ and show its graphical representation.

7. Given $x(t) = u(t)$ derive Fourier transformation $X(\omega)$ and show its graphical representation.

8. Given $x(t) = \text{sinc}(t)$ derive Fourier transformation $X(\omega)$ and show its graphical representation.

19.2 **   Fourier Integral, Exercises

1. Given function $x(t)$ in Fig. 19.2 derive $X(\omega)$.

Fig. 19.2 Example 19.2-1

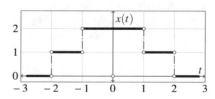

2. Given function $x(t)$ in Fig. 19.3 derive $X(\omega)$.

Fig. 19.3 Example 19.2-1

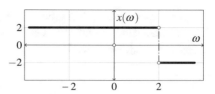

3. Given RF pulse, Fig. 19.4, where f_c is the RF waveform frequency, and T is the duration of RF pulse, derive its Fourier transform.

Fig. 19.4 Example 19.2-1

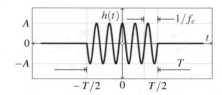

Solutions

Exercise 19.1, page 414

1. Continuous functions are transformed by Fourier integral as follows.

   (a) Dirac function equals non-zero only in one point, and its Fourier transformation is found by definition.

$$X(\omega) \overset{\text{def}}{=} \int_{-\infty}^{\infty} x(t)\, e^{-j\omega t}\, dt = \left\{ \delta(t) \neq 0 \Rightarrow t = 0 \quad \therefore \quad e^{j\omega\, 0} = 1 \right\}$$

$$= \int_{-\infty}^{\infty} \delta(t)\, e^{-j\omega t}{}^{\!\!1}\, dt = \int_{-\infty}^{\infty} \delta(t)\, dt \overset{\text{def}}{=} 1$$

We keep in mind this bidirectional relationship between Dirac function $x(t)$ and its Fourier transformation of the constant function $X(\omega)$, Fig. 19.5. That is to say, the two functions are said to be "dual" because Fourier transform of Dirac function is a constant function, as well as the opposite, i.e. Fourier transform of a constant function is Dirac function. Physical interpretation is that in order to synthesize Dirac function it is necessary to add all possible $\sin(\omega t)$ whose frequencies are in the range $\omega \in [-\infty, \infty]$.

Fig. 19.5 Example 19.1-1(a)

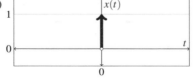
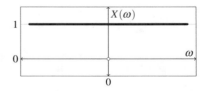

(b) Fourier transformation of a constant function may be derived, for example, by using "reverse engineering" technique, i.e. by using the inverse Fourier transformation integral. Given that $x(t) = a_0$, we write

$$\mathscr{F}^{-1}(X(\omega)) = x(t) \overset{\text{def}}{=} \frac{1}{2\pi} \int_{-\infty}^{\infty} X(\omega)\, e^{j\omega t}\, d\omega$$

$$\therefore$$

$$a_0 \overset{\text{def}}{=} \frac{1}{2\pi} \int_{-\infty}^{\infty} \underline{X(\omega)\, e^{j\omega t}}\, d\omega$$

The idea is to gradually deduce the form of $X(\omega)$ until the inverse Fourier equality satisfied, i.e. the right side of the equation is reduced to a_0 and becomes equal to the left side. To start with, $X(\omega)$ must include a 2π term, so that it cancels the already existing 2π term, as

$$a_0 \overset{\text{def}}{=} \frac{1}{\cancel{2\pi}} \int_{-\infty}^{\infty} \underline{\cancel{2\pi}\times?}\ e^{j\omega t}\, d\omega$$

Then, there must be the a_0 constant on the right side somewhere; thus we further deduce that $X(\omega) = 2\pi\, a_0 \times ?$, i.e.

$$a_0 \overset{\text{def}}{=} \int_{-\infty}^{\infty} \underline{a_0 \times ?}\ e^{j\omega t}\, d\omega$$

$$\therefore$$

$$a_0 \overset{\text{def}}{=} a_0 \underbrace{\int_{-\infty}^{\infty} \underline{\quad ? \quad}\ e^{j\omega t}\, d\omega}_{=1}$$

In order for the last equality to be valid, the right side integral must be reduced to one. One possible solution would be that the exponential term equals to one. As already seen in Example 19.1-1(a), we can use the sampling property of Dirac function because

$$\int_{-\infty}^{\infty} \delta(\omega)\, \cancel{e^{-j\omega \tau}}^{1}\, d\omega \overset{\text{def}}{=} 1$$

which leads to the following conclusion,

$$a_0 \overset{\text{def}}{=} a_0 \underbrace{\int_{-\infty}^{\infty} \underline{\delta(\omega)}\ e^{j\omega t}\, d\omega}_{=1}$$

In summary, we write

$$a_0 \overset{\text{def}}{=} \frac{1}{2\pi} \int_{-\infty}^{\infty} \underline{X(\omega)\, e^{j\omega t}}\, d\omega$$

$$\therefore$$

$$a_0 = \frac{1}{2\pi} \int_{-\infty}^{\infty} \underline{2\pi a_0 \, \delta(\omega)} \, e^{j\omega t} \, d\omega$$

$$\therefore$$

$$\underline{X(\omega) = 2\pi a_0 \, \delta(\omega)}$$

Which is to say that Fourier transformation of a constant function is indeed Dirac function (the $2\pi a_0$ term is the multiplying constant), as stated in Example 19.1-1(a) and illustrated in Fig. 19.5.

(c) Delayed Dirac function $\delta(t - \tau)$ results in

$$X(\omega) \overset{\text{def}}{=} \int_{-\infty}^{\infty} x(t) \, e^{-j\omega t} \, dt = \left\{ \delta(t - \tau) \neq 0 \Rightarrow t = \tau \right\}$$

$$= \int_{-\infty}^{\infty} \delta(t - \tau) \, e^{-j\omega \tau} \, dt = e^{-j\omega\tau} \underbrace{\int_{-\infty}^{\infty} \delta(t - \tau) \, dt}_{\overset{\text{def}}{=} 1}$$

$$= (1) \, e^{-j\omega\tau}$$

that is, Fourier transformation of delayed Dirac function $\delta(t - \tau)$ is also the constant function, where the complex exponent term only adds to phase delay τ that does not change the form of the constant functions.

We note that this property of Fourier transform is general for any delayed function, i.e. Fourier of a delayed function equals to Fourier transformation of its non-delayed version multiplied by the phase delayed term,

$$F(f(t - \tau)) = F(f(t)) \, e^{-j\omega\tau}$$

(d) Transformation of complex exponential function $x(t) = e^{j\omega_0 t}$ (its amplitude is constant, i.e. equals to the radius of the unity circle) may be deduced using similar reasoning as in Example 19.1-1(b). Note that ω_0 is a constant as opposed to ω that is a variable.

$$\mathscr{F}^{-1}(X(\omega)) = x(t) \overset{\text{def}}{=} \frac{1}{2\pi} \int_{-\infty}^{\infty} X(\omega) \, e^{j\omega t} \, d\omega$$

$$\therefore$$

$$e^{j\omega_0 t} = \frac{1}{2\pi} \int_{-\infty}^{\infty} \underline{X(\omega)} \, e^{j\omega t} \, d\omega$$

$$\therefore$$

$$e^{j\omega_0 t} = \frac{1}{2\pi} \int_{-\infty}^{\infty} \underline{2\pi \times ?} \, e^{j\omega t} \, d\omega$$

We note that by sampling the exponential function on the right side at $\omega = \omega_0$, we create term equal to the exponent on the left side. That can be done by using delayed Dirac function $\delta(\omega - \omega_0)$, so that we write

$$e^{j\,\omega_0\,t} = \frac{1}{2\pi} \int_{-\infty}^{\infty} \underline{2\pi\,\delta(\omega - \omega_0)}\ e^{j\,\omega\,t}\,d\omega$$

$$\therefore$$

$$e^{j\,\omega_0\,t} = \frac{1}{2\pi}\,2\pi \int_{-\infty}^{\infty} \underline{\delta(\omega - \omega_0)}\ e^{j\,\omega_0\,t}\,d\omega$$

$$\therefore$$

$$e^{j\,\omega_0\,t} = e^{j\,\omega_0\,t} \underbrace{\int_{-\infty}^{\infty} \underline{\delta(\omega - \omega_0)}\ d\omega}_{\stackrel{\text{def}}{=}1}$$

$$\therefore$$

$$e^{j\,\omega_0\,t} = e^{j\,\omega_0\,t} \ \checkmark$$

In conclusion,

$$X(\omega) = 2\pi\,\delta(\omega - \omega_0)$$

2. Continuous sinusoidal functions are transformed by Fourier integral as follows.

(a)   Application of Euler formula transforms cosine function to the sum of complex exponential functions.

$$x(t) = \cos(\omega_0\,t) = \frac{e^{j\,\omega_0\,t} + e^{-j\,\omega_0\,t}}{2}$$

Therefore,

$$X(\omega) = \mathscr{F}(x(t)) = \mathscr{F}\left[\frac{e^{j\,\omega_0\,t} + e^{-j\,\omega_0\,t}}{2}\right] = \mathscr{F}\left(\frac{e^{j\,\omega_0\,t}}{2}\right) + \mathscr{F}\left(\frac{e^{-j\,\omega_0\,t}}{2}\right)$$

In Example 19.1-1(d) we derived Fourier transformation of complex exponential function; thus we conclude

$$\mathscr{F}\left(\frac{e^{j\,\omega_0\,t}}{2}\right) = 2\pi\,\delta(\omega - \omega_0)$$

$$\therefore$$

$$\mathscr{F}\left(\frac{e^{-j\,\omega_0\,t}}{2}\right) = 2\pi\,\delta(\omega + \omega_0)$$

In conclusion,

$$X(\omega) = \frac{1}{2}\left[\not{2}\pi\ \delta(\omega - \omega_0) + \not{2}\pi\ \delta(\omega + \omega_0)\right] = \pi\ [\delta(\omega + \omega_0) + \delta(\omega - \omega_0)]$$

That is to say, Fourier transform of $\cos(\omega_0\ t)$ function is the sum of two Dirac functions, one delayed and the other advanced by ω_0, Fig. 19.6.

Fig. 19.6 Example 19.1-1(a)

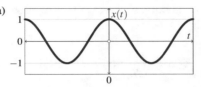
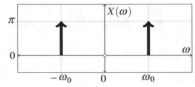

(b) Application of Euler formula transforms sine function to the difference of complex exponential functions.

$$x(t) = \sin(\omega_0\ t) = \frac{e^{j\omega_0 t} - e^{-j\omega_0 t}}{2j}$$

Therefore,

$$X(\omega) = \mathscr{F}(x(t)) = \mathscr{F}\left[\frac{e^{j\omega_0 t} - e^{-j\omega_0 t}}{2j}\right] = \mathscr{F}\left(\frac{e^{j\omega_0 t}}{2j}\right) - \mathscr{F}\left(\frac{e^{-j\omega_0 t}}{2j}\right)$$

In Example 19.1-1(d) we derived Fourier transformation of complex exponential function; thus we conclude

$$\mathscr{F}\left(\frac{e^{j\omega_0 t}}{2}\right) = 2\pi\ \delta(\omega - \omega_0)$$

$$\therefore$$

$$\mathscr{F}\left(\frac{e^{-j\omega_0 t}}{2}\right) = 2\pi\ \delta(\omega + \omega_0)$$

In conclusion,

$$X(\omega) = \frac{1}{2\not{j}}\left[\not{2}\pi\ \delta(\omega - \omega_0) - \not{2}\pi\ \delta(\omega + \omega_0)\right]$$

$$= \frac{\pi}{j}\ \frac{j}{j}\ [\delta(\omega - \omega_0) - \delta(\omega + \omega_0)]$$

$$= j\pi\ [\delta(\omega + \omega_0) - \delta(\omega - \omega_0)]$$

That is to say, Fourier transform of $\sin(\omega_0 t)$ function is also the sum of two Dirac functions, one delayed and the other advanced by ω_0; however, there is also phase rotation by $\pi/2$ (which is set by the j factor) as well as the negative orientation of the $\delta(\omega - \omega_0)$ term.

3. Exponential functions are transformed by Fourier integral as follows.

(a)   Product $x(t)$ of exponential function e^{-at} and $u(t)$ is highlighted in Fig. 19.7.
By definition, Fourier integral is

$$X(\omega) \stackrel{\text{def}}{=} \int_{-\infty}^{\infty} x(t)\, e^{-j\omega t}\, dt$$

$$= \int_{-\infty}^{\infty} e^{-at} u(t)\, e^{-j\omega t}\, dt$$

$$= \int_{0}^{\infty} e^{-at}\, e^{-j\omega t}\, dt$$

$$= \int_{0}^{\infty} e^{-(a+j\omega)t}\, dt$$

$$= \{\text{change of variable}\}$$

$$= -\frac{1}{a + j\omega}\, e^{-(a+j\omega)t}\, \Big|_{0}^{\infty}$$

$$= -\frac{1}{a + j\omega}\, (0 - \underbrace{e^{0}}_{1})$$

$$\therefore$$

$$X(\omega) = \frac{1}{a + j\omega}$$

Fig. 19.7 Example 19.1-3(a)

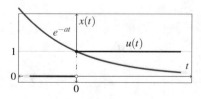

Fig. 19.8 Example 19.1-3(a)

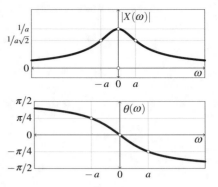

Modulus $|X(\omega)|$ and argument $\theta(\omega) = \arg X(\omega)$ of complex function $X(\omega)$ are calculated as usual for complex numbers,

$$|X(\omega)| = \left| \frac{1}{a + j\omega} \; \frac{a - j\omega}{a - j\omega} \right| = \left| \frac{a - j\omega}{a^2 + \omega^2} \right| = \left| \frac{a}{a^2 + \omega^2} - j \frac{\omega}{a^2 + \omega^2} \right|$$

$$= \sqrt{\frac{a^2}{(a^2 + \omega^2)^2} + \frac{\omega^2}{(a^2 + \omega^2)^2}} = \sqrt{\frac{a^2 + \omega^2}{(a^2 + \omega^2)^2}} = \frac{1}{\sqrt{a^2 + \omega^2}}$$

or, even faster calculation

$$|X(\omega)| = \left| \frac{1}{a + j\omega} \right| = \frac{1}{|a + j\omega|} = \frac{1}{\sqrt{a^2 + \omega^2}}$$

This even function may be sufficiently evaluated only in a few important points as

$$\omega = 0 \quad \therefore \quad |X(\omega)| = \frac{1}{a}$$

$$\omega = \pm a \quad \therefore \quad |X(\omega)| = \frac{1}{a\sqrt{2}}$$

$$\lim_{\omega \to \pm \infty} |X(\omega)| = 0$$

and then, its graph is sketched in Fig. 19.8 (top).
However, we pay attention to the signs of $\Re(X(\omega))$ and $\Im(X(\omega))$, so that the argument is calculated in the correct quadrant of the unity circle. The $\Re(X(\omega)) > 0$ and $\Im(X(\omega)) < 0$; therefore, the argument $\theta(\omega)$ is in the fourth quadrant thus negative.

$$\theta(\omega) \stackrel{\text{def}}{=} \arctan \frac{\Im(X(\omega))}{\Re(X(\omega))} = \arctan \frac{-\dfrac{\omega}{a^2 + \omega^2}}{\dfrac{a}{a^2 + \omega^2}} = \arctan \frac{-\omega}{a} = -\arctan \frac{\omega}{a}$$

This odd function may be sufficiently evaluated only in a few important points as

$$\omega = 0 \quad \therefore \quad \theta(0) = 0$$

$$\omega = +a \quad \therefore \quad \theta(a) = -\arctan(1) = -\frac{\pi}{4}$$

$$\omega = -a \quad \therefore \quad \theta(-a) = -\arctan(-1) = \frac{\pi}{4}$$

$$\lim_{\omega \to \infty} \theta(\omega) = -\frac{\pi}{2}$$

$$\lim_{\omega \to -\infty} \theta(\omega) = \frac{\pi}{2}$$

and then, its graph is sketched in Fig. 19.8 (bottom).

(b) $x(t) = e^{at} u(-t)$ Product $x(t)$ of exponential function e^{at} and $u(-t)$ is highlighted in Fig. 19.9. By definition, Fourier integral is

$$X(\omega) \stackrel{\text{def}}{=} \int_{-\infty}^{\infty} x(t) \, e^{-j\omega t} \, dt$$

$$= \int_{-\infty}^{\infty} e^{at} u(-t) \, e^{-j\omega t} \, dt$$

$$= \int_{-\infty}^{0} e^{at} \, e^{-j\omega t} \, dt$$

$$= \int_{-\infty}^{0} e^{(a-j\omega)t} \, dt$$

$$= \{\text{change of variable}\}$$

$$= \frac{1}{a - j\omega} \, e^{(a-j\omega)t} \Big|_{-\infty}^{0}$$

$$= \frac{1}{a - j\omega} (\underbrace{e^{0}}_{1} - 0)$$

$$\therefore$$

$$X(\omega) = \frac{1}{a - j\omega}$$

Fig. 19.9 Example 19.1-3(b)

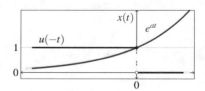

Fig. 19.10
Example 19.1-3(b)

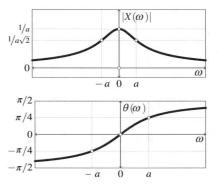

Modulus $|X(\omega)|$ and argument $\theta(\omega) = \arg X(\omega)$ of complex function $X(\omega)$ are calculated as usual for complex numbers, Fig. 19.10

$$|X(\omega)| = \left| \frac{1}{a - j\omega} \; \frac{a + j\omega}{a + j\omega} \right| = \left| \frac{a + j\omega}{a^2 + \omega^2} \right| = \left| \frac{a}{a^2 + \omega^2} + j \; \frac{\omega}{a^2 + \omega^2} \right|$$

$$= \sqrt{\frac{a^2}{(a^2 + \omega^2)^2} + \frac{\omega^2}{(a^2 + \omega^2)^2}} = \sqrt{\frac{a^2 + \omega^2}{(a^2 + \omega^2)^2}} = \frac{1}{\sqrt{a^2 + \omega^2}}$$

or, even faster calculation

$$|X(\omega)| = \left| \frac{1}{a - j\omega} \right| = \frac{1}{|a - j\omega|} = \frac{1}{\sqrt{a^2 + \omega^2}}$$

We note that this $|X(\omega)|$ is identical to the one in Example 19.1-3(a).
However, we pay attention to the signs of $\Re(X(\omega))$ and $\Im(X(\omega))$, so that the argument is calculated in the correct quadrant of the unity circle. The $\Re(X(\omega)) > 0$ and $\Im(X(\omega)) > 0$; therefore, the argument $\theta(\omega)$ is in the first quadrant thus positive.

$$\theta(\omega) \overset{\text{def}}{=} \arctan \frac{\Im(X(\omega))}{\Re(X(\omega))} = \arctan \frac{\dfrac{\omega}{a^2 + \omega^2}}{\dfrac{a}{a^2 + \omega^2}} = \arctan \frac{\omega}{a}$$

Similarly to Example 19.1-3(a), this odd function may be sufficiently evaluated only in a few important points as

$$\omega = 0 \quad \therefore \quad \theta(0) = 0$$

$$\omega = +a \quad \therefore \quad \theta(a) = \arctan(1) = \frac{\pi}{4}$$

$$\omega = -a \quad \therefore \quad \theta(-a) = \arctan(-1) = -\frac{\pi}{4}$$

$$\lim_{\omega \to \infty} \theta(\omega) = \frac{\pi}{2}$$

$$\lim_{\omega \to -\infty} \theta(\omega) = -\frac{\pi}{2}$$

and then, its graph is sketched in Fig. 19.9 (bottom).

(c)   Given function $x(t) = e^{-a|t|}$ may be written as

$$x(t) = \begin{cases} e^{-at}; & t > 0 \\ 1; & t = 0 \\ e^{at}; & t < 0 \end{cases}$$

Sketches of $x(t)$ and $X(\omega)$ are shown in Fig. 19.11. Being a real function $X(\omega)$, consequently its argument $\theta(\omega) = \arg X(\omega) = 0°$, i.e. it is a constant.

Fig. 19.11
Example 19.1-3(c)

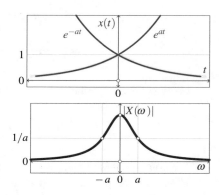

Thus, Fourier transformation may be written as

$$X(\omega) \overset{\text{def}}{=} \int_{-\infty}^{\infty} x(t)\, e^{-j\omega t}\, dt = \int_{-\infty}^{\infty} e^{-a|t|}\, e^{-j\omega t}\, dt$$

$$= \int_{-\infty}^{0} e^{at}\, e^{-j\omega t}\, dt + \int_{0}^{\infty} e^{-at}\, e^{-j\omega t}\, dt$$

$$= \big\{ \text{see Example 19.1-3(a) and (b)} \big\}$$

$$= \frac{1}{a - j\omega} + \frac{1}{a + j\omega} = \frac{a - j\cancel{\omega} + a + j\cancel{\omega}}{a^2 + \omega^2}$$

$$\therefore$$

$$X(\omega) = \frac{2}{a^2 + \omega^2} \quad \therefore \quad X(\omega) \in \mathbb{R} \quad \therefore \quad \theta(\omega) = 0°$$

4. The port function $x(t) = a_0 \Pi (t/\tau)$ where $t \in [-\tau/2, \tau/2]$ is transformed by Fourier integral as follows.

$$X(\omega) \overset{\text{def}}{=} \int_{-\infty}^{\infty} x(t)\, e^{-j\omega t}\, dt = \int_{-\infty}^{\infty} \Pi \left(\frac{t}{\tau} \right) e^{-j\omega t}\, dt = \int_{-\tau/2}^{\tau/2} a_0\, e^{-j\omega t}\, dt$$

$$= \{ \text{change of variable} \}$$

$$= -\frac{a_0}{j\omega}\, e^{-j\omega t} \bigg|_{-\tau/2}^{\tau/2} = -\frac{a_0}{j\omega} \left[e^{-j\omega\tau/2} - e^{j\omega\tau/2} \right]$$

$$= \frac{a_0}{j\omega} \left[e^{j\omega\tau/2} - e^{-j\omega\tau/2} \right] = \frac{2a_0}{\omega}\, \frac{e^{j\omega\tau/2} - e^{-j\omega\tau/2}}{2j}$$

$$= \frac{2a_0}{\omega}\, \frac{\tau/2}{\tau/2}\, \sin \left(\frac{\omega\tau}{2} \right) = 2a_0\, \frac{\tau}{2}\, \frac{\sin \left(\dfrac{\omega\tau}{2} \right)}{\dfrac{\omega\tau}{2}}$$

$$= a_0 \tau\, \text{sinc} \left(\frac{\omega\tau}{2} \right)$$

Fig. 19.12
Example 19.1-3(a)

 Π

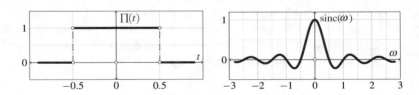

In conclusion, the $\Pi (t)$ and sinc (ω) are dual functions with respect to Fourier transformation, Fig. 19.12.

5. Fourier transform of the triangle function $x(t) = \Lambda (t/\tau)$ where $t \in [-\tau/2, \tau/2]$ may be performed by exploring properties of Fourier transform of function derivatives.
Given function $f(t)$, there are the following identities:

$$\mathscr{F}\left(f'(x) \right) = j\omega\, \mathscr{F}(\omega)$$

$$\mathscr{F}\left(f''(x) \right) = (j\omega)^2\, \mathscr{F}(\omega)$$

$$\mathscr{F}\left(f^{(3)}(x) \right) = (j\omega)^3\, \mathscr{F}(\omega)$$

$$\cdots$$

Derivatives of triangle function may be deduced by inspection of graphs.

First derivative of a piecewise linear triangle function may be deduced for each interval separately, Fig. 19.13. Horizontal parts of $x(t)$ are constant; thus derivative equals zero. Two slopes of the triangle are constant and found by definition of tangent in a right angle triangle, as a/τ and $a/-\tau$, respectively.

$$\frac{dx}{dt} = \frac{a}{\tau}; \quad x \in [-\tau, 0]$$

$$\frac{dx}{dt} = -\frac{a}{\tau}; \quad x \in [0, \tau]$$

Fig. 19.13
Example 19.1-3(a)

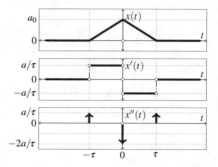

Second derivatives are found at points where $x'(t)$ change, i.e. $t = (-\tau, 0, \tau)$, where we find Dirac functions. Sharp step from zero to a/τ at $(t = -\tau)$ generates positive Delta function, sharp step from zero to $-a/\tau$ at $t = \tau$ generates negative Delta function, and sharp step from a/τ to $-a/\tau$ at $t = 0$ generates negative $-2a/\tau$ Delta function. In Example 19.1-1(a) and (b) the duality between delayed Delta and constant functions is derived; thus in combination with the Fourier transformations of function derivatives we find

$$\frac{d^2 x}{dt^2} = \frac{a}{\tau}\,\delta(t+\tau) - \frac{2a}{\tau}\,\delta(t) + \frac{a}{\tau}\,\delta(t-\tau)$$

$$= \frac{a}{\tau}\,[\delta(t+\tau) - 2\,\delta(t) + \delta(t-\tau)]$$

$$\xrightarrow{F} \left\{\text{Fourier transformations of function derivatives}\right\}$$

$$(j\omega)^2\, X(\omega) = \frac{a}{\tau}\left[e^{j\omega\tau} - 2 + e^{-j\omega\tau}\right]$$

$$\therefore$$

$$X(\omega) = \frac{a}{-\omega^2\tau}\Big[\underbrace{e^{j\omega\tau} + e^{-j\omega\tau}}_{2\cos(\omega\tau)} - 2\Big] = \frac{2a}{\omega^2\tau}\Big[\underbrace{1 - \cos\omega\tau}_{2\sin^2(\omega\tau/2)}\Big]$$

$$= \frac{\tau}{\tau}\,\frac{2a}{\omega\omega\tau}\,2\sin^2\left(\frac{\omega\tau}{2}\right) = a\tau\,\frac{\sin(\omega\tau/2)}{\omega\tau/2}\,\frac{\sin(\omega\tau/2)}{\omega\tau/2}$$

$$= a\tau\,\operatorname{sinc}\left(\frac{\omega\tau}{2}\right)\,\operatorname{sinc}\left(\frac{\omega\tau}{2}\right)$$

$$= a\tau \ \text{sinc}^2 \left(\frac{\omega\tau}{2} \right)$$

In conclusion, $\Lambda(t)$ and $\text{sinc}^2[\omega]$ are dual functions with respect to Fourier transformation, Fig. 19.14.

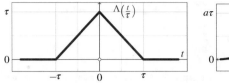 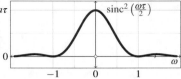

Fig. 19.14 Example 19.1-5

6. Sign function sign (t) does not converge when $t \to \infty$; thus direct application of Fourier integral also produces non-converging result. One possible method to derive Fourier transformation is to express sign (t) in terms of $u(t)$ and then to use limit of exponential function. Specifically, we write

$$\text{sign} (t) = u(t) - u(-t)$$

In addition, we can interpret the number one as the limit of exponential function when $a = 0$ (ie. $e^0 = 1$), highlighted function in Fig. 19.15, as

$$u(t) = \lim_{a=0} e^{-at} u(t)$$

$$-u(-t) = - \lim_{a=0} e^{at} u(-t)$$

By doing so, we write

Fig. 19.15 Example 19.1-6

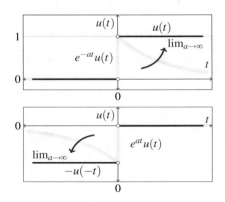

$$x(t) = \text{sign}(t) = u(t) - u(-t) = \lim_{a \to 0}\left[e^{-at}\,u(t) - e^{at}\,u(-t)\right]$$
$$= \{\text{see Example 19.1-3(a) and (b)}\}$$

$$\therefore$$

$$X(\omega) = \lim_{a \to 0}\left[\frac{1}{a^0 + j\omega} - \frac{1}{a^0 - j\omega}\right] = \frac{1}{j\omega} - \frac{1}{j\omega}$$

$$= \frac{2}{j\omega}$$

Second, and much faster, method to derive this Fourier transform function is based on the same idea as in Example 19.1-5, that is to say to exploit the derivative property of Fourier transform as

$$\frac{dx(t)}{dt} = 2\,\delta(t) \quad\overset{F}{\to}\quad j\omega\,X(\omega) = 2 \quad \therefore \quad X(\omega) = \frac{2}{j\omega}$$

7. Step function $u(t)$ does not converge when $t \to \infty$; thus direct application of Fourier integral also produces non-converging result. We can use the same technique as in Example 19.1-6 to write $u(t)$ in terms of sign (t) that is already derived in Example 16.2-2(c). Then

$$x(t) = u(t) = \frac{1 + \text{sign}(t)}{2}$$

$$\therefore$$

$$X(\omega) = \mathscr{F}\left[\frac{1}{2} + \frac{\text{sign}(t)}{2}\right] = \{\mathscr{F}(\text{constant}) + \mathscr{F}(\text{sign}(t))\}$$

$$= \mathscr{F}\left(\frac{1}{2}\right) + \frac{1}{2}\mathscr{F}(\text{sign}(t)) = 2\pi\,\frac{1}{2}\,\delta(\omega) + \frac{1}{2}\,\frac{2}{j\omega}$$

$$= \pi\,\delta(t) + \frac{1}{j\omega}$$

8. Formal derivation of Fourier transformation of cardinal sine function sinc (t) may be used a study case to illustrate various techniques in algebra and calculus working together.

First, let us do some preparatory work and quick reviews:

(a)   Reminder of basic trigonometric identities

$$\sin x\,\cos y = \frac{1}{2}\left[\sin(x + y) + \sin(x - y)\right]$$

$$\sin x\,\sin y = \frac{1}{2}\left[\cos(x - y) - \cos(x + y)\right]$$

$$\cos(-x) = \cos(x)$$

$$\sin(-x) = -\sin(x)$$

(b) Reminder of Euler formula

$$e^{\pm jx} = \cos x \pm j \sin x$$

(c) Reminder of special functions

$$\text{sign}(x) = \frac{|x|}{x} = \frac{x}{|x|} = \begin{cases} -1; & x < 1 \\ 0; & x = 0 \\ 1; & x > 0 \end{cases} \quad \text{and,} \quad |x| = \begin{cases} x; & x \geq 0 \\ -x; & x < 0 \end{cases}$$

(d) Reminder of partial integration

$$\int u \, dv = uv - \int v \, du$$

(e) Reminder of basic integrals

$$\int \frac{1}{x} \, dx = \ln|x| + C$$

$$\int \frac{1}{1+x^2} \, dx = \arctan x + C$$

$$\int e^x \, dx = e^x + C$$

(f) Solve the following integral:

$$\int_{\infty}^{\infty} \frac{a}{a^2 + x^2} \, dx = \frac{\not{a}}{\not{a}^2} \int_{\infty}^{\infty} \frac{1}{1 + \left(\dfrac{x}{a}\right)^2} \, dx = \left\{ \frac{x}{a} = t \quad \therefore \quad dx = |a| \, dt \right\}$$

$$= \frac{|a|}{a} \int_{\infty}^{\infty} \frac{1}{1+t^2} \, dt = \frac{|a|}{a} \left. \arctan(t) \right|_{-\infty}^{\infty}$$

$$= \frac{|a|}{a} \left[\frac{\pi}{2} - \left(-\frac{\pi}{2}\right) \right]$$

$$= \pi \, \text{sign}(a)$$

(g) Solve the following integral

$$\int_0^\infty e^{-\omega_0 t y} \, dy =$$

$$\left\{ \begin{array}{l} \text{change variable:} \quad -\omega_0 t y = x \quad \therefore \quad dy = -\dfrac{1}{\omega_0 t} \, dx \\[3mm] \qquad\qquad\qquad y = 0 \Rightarrow x = 0, \quad \text{and} \quad y = \infty \Rightarrow x = \infty \end{array} \right\}$$

$$= -\frac{1}{\omega_0 t} \int_0^\infty e^x \, dx = -\frac{1}{\omega_0 t} \, e^x \Big|_0^\infty = -\frac{1}{\omega_0 t} \left[e^{-\infty \nearrow 0} - e^{\nearrow 1} \right]$$

$$= \frac{1}{\omega_0 t}$$

(h) Solve the following integral where $x \in [0, \infty]$; first we solve

$$\int e^{-bx} \sin(ax) \, dx =$$

$$\left\{ \begin{array}{l} \text{partial integration:} \quad u = e^{-bx} \quad \therefore \quad du = -b\,e^{-bx} \, dx \\[3mm] \qquad\qquad\qquad dv = \sin(ax) \, dx \quad \therefore \quad v = -\dfrac{1}{a} \cos(ax) \end{array} \right\}$$

$$= e^{-bx} \left(-\frac{1}{a} \cos(ax) \right) - \int \left(-\frac{1}{a} \cos(ax) \right) \left(-b\,e^{-bx} \right) \, dx$$

$$= -\frac{1}{a} e^{-bx} \cos(ax) - \frac{b}{a} \underbrace{\int e^{-bx} \cos(ax) \, dx}_{\text{partial integration}}$$

$$\left\{ \begin{array}{l} \text{partial integration:} \quad u = e^{-bx} \quad \therefore \quad du = -b\,e^{-bx} \, dx \\[3mm] \qquad\qquad\qquad dv = \cos(ax) \, dx \quad \therefore \quad v = \dfrac{1}{a} \sin(ax) \end{array} \right\}$$

$$= -\frac{1}{a} e^{-bx} \cos(ax) - \frac{b}{a} \left[\frac{1}{a} e^{-bx} \sin(ax) + \frac{b}{a} \underline{\int e^{-bx} \sin(ax) \, dx} \right]$$

(h) *(cont.)* We note that after two partial integrations one of the terms on the right side is identical to the original integral on the left side. Thus further partial integration does not advance the solution. However, like any other common term in an algebra equation, this integral may be factored out as

$$\left[1 + \left(\frac{b}{a}\right)^2\right] \underline{\int e^{-bx} \sin(ax)\, dx} = -\frac{1}{a} e^{-bx} \cos(ax) - \frac{b}{a}\frac{1}{a} e^{-bx} \sin(ax)$$

$$\therefore$$

$$\left[\frac{a^2 + b^2}{a^2}\right] \int e^{-bx} \sin(ax)\, dx = -\frac{1}{a} e^{-bx} \left[\frac{a\cos(ax) + b\sin(ax)}{a}\right]$$

which gives

$$\int e^{-bx} \sin(ax)\, dx = \frac{\cancel{a^2}}{a^2 + b^2}\left(-\frac{1}{\cancel{a}}\right)\frac{1}{\cancel{a}} e^{-bx}\left[a\cos(ax) + b\sin(ax)\right] + C$$

$$= -\frac{1}{a^2 + b^2} e^{-bx}\left[a\cos(ax) + b\sin(ax)\right] + C$$

Now, we calculate definite integral (the integration constants cancel) as

$$\int_0^\infty e^{-bx} \sin(ax) = -\frac{1}{a^2 + b^2} e^{-bx}\left[a\cos(ax) + b\sin(ax)\right] \Bigg|_0^\infty$$

$$= -\frac{1}{a^2 + b^2}\left\{ \underbrace{e^{-b\infty}}_{\to 0}\Big[a\underbrace{\cos(a\,\infty)}_{\leq 1} + b\underbrace{\sin(a\,\infty)}_{\leq 1}\Big]_{}\right.$$
$$\underbrace{\phantom{-\frac{1}{a^2 + b^2}\left\{ e^{-b\infty}\Big[a\cos(a\,\infty) + b\sin(a\,\infty)\Big]}}_{=0}$$
$$\left. - \underbrace{e^{-b\cdot 0}}_{\to 1}\Big[a\,\underbrace{\cos(a\cdot 0)}_{=1} + b\,\underbrace{\sin(a\cdot 0)}_{=0}\Big]\right\}$$

$$= \frac{a}{a^2 + b^2}$$

Therefore,

$$\int_0^\infty e^{-bx} \sin(ax) = \frac{a}{a^2 + b^2}$$

By repeating the same idea, we find

$$\int_0^\infty e^{-bx} \cos(ax) = \frac{b}{a^2 + b^2}$$

After this preparatory work, we derive Fourier transform of sinc (t) function as follows:

$$X(\omega) \stackrel{\text{def}}{=} \int_{-\infty}^{\infty} x(t)\, e^{-j\omega t}\, dt = \int_{-\infty}^{\infty} \frac{\sin(\omega_0 t)}{\omega_0 t}\, e^{-j\omega t}\, dt$$

$$= \int_{-\infty}^{\infty} \frac{1}{\omega_0 t}\, \sin(\omega_0 t) \left[\cos(\omega t) - j \sin(\omega t) \right]\, dt$$

$$= \int_{-\infty}^{\infty} \frac{1}{\omega_0 t}\, \sin(\omega_0 t)\, \cos(\omega t)\, dt - j \int_{-\infty}^{\infty} \frac{1}{\omega_0 t}\, \sin(\omega_0 t)\, \sin(\omega t)\, dt$$

$$= I_1 - j\, I_2$$

In Example 19.1-8(d) we solved integral

$$\int_{0}^{\infty} e^{-\omega_0 t y}\, dy = \frac{1}{\omega_0 t}$$

whose solution appears as the leading term in both I_1 and I_2. Thus, that term may be replaced with the integral itself. In addition, we keep in mind that ω_0 is a constant, while ω is the integration variable and that integration is commutative. By doing so, after applying the trigonometric identities, the two integrals I_1 and I_2 are solved separately as follows. First, solution to I_1

$$I_1 = \int_{-\infty}^{\infty} \frac{1}{\omega_0 t}\, \sin(\omega_0 t)\, \cos(\omega t)\, dt$$

$$= \frac{1}{2} \int_{-\infty}^{\infty} \int_{0}^{\infty} e^{-\omega_0 t y}\, dy \left[\sin[(\omega_0 + \omega)\, t] + \sin[(\omega_0 - \omega)\, t] \right]\, dt$$

$$= \frac{1}{2} \int_{-\infty}^{\infty} \int_{0}^{\infty} e^{-\omega_0 t y}\, dy \left[\sin[(\omega + \omega_0)\, t] - \sin[(\omega - \omega_0)\, t] \right]\, dt$$

$$= \frac{1}{2} \left\{ \int_{-\infty}^{\infty} dy \int_{0}^{\infty} e^{-\omega_0 y t} \sin[(\omega + \omega_0)\, t]\, dt \right.$$

$$\left. - \int_{-\infty}^{\infty} dy \int_{0}^{\infty} e^{-\omega_0 y t} \sin[(\omega - \omega_0)\, t]\, dt \right\}$$

$\{$see: Example 19.1-8(h)$\}$

$$= \frac{1}{2} \int_{-\infty}^{\infty} \left[\frac{\omega + \omega_0}{(\omega + \omega_0)^2 + (\omega_0\, y)^2} \right]\, dy - \frac{1}{2} \int_{-\infty}^{\infty} \left[\frac{\omega - \omega_0}{(\omega - \omega_0)^2 + (\omega_0\, y)^2} \right]\, dy$$

$\{$see: Example 19.1-8(f)$\}$

$$= \frac{\pi}{2} \left[\text{sign}\, (\omega + \omega_0) - \text{sign}\, (\omega - \omega_0) \right]$$

The last result defines $\Pi\, (\omega/2\omega_0)$ as illustrated in Fig. 19.16 (a) to (d), which is well known dual function of sinc (t) function. That being the case, the question is: what is then I_2?

Fig. 19.16 Example 19.1-8

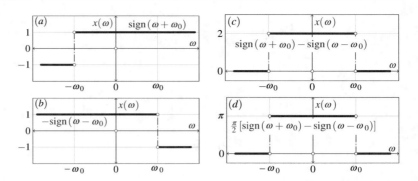

Solution to I_2 follows the same idea as for I_1; however, there are slight differences between the two paths to solutions.

$$I_2 = \int_{-\infty}^{\infty} \frac{1}{\omega_0 \, t} \, \sin(\omega_0 \, t) \, \sin(\omega \, t) \, dt$$

$$= \frac{1}{2} \int_{-\infty}^{\infty} \int_0^{\infty} e^{-\omega_0 \, t \, y} \, dy \, \Big[\cos[(\omega - \omega_0) \, t] - \cos[(\omega + \omega_0) \, t] \Big] \, dt$$

$$= \frac{1}{2} \left\{ \int_{-\infty}^{\infty} dy \int_0^{\infty} e^{-\omega_0 \, y \, t} \cos[(\omega - \omega_0) \, t] \, dt \right.$$

$$\left. - \int_{-\infty}^{\infty} dy \int_0^{\infty} e^{-\omega_0 \, y \, t} \cos[(\omega + \omega_0) \, t] \, dt \right\}$$

$$= \frac{1}{2} \int_{-\infty}^{\infty} \left[\frac{\omega_0 \, y}{(\omega - \omega_0)^2 + (\omega_0 \, y)^2} \right] dy - \frac{1}{2} \int_{-\infty}^{\infty} \left[\frac{\omega_0 \, y}{(\omega + \omega_0)^2 + (\omega_0 \, y)^2} \right] dy$$

$$\begin{cases} x & = (\omega \pm \omega_0)^2 + (\omega_0 \, y)^2 \quad \therefore \quad dx = 2\omega_0 \, y \, dy \quad \therefore \quad \omega_0 \, y \, dy = \frac{1}{2} \, dx \\ \\ \quad \therefore \quad \frac{1}{2} \int_{-\infty}^{\infty} \frac{dx}{x} = \frac{1}{2} \, \ln x \, \Big|_{-\infty}^{\infty} \end{cases}$$

$$= \frac{1}{4} \left[\ln \Big[(\omega - \omega_0)^2 + (\omega_0 \, y)^2 \Big] \, \Big|_{-\infty}^{\infty} - \ln \Big[(\omega + \omega_0)^2 + (\omega_0 \, y)^2 \Big] \, \Big|_{-\infty}^{\infty} \right]$$

$$= \frac{1}{4} \left[\ln \Big[(\omega - \omega_0)^2 + (\omega_0 \, y)^2 \Big] \, \Big|^{\infty} - \ln \Big[(\omega - \omega_0)^2 + (\omega_0 \, y)^2 \Big] \, \Big|_{-\infty} \right.$$

$$\left. - \ln \Big[(\omega + \omega_0)^2 + (\omega_0 \, y)^2 \Big] \, \Big|^{\infty} + \ln \Big[(\omega + \omega_0)^2 + (\omega_0 \, y)^2 \Big] \, \Big|_{-\infty} \right]$$

$$\left\{ \ln a + \ln b = \ln(ab); \quad \ln a - \ln b = \ln \frac{a}{b}; \quad \ln a^b = b \ln a \right\}$$

$$= \frac{1}{4} \left[\ln \frac{(\omega - \omega_0)^2 + (\omega_0 \, y)^2}{(\omega + \omega_0)^2 + (\omega_0 \, y)^2} \, \Big|^{\infty} - \ln \frac{(\omega - \omega_0)^2 + (\omega_0 \, y)^2}{(\omega + \omega_0)^2 + (\omega_0 \, y)^2} \, \Big|_{-\infty} \right]$$

$$= \frac{1}{4} \left\{ \lim_{y \to \infty} \left[\ln \frac{(\omega - \omega_0)^2 + (\omega_0 \, y)^2}{(\omega + \omega_0)^2 + (\omega_0 \, y)^2} \right] - \lim_{y \to -\infty} \left[\ln \frac{(\omega - \omega_0)^2 + (\omega_0 \, y)^2}{(\omega + \omega_0)^2 + (\omega_0 \, y)^2} \right] \right\}$$

$$= \frac{1}{4} \left\{ \ln \left[\lim_{y \to \infty} \frac{(\omega - \omega_0)^2 + (\omega_0 \, y)^2}{(\omega + \omega_0)^2 + (\omega_0 \, y)^2} \right] - \ln \left[\lim_{y \to -\infty} \frac{(\omega - \omega_0)^2 + (\omega_0 \, y)^2}{(\omega + \omega_0)^2 + (\omega_0 \, y)^2} \right] \right\}$$

$$= \left(\frac{\infty}{\infty} \right)$$

$$\overset{\text{L.H.}}{=} \frac{1}{4} \left\{ \ln \left[\lim_{y \to \infty} \frac{2\omega_0 \, y}{2\omega_0 \, y} \right] - \ln \left[\lim_{y \to -\infty} \frac{2\omega_0 \, y}{2\omega_0 \, y} \right] \right\}$$

$$= \frac{1}{4} \left[\ln(1) - \ln(1) \right] = 0$$

In conclusion,

$$X(\omega) = \int_{-\infty}^{\infty} \frac{\sin(\omega_0 \, t)}{\omega_0 \, t} \, e^{-j \omega t} \, dt = I_1 + j \overset{0}{I_2} = I_1$$

$$= \frac{\pi}{2} \left[\text{sign} \, (\omega + \omega_0) - \text{sign} \, (\omega - \omega_0) \right]$$

Exercise 19.2, page 414

1. Given $x(t)$ may be decomposed into the sum of two $\Pi \, (t)$ functions, $x_1(t)$ and $x_2(t)$ as illustrated in Fig. 19.17, where

$$x_1(t) = \Pi \left(\frac{t}{4} \right) \quad \text{and} \quad x_2(t) = \Pi \left(\frac{t}{2} \right)$$

Therefore,

$$x(t) = \Pi \left(\frac{t}{4} \right) + \Pi \left(\frac{t}{2} \right)$$

Fig. 19.17 Example 19.2-1

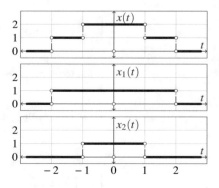

Fourier transformation is therefore the sum of two transformations as

$$X(\omega) = \mathscr{F}\left(\Pi\left(\frac{t}{4}\right)\right) + \mathscr{F}\left(\Pi\left(\frac{t}{2}\right)\right) = 4\mathrm{sinc}\,(2\omega) + 2\mathrm{sinc}\,(\omega)$$

2. Using, for example, properties of Fourier transformation with regard to derivatives given $x(t)$ we write

$$x(t) = -2\mathrm{sign}\,(t-2) \quad \therefore \quad x'(t) = -4\delta(t-2) \quad \therefore \quad j\omega\, X(\omega) = -4e^{-2j\omega}$$

$$\therefore$$

$$X(\omega) = -\frac{4}{j\omega}\, e^{-2j\omega}$$

3. Given RF pulse $h(t)$ may be decomposed into the product of $x(t) = \cos(t)$ and $\Pi(t)$ functions, Fig. 19.18, i.e.

$$h(t) = A\cos(2\pi f_c\, t)\, \Pi\left(\frac{t}{T}\right)$$

where f_c is the RF waveform frequency, and T is the duration of RF pulse.

Fig. 19.18 Example 19.2-3

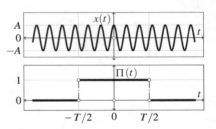

By using Euler's formula, we write

$$h(t) = \cos(2\pi f_s\, t) = \frac{1}{2}\left[\exp(j\,2\pi f_c\, t) + \exp(-j\,2\pi f_c\, t)\right]$$

Fourier transformation is therefore derived with the help of frequency shifting property, as

$$h(t) = \frac{1}{2}\Pi\left(\frac{t}{T}\right)\exp(j\,2\pi f_c\, t) + \frac{1}{2}\Pi\left(\frac{t}{T}\right)\exp(-j\,2\pi f_c\, t)$$

$$\therefore$$

$$\mathscr{F}(h(t)) = \frac{AT}{2}\left[\mathrm{sinc}\,(T(f - f_c)) + \mathrm{sinc}\,(T(f + f_c))\right]$$

That is to say, if $f_c T \gg 1$, we write, see Fig. 19.19

$$\mathscr{F}(h(t)) = \begin{cases} \dfrac{AT}{2}\,\text{sinc}\,(T(f - f_c)) & ; \quad f > 0 \\ 0 & ; \quad f = 0 \\ \dfrac{AT}{2}\,\text{sinc}\,(T(f + f_c)) & ; \quad f < 0 \end{cases}$$

Fig. 19.19 Example 19.2-3

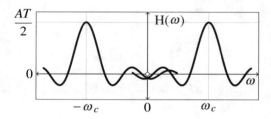

Discrete Fourier Transformation

20

Important to Know

Fourier sum: is defined as

$$\mathscr{F}(x_n) = X_k \overset{\text{def}}{=} \sum_{n=0}^{N-1} x_n\, e^{-j\frac{2\pi\, k}{N}\, n}$$

$$\mathscr{F}^{-1}(X[k]) = x[n] = \frac{1}{T} \sum_{-\infty}^{\infty} X[k]\, \exp\left(j\frac{2\pi k}{N} n \right)$$

Fourier series: is defined as

$$x(t) = \sum_{k=-\infty}^{\infty} C_k\, e^{j\,\omega_0\, t}, \quad \omega_0 = \frac{2\pi}{T}$$

$$C_k = \frac{1}{T} \int_{t_0}^{t_0+T} x(t)\, e^{-j\,\omega_0\, t}\, dt$$

20.1 Exercises

20.1 ** Discrete Fourier Sum

---

1. Derive discrete Fourier transformations $X[k]$ of the following discrete series $x[n]$.

   (a) $x[n] = \{1, 2, 3, 4\}$

   (b) $x[n] = \{2, 3, -1, 1\}$

   (c) $x[n] = \delta[n], \quad N = 16$

   (d) $x[n] = \delta[n - n_0]$

   (e) $x[n] = 1, \quad N = 16$

   (f) $x[n] = \alpha^n\, u[n-1], \quad (|\alpha| < 1)$

2. Derive inverse discrete Fourier transformations $x[n]$ of the following discrete series $X[k]$.

   (a) $X[k] = \{10, -2 + 2j, -2, -2 - 2j\}$

   (b) $X[k] = \{5, 3 - j2, -3, 3 + j2\}$

© Springer Nature Switzerland AG 2021
R. Sobot, *Engineering Mathematics by Example*,
https://doi.org/10.1007/978-3-030-79545-0_20

3. Derive discrete Fourier transformations $X[k]$ of the following functions $x(t)$.

(a) $x(t) = \sin(1\,\text{Hz})$, sampling frequency $f_s = 8\,\text{Hz}$.

20.2 **   Fourier Series

1. Given square waveform in Fig. 20.1 (left) derive Fourier series assuming

(a) $a = T/4$ (b) $a = T/8$

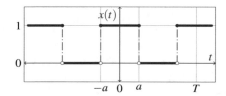 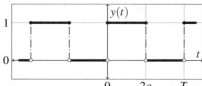

(a) $a = T/4$ (b) $a = T/8$

Fig. 20.1 Example 20.2-1 (left) and 2 (right)

2. Further to result derived in Example 20.2-1, given square waveform in Fig. 20.1 (right), derive Fourier series by applying the time-shifting property.

3. Given $x(t)$ waveform in Fig. 20.2, derive its Fourier series.

Fig. 20.2 Example 20.2-3

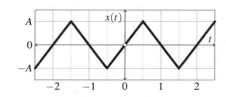

4. Given $f(t) = 1 - (t/\pi)$, derive its Fourier series if $t \in [-\pi, \pi]$.

20.3 **   Fast Fourier Transformation

1. Assuming the audio spectrum signal, [20 Hz, 20 kHz], estimate the time needed to calculate discrete Fourier transformation (DFF) by hands, if the desired resolution is 1 Hz and it takes 1 min to perform one calculation (extremely fast, but for the sake of argument). Then, compare with the time needed to do the same calculations by using Fast Fourier Transformation (FFT).

2. Given a general discrete series $x_n = x[n]$, develop transformation equations and illustrate the associated graph diagram of FFT algorithm if

(a) $N = 2$ (b) $N = 4$

3. By using graph diagram of FFT in Fig. 20.15, calculate Fourier transformation of $x[n] = \{1, 2, 3, 4\}$ (already solved using DFT in Example 20.1-1(a)).

4. Following the logic developed in Example 20.3-2,

    (a)   Deduce FFT graph diagram if the number of transformation points is $N = 8$.

    (b)   Using FFT graph diagram calculate $\mathscr{F}(x_n)$ if $x_n = [1, 1, -1, -1, 1, 1, -1, -1]$.

Solutions

Exercise 20.1, page 437

1. Fourier sum for a simply defined function is derived by definition.

    (a)   Given a simply defined function $x[n] = \{1, 2, 3, 4\}$, we deduce $N = 4$ and we write

$$X[k] = \sum_{n=0}^{N-1} x[n] \exp\left(-j\frac{2\pi k}{N}n\right) = \sum_{n=0}^{3} x[n] \exp\left(-j\frac{2\pi k}{4}n\right)$$

Calculations for each k are as follows.

$$k = 0: \quad X[0] = \sum_{n=0}^{3} x[n] \exp\left(-j\frac{2\pi\,0}{4}n\right) = \sum_{n=0}^{3} x[n]\,\underset{1}{e^0}$$

$$= x[0] + x[1] + x[2] + x[3] = 1 + 2 + 3 + 4$$

$$= 10$$

$$k = 1: \quad X[1] = \sum_{n=0}^{3} x[n] \exp\left(-j\frac{\cancel{2}^{1}\,\pi\,\cancel{1}}{\cancel{4}^{2}}n\right) = \sum_{n=0}^{3} x[n]\,\underbrace{\exp\left(-j\frac{\pi}{2}n\right)}_{(-j)^n}$$

$$\left\{\text{see the examples for calculating powers of complex numbers}\right\}$$

$$= x[0]\,(-j)^0 + x[1]\,(-j)^1 + x[2]\,(-j)^2 + x[3]\,(-j)^3$$

$$= 1(1) + 2(-j) + 3(-1) + 4(j) = 1 - 2j - 3 + 4j$$

$$= -2 + 2j$$

$$k = 2: \quad X[2] = \sum_{n=0}^{3} x[n] \exp\left(-j\frac{\cancel{2}\,\pi\,\cancel{2}}{\cancel{4}\,1}n\right) = \sum_{n=0}^{3} x[n]\,\underbrace{\exp\left(-j\pi\,n\right)}_{(-1)^n}$$

$$= x[0]\,(-1)^0 + x[1]\,(-1)^1 + x[2]\,(-1)^2 + x[3]\,(-1)^3$$

$$= 1(1) + 2(-1) + 3(1) + 4(-1) = 1 - 2 + 3 - 4$$

$$= -2$$

(a)   *(cont.)*

$$k = 3: \quad X[3] = \sum_{n=0}^{3} x[n] \exp\left(-j\frac{\cancel{2}\,1}{\cancel{4}\,2}\frac{\pi\,3}{}n\right) = \sum_{n=0}^{3} x[n] \underbrace{\exp\left(-j\frac{3\pi}{2}n\right)}_{(j)^n}$$

$$= x[0]\,(j)^0 + x[1]\,(j)^1 + x[2]\,(j)^2 + x[3]\,(j)^3$$

$$= 1(1) + 2(j) + 3(-1) + 4(-j) = 1 + 2j - 3 - 4j$$

$$= -2 - 2j = x[1]^*$$

Therefore, after recalling that $\theta = -\pi \equiv \pi$, and $\theta = -(3\pi/4) \equiv (5\pi/4)$, we write its graphical representation in Fig. 20.3 and the complete solution as

$$X[k] = \{10, -2 + 2j, -2, -2 - 2j\}$$

$$|X[k]| = \left\{10, 2\sqrt{2}, 2, 2\sqrt{2}\right\}$$

$$\theta_k = \left\{0, \frac{3\pi}{4}, \pi, \frac{5\pi}{4}\right\}$$

Fig. 20.3 Example 20.2-1

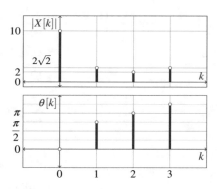

(b)   Given function $x[n] = \{2, 3, -1, 1\}$ we conclude $N = 4$; by definition we write

$$X[k] = \sum_{n=0}^{N-1} x[n] \exp\left(-j\frac{2\pi k}{N}n\right) = \sum_{n=0}^{3} x[n] \exp\left(-j\frac{\cancel{2}}{\cancel{4}\,2}\frac{\pi k}{}n\right)$$

Thus,

$$k = 0: \quad X[0] = \sum_{n=0}^{3} x[n] \exp\left(-j\frac{\pi}{2}\frac{0}{n}\right) = \sum_{n=0}^{3} x[n] \underbrace{\exp(0)}_{1} = \sum_{n=0}^{3} x[n]$$

$$= 2 + 3 - 1 + 1 = 5$$

$$k = 1: \quad X[1] = \sum_{n=0}^{3} x[n] \exp\left(-j\frac{\pi}{2}\frac{1}{n}\right)$$

$$= 2\exp\left(-j\frac{0\pi}{2}\right) + 3\exp\left(-j\frac{1\pi}{2}\right) - 1\exp\left(-j\frac{2\pi}{2}\right) + 1\exp\left(-j\frac{3\pi}{2}\right)$$

$$= 2 + 3(-j) - 1(-1) + 1(j)$$

$$= 3 - 2j$$

$$k = 2: \quad X[2] = \sum_{n=0}^{3} x[n] \exp\left(-j\frac{\pi}{2}\frac{2}{n}\right)$$

$$= 2\exp\left(-j(0\pi)\right) + 3\exp\left(-j(1\pi)\right) - 1\exp\left(-j(2\pi)\right) + 1\exp\left(-j(3\pi)\right)$$

$$= 2 + 3(-1) - 1(1) + 1(-1)$$

$$= -3$$

$$k = 3: \quad X[3] = \sum_{n=0}^{3} x[n] \exp\left(-j\frac{\pi}{2}\frac{3}{n}\right)$$

$$= 2\exp\left(-j\frac{\pi}{2}\frac{3}{0}\right) + 3\exp\left(-j\frac{\pi}{2}\frac{3}{1}\right) - 1\exp\left(-j\frac{\pi}{2}\frac{3}{2}\right) + 1\exp\left(-j\frac{\pi}{2}\frac{3}{3}\right)$$

$$= 2 + 3(j) - 1(-1) + 1(-j)$$

$$= 3 + 2j$$

Therefore,

$$X[k] = \{5, 3 - 2j, -3, 3 + 2j\}$$

$$|X[k]| = \left\{5, \sqrt{13}, 3, \sqrt{13}\right\}$$

$$\theta[k] = \{0, -33.7°, \pi, 33.7°\}$$

(c)   Given Delta function $x[n] = \delta[n]$ in $N = 16$ points, that is to say $x[0] = 1$ otherwise $x[n] = 0, n \neq 0$, or

$$x[n] = \{1, 0, 0, 0, 0, 0, 0, 0, 0, 0, 0, 0, 0, 0, 0, 0\}$$

By definition we write

$$X[k] = \sum_{n=0}^{N-1} x[n] \exp\left(-j\frac{2\pi k}{N}n\right) = \sum_{n=0}^{15} x[n] \exp\left(-j\frac{2\pi k}{\cancel{16}\ 8}n\right)$$

Calculations for each k are as follows.

$$\underline{k=0:}\quad X[0] = \sum_{n=0}^{15} x[n] \exp\left(-j\frac{\pi}{8}0\ n\right) = \sum_{n=0}^{15} x[n] \underbrace{\exp(0)}_{1}$$

$$= x[0] + x[1] + x[2] + \cdots + x[15] = 1 + 0 + 0 + \cdots + 0$$

$$= 1$$

$$\underline{k=1:}\quad X[1] = \sum_{n=0}^{15} x[n] \exp\left(-j\frac{\pi}{8}1\ n\right) = \sum_{n=0}^{15} x[n] \exp\left(-j\frac{\pi}{8}n\right)$$

$$= x[0] \exp\left(-j\frac{\pi}{8}0\right) + 0 + 0 + \cdots + 0$$

$$= 1$$

$$\underline{k=2:}\quad X[2] = \sum_{n=0}^{15} x[n] \exp\left(-j\frac{\pi}{8}2\ n\right) = \sum_{n=0}^{15} x[n] \exp\left(-j\frac{2\pi}{8}n\right)$$

$$= x[0] \exp\left(-j\frac{2\pi}{8}0\right) + 0 + 0 + \cdots + 0$$

$$= 1$$

$$\vdots$$

etc.

In other words $X[k] = 1$, $\forall k$, which is, in general, a constant function that is known to be dual function of impulse.

(d)   DFT of delayed impulse $x[n] = \delta[n - n_0]$ we derive as

$$X[k] = \sum_{n=-\infty}^{\infty} x[n] \exp\left(-j\frac{2\pi k}{N}n\right) = \sum_{n=-\infty}^{\infty} \delta[n - n_0] \exp\left(-j\frac{2\pi k}{N}n\right)$$

$$= \cdots + 0 + 0 + \underbrace{\delta[n_0 - n_0] \exp\left(-j\frac{2\pi k}{N}n_0\right)}_{n=n_0} + 0 + 0 + \cdots$$

$$= 1 \exp\left(-j\frac{2\pi k}{N}n_0\right)$$

Therefore, $|X[k]| = 1, \forall k$, and the argument is $(2\pi k\, n_0/N)$, which illustrates the property of delayed functions in general where the delay transforms into the equivalent argument (i.e. angle or phase).

(e)   A constant function $x[n] = 1$ defined in $N = 16$ points is transformed as

$$X[k] = \sum_{n=0}^{N-1} x[n] \exp\left(-j\frac{2\pi k}{N}n\right) = \sum_{n=0}^{15} 1 \exp\left(-j\frac{2\pi k}{\cancel{16}\,8}n\right)$$

Calculations for each k are as follows.

$$\underline{k = 0}: \quad X[0] = \sum_{n=0}^{15} \exp\left(-j\frac{\pi}{8}\frac{0}{}n\right) = \sum_{n=0}^{15} \cancel{\exp(0)}^{1} = \sum_{n=0}^{15} 1_n$$

$$= 16$$

$$\underline{k = 1}: \quad X[1] = \sum_{n=0}^{15} \exp\left(-j\frac{\pi}{8}\frac{1}{}n\right) = \sum_{n=0}^{15} \exp\left(-j\frac{\pi}{8}n\right)$$

$$= \exp\left(-j\frac{\pi}{8}0\right) + \exp\left(-j\frac{\pi}{8}1\right) + \cdots + \exp\left(-j\frac{\pi}{8}15\right)$$

The question is: how to do summation of the last sum? First, modules of all terms equal to one. Second, each term in the sum is a complex number that is shifted along the circle at the multiples of $\pi/8$, that is to say all sixteen terms are uniformly distributed on the unity circle in the complex plain, see Fig. 20.4. We note that, for example, $x[6]$ and $x[14]$ are at the opposite sides; therefore, their sum equals zero. Formally, we can confirm that statement as

Fig. 20.4 Example 20.1-1

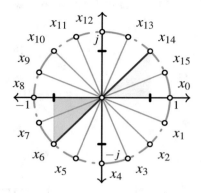

$$x[6] + x[14] = \exp\left(-j\frac{\pi}{8}6\right) + \exp\left(-j\frac{\pi}{8}14\right)$$

$$= \exp\left(-j\frac{\cancel{2}\,\pi}{\cancel{8}\,4}3\right) + \exp\left(-j\frac{\cancel{2}\,\pi}{\cancel{8}\,4}7\right)$$

$$= \exp\left(-j\frac{3\pi}{4}\right) + \exp\left(j\frac{\pi}{4}\right)$$

$$= \cos\left(-\frac{3\pi}{4}\right) + j\sin\left(-\frac{3\pi}{4}\right) + \cos\left(\frac{\pi}{4}\right) + j\sin\left(\frac{\pi}{4}\right)$$

$$= -\frac{\sqrt{2}}{2} - j\frac{\sqrt{2}}{2} + \frac{\sqrt{2}}{2} + j\frac{\sqrt{2}}{2}$$

$$= 0$$

Similarly, we show that all other sums of the opposite pairs also equal zero, i.e.

$$x[0] + x[8] = 0 \qquad\qquad x[4] + x[12] = 0$$
$$x[1] + x[9] = 0 \qquad\qquad x[5] + x[13] = 0$$
$$x[2] + x[10] = 0 \qquad\qquad x[6] + x[14] = 0$$
$$x[3] + x[11] = 0 \qquad\qquad x[7] + x[15] = 0$$

Same calculation is valid for all the other values of k; therefore, we conclude

$$X[0] = 16$$

$$X[1] = \exp\left(-j\frac{\pi}{8}0\right) + \exp\left(-j\frac{\pi}{8}1\right) + \cdots + \exp\left(-j\frac{\pi}{8}15\right) = 0$$

$$X[2] = X[3] = \cdots = X[15] = 0$$

In summary, given $N = 16$

$$\mathscr{F}(x[n] = 1) = 16\,\delta[n]$$

and $X[k]$ being real function, the argument $\theta[k] = 0$.

(f)   Given infinitely long series $x[n] = \alpha^n\, u[n-1], \quad (|\alpha| < 1)$ we apply Fourier sum as follows:

$$X[k] = \sum_{n=-\infty}^{\infty} x[n]\, \exp\left(-j\frac{2\pi k}{N}n\right) = \sum_{n=-\infty}^{\infty} \alpha^n\, u[n-1]\, \exp\left(-j\frac{2\pi k}{N}n\right)$$

$$\left\{ \text{delayed step function } u[n-1] = 0, \ \forall n < 1 \quad \therefore \quad u[n-1] = 1, \ \forall n \geq 1 \right\}$$

$$= \sum_{n=1}^{\infty} \alpha^n\, (1)\, \exp\left(-j\frac{2\pi k}{N}n\right) = \sum_{n=1}^{\infty} \left[\alpha\, \exp\left(-j\frac{2\pi k}{N}\right)\right]^n$$

$$\left\{ x^a\, y^a = (xy)^a; \quad \text{as well as} \quad x^{ab} = \left(x^a\right)^b \right\}$$

$$\left\{ \sum_{k=m}^{N} a^k = \frac{a^{N+1} - a^m}{a - 1}, \quad (|a| < 1) \right\}$$

$$= \frac{\left[\alpha\, \exp\left(-j\frac{2\pi k}{N}\right)\right]^{\infty + 1^{\,0}} - \left[\alpha\, \exp\left(-j\frac{2\pi k}{N}\right)\right]^1}{\alpha\, \exp\left(-j\frac{2\pi k}{N}\right) - 1}$$

$$\left\{ \left|\exp\left(-j\frac{2\pi k}{N}\right)\right| = 1, \quad \alpha < 1 \quad \therefore \quad \alpha\, \exp\left(-j\frac{2\pi k}{N}\right) < 1 \right\}$$

$$= \frac{\alpha\, \exp\left(-j\frac{2\pi k}{N}\right)}{1 - \alpha\, \exp\left(-j\frac{2\pi k}{N}\right)}$$

2. Inverse Fourier sum is derived by definition.

(a)   Given a simply defined function $X[k] = \{10, -2+2j, -2, -2-2j\}$, we deduce $N = 4$ and we write

$$x[n] = \mathscr{F}^{-1}(X[k]) = \frac{1}{T}\sum_{-\infty}^{\infty} X[k]\exp\left(j\frac{2\pi k}{N}n\right) = \frac{1}{4}\sum_{n=0}^{3} X[k]\exp\left(j\frac{2\pi k}{4}n\right)$$

$$= \frac{1}{4}\left[10\exp\left(j\frac{2\pi\,0}{4}n\right)^{\!1} + (-2+j2)\exp\left(j\frac{2\,\pi\,1}{4\,2}n\right)\right.$$

$$\left.-2\exp\left(j\frac{2\,\pi\,2}{4}n\right) + (-2-j2)\exp\left(j\frac{2\,\pi\,3}{4\,2}n\right)\right]$$

$\underline{n = 0:}$ $x[0] = \dfrac{1}{4}\left[10 + (-2+j2)\exp\left(j\dfrac{\pi}{2}0\right)\right.$

$$\left.-2\exp\left(j\pi\,0\right) + (-2-j2)\exp\left(j\dfrac{3\pi}{2}0\right)\right]$$

$$= \frac{1}{4}\left[10 - 2 + j2 - 2 - 2 - j2\right]$$

$$= 1$$

Therefore,

$\underline{n = 1:}$ $x[1] = \dfrac{1}{4}\left[10 + (-2+j2)\exp\left(j\dfrac{\pi}{2}1\right)^{\!j}\right.$

$$\left.-2\exp\left(j\pi\,1\right)^{-1} + (-2-j2)\exp\left(j\dfrac{3\pi}{2}1\right)^{-j}\right]$$

$$= \frac{1}{4}\left[10 + (-2+j2)(j) - 2(-1) + (-2-j2)(-j)\right]$$

$$= \frac{1}{4}\left[10 + 2j - 2 + 2 + 2j - 2\right]$$

$$= 2$$

$\underline{n = 2:}$ $x[2] = \dfrac{1}{4}\left[10 + (-2+j2)\exp\left(j\dfrac{\pi}{2}2\right)^{-1}\right.$

$$\left.-2\exp\left(j\pi\,2\right)^{1} + (-2-j2)\exp\left(j\dfrac{3\pi}{2}2\right)^{-1}\right]$$

$$= \frac{1}{4}\left[10 + (-2+j2)(-1) - 2(1) + (-2-j2)(-1)\right]$$

$$= \frac{1}{4}\left[10 + 2 - 2j - 2 + 2 + 2j\right]$$

$$= 3$$

(a)   *(cont.)*

$$n = 3: \quad x[3] = \frac{1}{4}\left[10 + (-2 + j2) \, \exp\left(j\frac{\pi}{2}3\right)^{-j} \right.$$

$$\left. - 2 \, \exp\left(j\pi \, 3\right)^{-1} + (-2 - j2) \, \exp\left(j\frac{3\pi}{2}3\right)^{j} \right]$$

$$= \frac{1}{4}\left[10 + (-2 + j2)(-j) - 2(-1) + (-2 - j2)(j) \right]$$

$$= \frac{1}{4}\left[10 + 2j + 2 + 2 - 2j + 2 \right]$$

$$= 4$$

In conclusion, $x[n] = \{1, 2, 3, 4\}$, as same as in Example 20.1-1(a).

(b)   Given a simply defined function $X[k] = \{5, 3 - j2, -3, 3 + j2\}$, we deduce $N = 4$ and we write

$$x[n] = \mathscr{F}^{-1}(X[k]) = \frac{1}{T}\sum_{-\infty}^{\infty} X[k] \, \exp\left(j\frac{2\pi k}{N}n\right) = \frac{1}{4}\sum_{n=0}^{3} X[k] \, \exp\left(j\frac{2\pi k}{4}n\right)$$

$$= \frac{1}{4}\left[5 \, \exp\left(j\frac{2\pi \, 0}{4}n\right)^{1} + (3 - j2) \, \exp\left(j\frac{2\pi \, 1}{4 \, 2}n\right) \right.$$

$$\left. - 3 \, \exp\left(j\frac{2\pi \, 2}{4}n\right) + (3 + j2) \, \exp\left(j\frac{2\pi \, 3}{4 \, 2}n\right) \right]$$

Therefore,

$$n = 0: \quad x[0] = \frac{1}{4}\left[5 + (3 - j2) \, \exp\left(j\frac{\pi}{2}0\right) \right.$$

$$\left. - 3 \, \exp\left(j\pi \, 0\right) + (3 + j2) \, \exp\left(j\frac{3\pi}{2}0\right) \right]$$

$$= \frac{1}{4}\left[5 + 3 - j2 - 3 + 3 + j2 \right]$$

$$= 2$$

(b) *(cont.)*

$$\underline{n=1}: \quad x[1] = \frac{1}{4}\left[5 + (3-j2)\,\exp\left(j\frac{\pi}{2}1\right)^{j}\right.$$

$$-3\,\exp(j\pi 1)^{-1} + (3+j2)\,\exp\left(j\frac{3\pi}{2}1\right)^{-j}\Bigg]$$

$$= \frac{1}{4}\left[5 + (3-j2)(j) - 3(-1) + (3+j2)(-j)\right]$$

$$= \frac{1}{4}\left[5 + 3j + 2 + 3 - 3j + 2\right]$$

$$= 3$$

$$\underline{n=2}: \quad x[2] = \frac{1}{4}\left[5 + (3-j2)\,\exp\left(j\frac{\pi}{2}2\right)^{-1}\right.$$

$$-3\,\exp(j\pi 2)^{1} + (3+j2)\,\exp\left(j\frac{3\pi}{2}2\right)^{-1}\Bigg]$$

$$= \frac{1}{4}\left[5 + (3-j2)(-1) - 3(1) + (3+j2)(-1)\right]$$

$$= \frac{1}{4}\left[5 - 3 + 2j - 3 - 3 - 2j\right]$$

$$= -1$$

$$\underline{n=3}: \quad x[3] = \frac{1}{4}\left[5 + (3-j2)\,\exp\left(j\frac{\pi}{2}3\right)^{-j}\right.$$

$$-3\,\exp(j\pi 3)^{-1} + (3+j2)\,\exp\left(j\frac{3\pi}{2}3\right)^{j}\Bigg]$$

$$= \frac{1}{4}\left[5 + (3-j2)(-j) - 3(-1) + (3+j2)(j)\right]$$

$$= \frac{1}{4}\left[5 - 3j - 2 + 3 + 3j - 2\right]$$

$$= 1$$

In conclusion, $x[n] = \{2, 3, -1, 1\}$, as same as in Example 20.1-1(b).

3. Given $x(t) = \sin(1\,\text{Hz})$, in the first step it is necessary to convert continuous function $x(t)$ into its discrete version sampled at $N = 8$ points. Given the sampling frequency $f_s = 8\,\text{Hz}$ we normalize period $T(1\,\text{Hz}) = 1\,\text{s}$ as

$$f = 1\text{Hz} \Rightarrow T = \frac{1}{f} = 1\,\text{s} \Leftrightarrow 2\pi$$

$$f_s = 8\,\text{Hz} \Rightarrow N = 8$$

$$\therefore$$

$$\frac{1}{8}\,\text{s} = \frac{2\pi}{8} = \frac{\pi}{4}$$

that is to say, each sample is found at the multiples of $\pi/4$, see Fig. 20.5.
We calculate $N = 8$ samples as

Fig. 20.5 Example 20.1-3

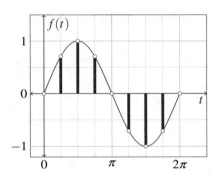

$$x[0] = (1)\sin(0) = 0$$

$$x[1] = (1)\sin\left(\frac{\pi}{4}\right) = \frac{\sqrt{2}}{2}$$

$$x[2] = (1)\sin\left(\frac{\pi}{4}\,2\right) = 1$$

$$x[3] = (1)\sin\left(\frac{\pi}{4}\,3\right) = \frac{\sqrt{2}}{2}$$

$$x[4] = (1)\sin\left(\frac{\pi}{4}\,4\right) = 0$$

$$x[5] = (1)\sin\left(\frac{\pi}{4}\,5\right) = -\frac{\sqrt{2}}{2}$$

$$x[6] = (1)\sin\left(\frac{\pi}{4}\,6\right) = -1$$

$$x[7] = (1)\sin\left(\frac{\pi}{4}\,7\right) = -\frac{\sqrt{2}}{2}$$

Therefore,

$$x[n] = \left\{0, \frac{\sqrt{2}}{2}, 1, \frac{\sqrt{2}}{2}, 0, -\frac{\sqrt{2}}{2}, -1, -\frac{\sqrt{2}}{2}\right\}$$

We note

$$\sum_{n=1}^{7} x[n] = 0 + \frac{\sqrt{2}}{2} + 1 + \frac{\sqrt{2}}{2} + 0 - \frac{\sqrt{2}}{2} - 1 - \frac{\sqrt{2}}{2} = 0$$

By definition, we write

$$X[k] = \sum_{n=0}^{N-1} x[n] \exp\left(-j\frac{2\pi k}{N}n\right) = \sum_{n=0}^{7} x[n] \exp\left(-j\frac{2\,\pi k}{8\;4}n\right)$$

Calculations for each k are as follows.

$\underline{k=0:}$ $X[0] = \sum_{n=0}^{7} x[n] \exp\left(-j\frac{\pi\,0}{4}n\right) = \sum_{n=0}^{7} x[n]\, e^{0\,1} = \sum_{n=0}^{7} x[n] = 0$

$\underline{k=1:}$ $X[1] = \sum_{n=0}^{7} x[n] \exp\left(-j\frac{\pi\,1}{4}n\right)$

$$= (0)\exp(0) + \frac{\sqrt{2}}{2}\exp\left(-j\frac{\pi}{4}1\right) + (1)\exp\left(-j\frac{\pi}{4}2\right)$$

$$+ \frac{\sqrt{2}}{2}\exp\left(-j\frac{\pi}{4}3\right) + (0)\exp\left(-j\frac{\pi}{4}4\right) - \frac{\sqrt{2}}{2}\exp\left(-j\frac{\pi}{4}5\right)$$

$$- (1)\exp\left(-j\frac{\pi}{4}6\right) - \frac{\sqrt{2}}{2}\exp\left(-j\frac{\pi}{4}7\right)$$

$$\{\cos(-x) = \cos(x) \quad \sin(-x) = -\sin(x)\}$$

$$= \frac{\sqrt{2}}{2}\left[\cos\left(\frac{\pi}{4}\right) - j\sin\left(\frac{\pi}{4}\right)\right] + (1)\left[\cos\left(\frac{2\pi}{4}\right) - j\sin\left(\frac{2\pi}{4}\right)\right] + \cdots$$

$$= (0.5 - j0.5) + (-j) + (-0.5 - j0.5) + (0.5 - j0.5) + (-j) + (-0.5 - j0.5)$$

$$= -4j$$

$\underline{k=2:}$ $X[2] = \sum_{n=0}^{7} x[n] \exp\left(-j\frac{\pi\,2}{4\;2}n\right)$

$$= (0)\exp(0) + \frac{\sqrt{2}}{2}\exp\left(-j\frac{\pi}{2}\right) + (1)\exp\left(-j\frac{\pi}{2}2\right)$$

$$+ \frac{\sqrt{2}}{2}\exp\left(-j\frac{\pi}{2}3\right) + (0)\exp\left(-j\frac{\pi}{2}4\right) - \frac{\sqrt{2}}{2}\exp\left(-j\frac{\pi}{2}5\right)$$

$$- (1)\exp\left(-j\frac{\pi}{2}6\right) - \frac{\sqrt{2}}{2}\exp\left(-j\frac{\pi}{2}7\right)$$

$$= \frac{\sqrt{2}}{2}\left[\cos\left(\frac{\pi}{2}\right) - j\sin\left(\frac{\pi}{2}\right)\right] + (1)\left[\cos\left(\frac{2\pi}{2}\right) - j\sin\left(\frac{2\pi}{2}\right)\right] + \cdots$$

$$= -j\frac{\sqrt{2}}{2} - 1 + j\frac{\sqrt{2}}{2} + j\frac{\sqrt{2}}{2} + 1 - j\frac{\sqrt{2}}{2}$$

$$= 0$$

Similarly,

$$X[2] = X[3] = X[4] = X[5] = X[6] = 0$$
$$X[7] = \cdots = 4j$$

In summary,

$$X[k] = \{0, -4j, 0, 0, 0, 0, 0, 4j\}$$
$$|X[k]| = \{0, 4, 0, 0, 0, 0, 0, 4\}$$

This Fourier transformation result is referred to as "bilateral" as illustrated in Fig. 20.6 (top). There are two impulses in $|X[k]|$; however, we know that sinusoidal function is dual with only impulse.

In order to show amplitude $|X[k]|$ relative to the units of frequency, i.e. [Hz], we must calculate resolution of each "bin" as

$$\frac{f_s}{N} = \frac{8\,\text{Hz}}{8} = 1\,\text{Hz/bin}$$

Fig. 20.6 Example 20.1-3

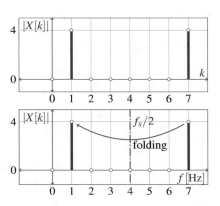

In this case, because N and sampling frequency f_s are numerically equal it happens that k maps directly into Hz. That is to say, by selecting different either N or f_s we can arbitrary set the resolution of Fourier transformation. In addition, impulses found above half of sampling frequency, known as Nyquist frequency, are folded over in "unilateral" view of Fourier transformation (as commonly shown by spectrum analyzers or simulated graphs), see Fig. 20.6 (bottom).

Consequently, in unilateral view amplitudes at each frequency are calculated as the sum of samples already in place and the ones "folded over".

For example amplitude of sample at $x(1\,\text{Hz})$ is increased by amplitude of $x(7\,\text{Hz})$ that after folding over falls on top, thus $|H[1\,\text{Hz}]| = 4 + 4 = 8$. This amplitude is the consequence of $N = 8$; thus in order to show correct amplitude, i.e. $|x(t)| = 1$, the unilateral graph must be recalibrated by dividing the amplitudes by N, see Fig. 20.7.

Fig. 20.7 Example 20.1-3

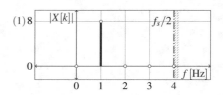

Exercise 20.2, page 438

1. Square pulse waveform that is even (i.e. symmetric around $x = 0$) may be decomposed on Fourier series as follows. By definition, coefficients of Fourier series for periodic function $x(t) = \sin(\omega t)$ are

$$x(t) = \sum_{k=-\infty}^{\infty} C_n \exp(j n\omega_0 t); \quad \text{where,} \quad \omega_0 = \frac{2\pi}{T} \quad \text{and}$$

$$C_n = \frac{1}{T} \int_{t_0}^{t_0+T} x(t) \exp(-j n\omega_0 t) \, dt$$

Case $n = 0$: note that $x(t) = 1, \quad t \in [-a, a]$, thus

$$C_0 = \frac{1}{T} \int_{t_0}^{t_0+T} x(t) \underbrace{\exp(-j\,0\omega_0 t)}_{1} \, dt = \frac{1}{T} \int_{-a}^{a} (1) \, dt = \frac{2a}{T}$$

otherwise,

$$C_n = \frac{2}{T} \int_{t_0}^{t_0+T} x(t) \exp(-j n\omega_0 t) \, dt = \frac{2}{T} \int_{-a}^{a} (1) \exp(-j n\omega_0 t) \, dt$$

$$= \frac{2}{-jn\omega_0} \frac{1}{T} \left[\exp(-j n\omega_0 a) - \exp(j n\omega_0 a) \right]$$

$$= \frac{2T}{n\pi} \frac{1}{T} \frac{1}{2j} \left[\exp(-j n\omega_0 a) - \exp(j n\omega_0 a) \right]$$

$$= \frac{2}{n\pi} \sin\left(\frac{2n\pi}{T} a \right)$$

Relative to the period T we calculate

(a) $\underline{a = T/4}$: therefore,

$$C_0 = \frac{2a}{T} = \frac{2\,\cancel{T}}{\cancel{T}\,4\,2} = \frac{1}{2}$$

$$C_n = \frac{2}{n\pi} \sin\left(\frac{2n\pi}{T} a\right) = \frac{2}{n\pi} \sin\left(\frac{2\,n\pi\,\cancel{T}}{\cancel{T}\,4\,2}\right) = \frac{\sin\,(n\pi/2)}{n\pi/2} = \text{sinc}\,(n\pi/2)$$

$$\therefore$$

$$|C_n| = |\text{sinc}\,(n\pi/2)\,|$$

We calculate a few samples,

$$C_1 = \frac{2}{\pi} \sin\left(\frac{\pi}{2}\right) = \frac{2}{\pi}$$

$$C_2 = \frac{2}{\pi} \sin\left(\frac{2\pi}{2}\right) = 0$$

$$C_3 = \frac{2}{\pi} \sin\left(\frac{3\pi}{2}\right) = -\frac{2}{3\pi}$$

$$C_4 = \frac{2}{\pi} \sin\left(\frac{4\pi}{2}\right) = 0$$

$$C_5 = \frac{2}{\pi} \sin\left(\frac{5\pi}{2}\right) = \frac{2}{5\pi}$$

$$\cdots$$

That is to say, function $x(t)$ can be synthesized as the sum of infinite sine series. Recall that $|C_n| = \sqrt{a_n^2 + b_n^2}$; thus

$$x(t) = \sum_{k=-\infty}^{\infty} C_n \, \exp\,(j\,n\omega_0 t) = a_0 + \sum_{n=1}^{\infty} a_n \cos(n\omega_0 t) + b_n \sin(n\omega_0 t)$$

$$\left\{\text{for even functions } b_n = 0 \quad \therefore \quad |C_n| = |a_n|\right\}$$

$$= \frac{1}{2} + \sum_{n=1}^{\infty} \frac{2}{n\pi} \sin\left(\frac{n\pi}{2}\right) \cos(n\,\omega_0 t)$$

$$= \frac{1}{2} + \frac{2}{\pi}\left[\cos\left(\frac{2\pi}{T} t\right) - \frac{1}{3}\cos\left(3 \times \frac{2\pi}{T} t\right) + \frac{1}{5}\cos\left(5 \times \frac{2\pi}{T} t\right) + \cdots\right]$$

As a comparison, this series is overlaid on $x(t)$, see Fig. 20.8, where only the first four terms are used, and spectrum is illustrated in Fig. 20.9 (left). Note that all even terms equal to zero.

Fig. 20.8 Example 20.2-1

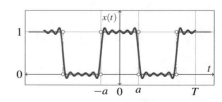

(b) $a = T/8$: therefore,

$$C_0 = \frac{2a}{T} = \frac{2\,\cancel{T}}{\cancel{T}84} = \frac{1}{4}$$

$$C_n = \frac{2}{n\pi} \sin\left(\frac{2n\pi}{T}a\right) = \frac{2}{2}\frac{2}{n\pi}\sin\left(\frac{2\,n\pi\,\cancel{T}}{\cancel{T}\,84}\right) = \frac{1}{2}\frac{\sin(n\pi/4)}{n\pi/4} = \frac{1}{2}\,\mathrm{sinc}\,(n\pi/4)$$

$$\therefore$$

$$|C_n| = \frac{1}{2}\,|\mathrm{sinc}\,(n\pi/4)\,|$$

We calculate a few samples,

$$C_1 = \frac{2}{\pi}\sin\left(\frac{\pi}{4}\right) = \frac{\sqrt{2}}{\pi}$$

$$C_2 = \frac{2}{2\pi}\sin\left(\frac{2\pi}{4}\right) = \frac{1}{\pi}$$

$$C_3 = \frac{2}{3\pi}\sin\left(\frac{3\pi}{4}\right) = \frac{\sqrt{2}}{3\pi}$$

$$C_4 = \frac{2}{4\pi}\sin\left(\frac{4\pi}{4}\right) = 0$$

$$C_5 = \frac{2}{5\pi}\sin\left(\frac{5\pi}{4}\right) = -\frac{\sqrt{2}}{5\pi}$$

$$\ldots$$

and spectrum is illustrated in Fig. 20.9 (right).

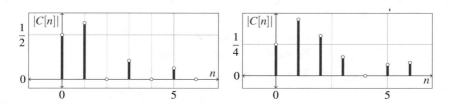

Fig. 20.9 Example 20.2-1

2. Relative to square pulse waveform in Example 20.2-1, shapes of $x(t)$ and $y(t)$ waveforms are same except for the time delay, that is to say $y(t) = x(t - a)$. Fourier transform of time delay is the exponential phase shift term; thus we write

$$C_n = \exp(-jn\omega_0 a)\,\frac{1}{n\pi}\sin\left(\frac{2n\pi}{T}a\right)$$

3. Given function $x(t)$ is odd, by graph inspection, we write

$$T_0 = 2 \quad \therefore \quad \omega_0 = \frac{2\pi}{T_0} = \frac{2\pi}{2} = \pi$$

$$x(t) = 2At \quad x \in \left[-\frac{T_0}{4}, \frac{T_0}{4}\right] = \left[-\frac{1}{2}, \frac{1}{2}\right]$$

Coefficients of Fourier series for an odd periodic function $x(t)$ may be calculated by definition as follows:

$$x(t) = \sum_{k=-\infty}^{\infty} C_n \exp(j\, n\omega_0 t) = a_0 + \sum_{n=1}^{\infty} \left[a_n \cos(n\omega_0 t) + b_n \sin(n\omega_0 t)\right]$$

where, in odd function all $a_0 = a_n = 0$, while

$$b_n = \frac{2}{T_0} \int_0^{T_0} x(t) \sin(n\omega_0 t)\ dt = \frac{4}{T_0} \int_0^{T_0/2} x(t) \sin(n\omega_0 t)\ dt$$

$$= \frac{4}{T_0} \int_{-T_0/4}^{T_0/4} x(t) \sin(n\omega_0 t)\ dt = \frac{4}{2} \int_{-1/2}^{1/2} 2At\ \sin(n\pi\, t)\ dt$$

$$= 4A \int_{-1/2}^{1/2} t\ \sin(n\pi\, t)\ dt$$

Integral in the form $\int x \sin(ax)\ dx$ is solved by partial integration as

$$\int x \sin(ax)\ dx = \begin{cases} u = x & \therefore\quad du = dx \\ dv = \sin(ax)\ dx & \therefore\quad v = -\dfrac{1}{a}\cos(ax) \end{cases}$$

$$= -\frac{1}{a}x \cos(ax) + \int \frac{1}{a}\cos(ax)\ dx = -\frac{1}{a}x \cos(ax) + \frac{1}{a^2}\sin(ax)$$

$$= \frac{1}{a^2}\left[\sin(ax) - ax\, \cos(ax)\right]$$

Therefore,

$$b_n = 4A \left[\frac{1}{n\pi^2}\left[\sin(n\pi\, t) - n\pi t\, \cos(n\pi\, t)\right]\right]\Bigg|_{-1/2}^{1/2}$$

$$= \frac{4A}{(n\pi)^2}\left[\sin\left(\frac{n\pi}{2}\right) + \sin\left(\frac{n\pi}{2}\right) - \left(\frac{n\pi}{2}\right)\cos\left(\frac{n\pi}{2}\right) + \left(\frac{n\pi}{2}\right)\cos\left(\frac{n\pi}{2}\right)\right]$$

$$= \frac{8A}{(n\pi)^2}\ \sin\left(\frac{n\pi}{2}\right)$$

$$
= \begin{cases} 0; & n \text{ even} \\ \dfrac{8A}{(n\pi)^2}; & n = 1, 5, 9, \ldots \\ -\dfrac{8A}{(n\pi)^2}; & n = 3, 7, 11, \ldots \end{cases}
$$

Fourier series is therefore

$$
x(t) = \sum_{n=1}^{\infty} b_n \sin{(n\pi\, t)} = \frac{8A}{\pi^2}\left[\, \sin{(\pi\, t)} - \frac{1}{9}\sin{(3\pi\, t)} + \frac{1}{25}\sin{(5\pi\, t)} + \cdots \right]
$$

As a comparison, this series is overlaid on $x(t)$, see Fig. 20.10, where only the first three terms are used, and amplitude spectrum is illustrated in Fig. 20.11 (left) and phase spectrum is in Fig. 20.11 (right). Note that all even terms equal to zero.

Fig. 20.10 Example 20.2-1

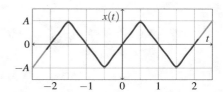

$$
|C_n| = \sqrt{a_n^2 + b_n^2} = \sqrt{b_n^2} = |b_n| = \frac{8A}{(n\pi)^2}
$$

$$
\theta_n = \arctan\left(\frac{-b_n}{a_n^{\nearrow 0}}\right) = \arctan(\pm\infty) = \begin{cases} 0; & n \text{ even} \\ -\dfrac{\pi}{2}; & n = 1, 5, 9, \ldots \\ \dfrac{\pi}{2}; & n = 3, 7, 11, \ldots \end{cases}
$$

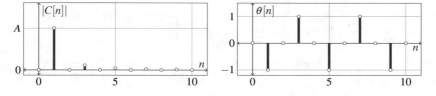

Fig. 20.11 Example 20.2-3

4. Given a linear function $f(t) = 1 - t/\pi$ its Fourier series within the interval $x \in [-\pi, \pi]$ is calculated as follows. By definition we write

$$f(t) = a_0 + \sum_{n=1}^{\infty} a_n \cos\left(\frac{n\pi t}{T}\right) + \sum_{n=1}^{\infty} b_n \sin\left(\frac{n\pi t}{T}\right)$$

where

$$a_0 = \frac{1}{2T} \int_{-T}^{T} f(t)\, dt$$

$$a_n = \frac{1}{T} \int_{-T}^{T} f(t) \cos\left(\frac{n\pi t}{T}\right) dt$$

$$b_n = \frac{1}{T} \int_{-T}^{T} f(t) \sin\left(\frac{n\pi t}{T}\right) dt$$

In addition, given interval it follows that $T = \pi$; therefore

$$a_0 = \frac{1}{2\pi} \int_{-\pi}^{\pi} \left(1 - \frac{t}{\pi}\right) dt = \frac{1}{2\pi}\left[t\,\Big|_{-\pi}^{\pi} - \frac{1}{2\pi}t^2\,\Big|_{-\pi}^{\pi}\right] = \frac{1}{2\pi}\left[2\pi - \frac{1}{2\pi}\left(\pi^2 - \pi^2\right)\right]$$

$$= 1$$

$$a_n = \frac{1}{\pi} \int_{-\pi}^{\pi} \left(1 - \frac{t}{\pi}\right) \cos\left(\frac{n\pi t}{\pi}\right) dt = \frac{1}{\pi} \int_{-\pi}^{\pi} \cos(nt)\, dt - \frac{1}{\pi^2}\left[\int_{-\pi}^{\pi} x \cos(nt)\, dt \nearrow^{0}\right]$$

$$\begin{cases} \text{for odd functions: } f(x) = x \cos(nx) = -f(-x) \quad \therefore \quad \int_{-a}^{a} f(x)\, dx = 0, \\[2mm] \text{or, use partial fraction integration to prove it} \end{cases}$$

$$= \frac{1}{n\pi} \sin(nt)\,\Big|_{-\pi}^{\pi} = \frac{1}{n\pi}\left[\sin(n\pi)\nearrow^{0} - \sin(-n\pi)\nearrow^{0}\right]$$

$$= 0$$

$$b_n = \frac{1}{\pi} \int_{-\pi}^{\pi} \left(1 - \frac{t}{\pi}\right) \sin\left(\frac{n\pi t}{\pi}\right) dt = \frac{1}{\pi} \underbrace{\int_{-\pi}^{\pi} \sin(nt)\, dt}_{\cos(n\pi) - \cos(n\pi) = 0}\nearrow^{0} - \frac{1}{\pi^2} \int_{-\pi}^{\pi} t \sin(nt)\, dt$$

$$\{\text{use partial fraction integration}\}$$

$$= -\frac{1}{\pi^2}\frac{1}{n}\left[-t\cos(nt) + \underbrace{\sin(nt)}_{\sin(n\pi)=0}\right]_{-\pi}^{\pi} = \frac{1}{\pi^2}\frac{1}{n}\left[\pi\cos(n\pi) + \pi\cos(n\pi)\right]$$

$$= \frac{1}{\pi^2}\frac{1}{n}\, 2\,\pi \underbrace{\cos(n\pi)}_{(-1)^n}$$

$$= (-1)^n \frac{2}{n\pi}$$

Therefore,

$$f(t) = 1 + \sum_{n=1}^{\infty} (-1)^n \frac{2}{n\pi} \sin(nt) = 1 - \frac{2}{\pi} \sin(t) + \frac{2}{2\pi} \sin(2t) - \frac{2}{3\pi} \sin(3t) + \cdots$$

This series is superimposed to $f(t)$, see Fig. 20.12, which illustrates the approximation. For "perfect" approximation, there should be an infinite number of terms. The largest discrepancy between the continuous function $f(t)$ and its approximation is at the end of given interval. Over wider interval Fourier series is a periodic function.

Fig. 20.12 Example 20.2-4

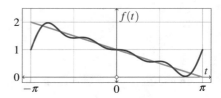

Exercise 20.3, page 438

1. Discrete Fourier transformation (DFT) of series $x[n]$ is by definition

$$\mathscr{F}_k(x[n]) \overset{\text{def}}{=} \sum_{n=0}^{N-1} x[n] \exp\left(-j\frac{2\pi k}{N} n\right) \quad \text{where,} \quad k = 0, 1, 2, \ldots$$

In order to simplify the long equations writing, often we use exp() to denote the necessary exponential term in the above definition. As n, k, N changes, we find that the number of necessary calculations increases as follows:

$\underline{N = 1}$ $\mathscr{F}_0 = x[0]$ exp() (one calculation)

$\underline{N = 2}$ $\mathscr{F}_0 = x[0]$ exp() $+ x[1]$ exp() (two calculations)

$\quad\quad\quad$ $\mathscr{F}_1 = x[0]$ exp() $+ x[1]$ exp() (two calculations)

$\quad\quad\quad\quad\quad\quad\quad\quad\quad\quad\quad\quad$ (total: four calculations)

$\underline{N = 3}$ $\mathscr{F}_0 = x[0]$ exp() $+ x[1]$ exp() $+ x[2]$ exp() (three calculations)

$\quad\quad\quad$ $\mathscr{F}_1 = x[0]$ exp() $+ x[1]$ exp() $+ x[2]$ exp() (three calculations)

$\quad\quad\quad$ $\mathscr{F}_2 = x[0]$ exp() $+ x[1]$ exp() $+ x[2]$ exp() (three calculations)

$\quad\quad\quad\quad\quad\quad\quad\quad\quad\quad\quad\quad$ (total: nine calculations)

$\quad\quad\quad\vdots$

$\quad\quad N$ $\quad\quad\quad\quad\quad\quad\quad\quad\quad\quad\quad\quad$ (total: N^2 calculations)

That is to say, for example, the resolution of $1\,\mathrm{Hz}$ within $20\,\mathrm{kHz}$ bandwidth means that it is necessary to do $N = 20\,000$ points in the transformation. Knowing that DFT calculation does N^2 operations, we calculate

$$t = N^2 \times 1\,\mathrm{min} = \left(20 \times 10^3\right)^2 \mathrm{min} = 400 \times 10^6\,\mathrm{min} \approx 278 \times 10^3\ \mathrm{days}$$

$$\approx 760\ \ \mathrm{years}$$

By comparison, FFT algorithm does $N \log_2 N$ calculations, that is to say

$$t = (N \log_2 N) \times 1\,\mathrm{min} = \left(N \frac{\log_2}{\ln 2}\right) \times 1\,\mathrm{min} = 286 \times 10^3\,\mathrm{min} \approx 198\ \ \mathrm{days}$$

The advantage of FFT over DFT increases as N becomes larger.

2. The main idea of FFT is to split the grand sum into two separate sums, in the first iteration to one with odd and one with even coefficients. In the subsequent iteration, again every second term is grouped as "odd" and the other half as "even" group, see Fig. 20.13. The process is repeated until each group consists of two terms only. This process is highly symmetric; as a consequence the binary form of aligned term indexes in the final iteration is binary inverted relative to the binary form of aligned term indexes at the beginning.

Fig. 20.13 Example 20.3-2

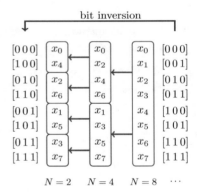

That is to say, we can write as follows.

$$\mathscr{F}_k(x[n]) \overset{\text{def}}{=} \sum_{n=0}^{N-1} x[n] \exp\left(-j\frac{2\pi k}{N}\,n\right)$$

$$= \underbrace{\sum_{m=0}^{N/2-1} x[2m] \exp\left(-j\frac{2\pi k}{N/2}\frac{2m}{2}\right)}_{\text{even index}} + \underbrace{\sum_{m=0}^{N/2-1} x[2m+1] \exp\left(-j\frac{2\pi k}{N/2}\frac{2m+1}{2}\right)}_{\text{odd index}}$$

$$\left\{C_k = \exp\left(-j\frac{2\pi k}{N}\right)\right\}$$

$$= \sum_{m=0}^{N/2-1} x[2m] \exp\left(-j\frac{2\pi k}{N/2}\,m\right) + C_k \sum_{m=0}^{N/2-1} x[2m+1] \exp\left(-j\frac{2\pi k}{N/2}\,m\right)$$

where

$$\exp\left(-j\frac{2\pi\,k}{N/2}\,m\right) = \cos\left(-\frac{2\pi\,k}{N/2}\,m\right) + j\sin\left(-\frac{2\pi\,k}{N/2}\,m\right);\quad (k=0,1,\ldots,N)$$

However, when $k > N/2$, that is to say $k = N/2 + r$ where $r = 1, 2, \ldots, N/2$, we can write

$$\cos\left(-\frac{2\pi\,(N/2+r)}{N/2}\,m\right) = \cos\left(\frac{-2\pi\,(N/2)\,m - 2\pi\,r\,m}{N/2}\right)$$

$$= \cos\left(-2\pi\,m - \frac{2\pi\,r\,m}{N/2}\right)$$

$$\{m = 0, 1, \ldots, N/2 \quad\therefore\quad -2\pi\,m = 0\}$$

$$= \cos\left(-\frac{2\pi\,r\,m}{N/2}\right) = \cos\left(-\frac{2\pi\,(N/2+r)\,m}{N/2}\right)$$

$$\{\text{known as "symmetry identity"}\}$$

This identity is valid for sine terms as well, which is to say that

$$\exp\left(-j\frac{2\pi\,k}{N/2}\,m\right) = \exp\left(-j\frac{2\pi\,(N/2+k)}{N/2}\,m\right);\quad (k=0,1,2,\ldots,N)$$

As a consequence, it is sufficient to carry out only $N/2$ calculations. However, each of the summing groups can be split again and again; thus the total number of calculations is $N\log_2 N$.

(a)   Given $N = 2$ and a general discrete series $x_n = x[n]$, we write Fourier transform $\mathcal{F}_k(x[n]) = X_k$ equations using the following syntax

$$X_k = \sum_{n=0}^{N-1} x_n \exp\left(-j\frac{2\pi}{N}\,n\,k\right) = \sum_{n=0}^{N-1} x_n\,w_N^{n\cdot k} = \{N = 2\}$$

$$= \sum_{n=0}^{1} x_n\,w_2^{n\cdot k}$$

This syntax simplifies the repetitive writing of complex exponential terms; we keep in mind that $w_N^{n\cdot k}$ represents the following calculations:

$$w_N \equiv \exp\left(-j\frac{2\pi}{N}\right) \quad\therefore\quad \exp\left(-j\frac{2\pi}{N}\,n\,k\right) = \left[\exp\left(-j\frac{2\pi}{N}\right)\right]^{n\cdot k} \stackrel{\text{def}}{=} w_N^{n\cdot k}$$

where the power term is calculated as the product nk, for example, if $N = 2$ it follows that

$$w_2^{0\cdot 0} = \left[\exp\left(-j\frac{2\pi}{2}\right)\right]^{0\cdot 0} = \left[\exp\left(-j\pi\right)\right]^{0\cdot 0} = (-1)^0 = 1$$

or, =

$$w_2^{1 \cdot 1} = \left[\exp\left(-j\frac{2\pi}{2}\right)\right]^{1 \cdot 1} = [\exp(-j\pi)]^{1 \cdot 1} = (-1)^1 = -1$$

For the moment, let us keep $n > 0$ exponential terms in place and also use the identity $w_2^{1 \cdot 1} \equiv -w_2^{0 \cdot 0}$, even if their values are as simple as $+1$ or -1. Reason for this choice becomes evident as $N = 4$, $N = 8$, etc. In this case,

<u>k=0:</u> $\quad X_0 = x_0 \, w_2^{0 \cdot 0} + x_1 \, w_2^{1 \cdot 0} = x_0 + x_1 \equiv x_0 + x_1 \, w_2^0$
<u>k=1:</u> $\quad X_1 = x_0 \, w_2^{0 \cdot 1} + x_1 \, w_2^{1 \cdot 1} = x_0 - x_1 \equiv x_0 - x_1 \, w_2^0$

This calculation process is illustrated by "butterfly diagram", Fig. 20.14. The solutions X_k on the right side are formed by the graph inspection. For example, x_0 is first multiplied by "1" and then added to $x_1 \, w_2^0$ at the summing node to produce $X_0 = x_0 + x_1 \, w_2^0$. Similarly, the graphs illustrate the formation of X_1 equation. This basic butterfly diagram is reused for any $N = 2$ Fast Fourier Transformation. As illustrated by the colour scheme, note paths that carry "$+1$" multiplication factor, paths that carry positive exponential factor, and paths that carry negative exponential factor.

Fig. 20.14
Example 20.3-2(a)

(b)   In the case of $N = 4$, we reorganize the four X_k DFT transformation into symmetric form that can be directly translated into butterfly diagram as follows.
As a reminder, we use shorthand syntax for $N = 4$ as

$$w_4 \equiv \exp\left(-j\frac{2\pi}{4}\right) = \exp\left(-j\frac{\pi}{2}\right) \quad \therefore \quad \exp\left(-j\frac{\pi}{2}\right)^{n k} \equiv w_4^{n \cdot k}$$

$$\therefore$$

$$w_4^0 = \exp\left(-j\frac{\pi}{2}\right)^0 = 1$$

$$w_4^1 = \exp\left(-j\frac{\pi}{2}\right)^1 = -j$$

$$w_4^2 = \exp\left(-j\frac{\pi}{2}\right)^2 = \exp(-j\pi) = -1$$

$$w_4^3 = \exp\left(-j\frac{\pi}{2}\right)^3 = \exp\left(-j\frac{3\pi}{2}\right) = \exp\left(j\frac{\pi}{2}\right) = j$$

so that we can write

$$\underline{k=0}: \; X_0 = x_0 \, w_4^{0 \cdot 0} + x_1 \, w_4^{1 \cdot 0} + x_2 \, w_4^{2 \cdot 0} + x_3 \, w_4^{3 \cdot 0}$$

$$= x_0 + x_1 + x_2 + x_3 \quad \text{(``DC component'')}$$

$$\underline{k=1}: \; X_1 = x_0 \, w_4^{0 \cdot 1} + x_1 \, w_4^{1 \cdot 1} + x_2 \, w_4^{2 \cdot 1} + x_3 \, w_4^{3 \cdot 1}$$

$$= x_0 + x_1 \, w_4^1 + x_2 \, w_4^2 + x_3 \, w_4^3 = \left\{ \frac{3\pi}{2} \equiv -\frac{\pi}{2} \right\}$$

$$= x_0 + x_1 \, w_4^1 - x_2 - x_3 \, w_4^1$$

$$\underline{k=2}: \; X_2 = x_0 \, w_4^{0 \cdot 2} + x_1 \, w_4^{1 \cdot 2} + x_2 \, w_4^{2 \cdot 2} + x_3 \, w_4^{3 \cdot 2}$$

$$= x_0 + x_1 \, w_4^2 + x_2 \, w_4^4 + x_3 \, w_4^6$$

$$= x_0 - x_1 + x_2 - x_3$$

$$\underline{k=3}: \; X_3 = x_0 \, w_4^{0 \cdot 3} + x_1 \, w_4^{1 \cdot 3} + x_2 \, w_4^{2 \cdot 3} + x_3 \, w_4^{3 \cdot 3}$$

$$= x_0 + x_1 \, w_4^3 + x_2 \, w_4^6 + x_3 \, w_4^9 = \left\{ \frac{9\pi}{2} \equiv -\frac{3\pi}{2} \right\}$$

$$= x_0 + x_1 \, w_4^1 - x_2 - x_3 \, w_4^1$$

This set of equation may be written in a bit symmetric form as

$$X_0 = (x_0 + x_2) + w_4^0 (x_1 + x_3)$$

$$X_1 = (x_0 - x_2) + w_4^1 (x_1 - x_3)$$

$$X_2 = (x_0 + x_2) + w_4^2 (x_1 + x_3)$$

$$X_3 = (x_0 - x_2) + w_4^3 (x_1 - x_3)$$

which can be further structured to explicitly include the form of two-bit butterfly (i.e. $N = 2$) as

$$X_0 = (x_0 + w_2^0 \, x_2) + w_4^0 (x_1 + w_2^0 \, x_3)$$

$$X_1 = (x_0 - w_2^0 \, x_2) + w_4^1 (x_1 - w_2^0 \, x_3)$$

$$X_2 = (x_0 + w_2^0 \, x_2) - w_4^0 (x_1 + w_2^0 \, x_3)$$

$$X_3 = (x_0 - w_2^0 \, x_2) - w_4^1 (x_1 - w_2^0 \, x_3)$$

Direct graph implementation of these equations is illustrated in Fig. 20.15, where starting from the left side of the graph, and following x_n terms along the associated paths (the addition operations are done at the merging nodes), we compose the four X_k equations simply by inspection. Note the index order of x_n and X_k indexes in this graph structure.

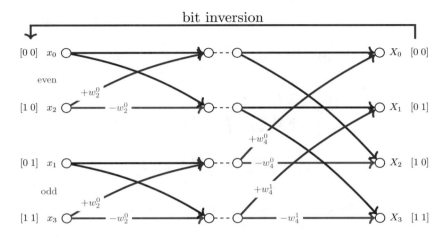

Fig. 20.15 Example 20.3-2(b)

3. In order to use FFT graph diagram, in the first step we reorganize x_n terms on the left side by using the index bit inversion relative to X_k, see Fig. 20.16. The intermediate sums are calculated using the two-bit grouping on the left side, and then the second group of addition/multiplication operations is done by following paths on the right side of the graph.

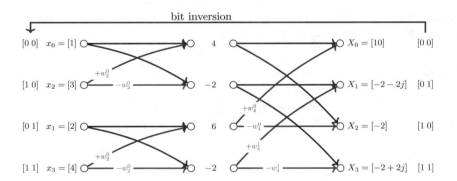

Fig. 20.16 Example 20.3-3

4. (a) In the case of $N = 8$, we use shorthand syntax as

$$w_8 \equiv \exp\left(-j\frac{2\pi}{8}\right) = \exp\left(-j\frac{\pi}{4}\right) \quad \therefore \quad \exp\left(-j\frac{\pi}{4}\right)^{n\,k} \equiv w_8^{n \cdot k}$$

$\therefore$

$$w_8^0 = \exp\left(-j\frac{\pi}{4}\right)^0 = 1$$

$$w_8^1 = \exp\left(-j\frac{\pi}{4}\right)^1 = \frac{\sqrt{2}}{2}(1 - j)$$

$$w_8^2 = \exp\left(-j\frac{\pi}{4}\right)^2 = \exp\left(-j\frac{\pi}{2}\right) = -j$$

$$w_8^3 = \exp\left(-j\frac{\pi}{4}\right)^3 = \exp\left(-j\frac{3\pi}{4}\right) = -\frac{\sqrt{2}}{2}(1 + j)$$

$$w_8^4 = \exp\left(-j\frac{\pi}{4}\right)^4 = \exp\left(-j\pi\right) = -1 \equiv -w_8^0$$

$$w_8^5 = \exp\left(-j\frac{\pi}{4}\right)^5 = \exp\left(-j\frac{5\pi}{4}\right) = -\frac{\sqrt{2}}{2}(1 - j) \equiv -w_8^1$$

$$w_8^6 = \exp\left(-j\frac{\pi}{4}\right)^6 = \exp\left(-j\frac{6\pi}{4}\right) = j \equiv -w_8^2$$

$$w_8^7 = \exp\left(-j\frac{\pi}{4}\right)^7 = \exp\left(-j\frac{7\pi}{4}\right) = \frac{\sqrt{2}}{2}(1 + j) \equiv -w_8^3$$

With the help of Fig. 20.13, we derive FFT graph flow for $N = 8$ as in Fig. 20.17, where horizontal red lines imply multiplication by "-1", dark lines imply multiplication by "1", and other multiplication factors are written explicitly.

(b) Given $x_n = [1, 1, -1 - 1, 1, 1, -1 - 1]$, by inspection of graph in Fig. 20.17, where horizontal red lines imply multiplication by "-1", we write

$$X_0 = x_0 + x_4 + x_2 + x_6 + x_1 + x_5 + x_3 + x_7 = 1 + 1 - 1 - 1 + 1 + 1 - 1 - 1$$

$$= 0$$

$$X_1 = x_0 - x_4 + (x_2 - x_6)(-j) + \left(x_1 - x_5 + (x_3 - x_7)(-j)\right)w_8^2$$

$$= 1 - 1 - 1 + 1 + (1 - 1 + (-1 + 1)(-j))w_8^2$$

$$= 0$$

$$X_2 = x_0 + x_4 + (x_2 + x_6)(-1) + \left(x_1 + x_5 + (x_3 + x_7)(-1)\right)(-j)$$

$$= 1 + 1 + 1 + 1 + (1 + 1 + 1 + 1)(-j)$$

$$= 4 - 4j$$

$$X_3 = x_0 - x_4 + (x_2 - x_6)(-w_8^2) + x_1 - x_5 + (x_3 - x_7)(-w_8^2)$$

$$= 1 - 1 + (-1 + 1)(-w_8^2) + 1 - 1 + (-1 + 1)(-w_8^2)$$

$$= 0$$

$$X_4 = x_0 + x_4 + x_2 + x_6 + (x_1 + x_5 + x_3 + x_7)(-1)$$

$$= 1 + 1 - 1 - 1 - 1 - 1 + 1 + 1$$

$$= 0$$

$$X_5 = x_0 - x_4 + (x_2 - x_6)(-j) + \left(x_1 - x_5 + (x_3 - x_7)(-j)\right)(-w_8^1)$$

$$= 1 - 1 + (-1 + 1)(-j) + \left(1 - 1 + (-1 + 1)(-j)\right)(-w_8^1)$$

$$= 0$$

$$X_6 = x_0 + x_4 + (x_2 + x_6)(-1) + \left(x_1 + x_5 + (x_3 + x_7)(-1)\right)(j)$$

$$= 1 + 1 + 1 + 1 + (1 + 1 + 1 + 1)(j)$$

$$= 4 + 4j$$

$$X_7 = x_0 + x_4 + (x_2 - x_6)(j) + \left(x_1 - x_5 + (x_3 - x_7)(j)\right)(-w_8^3)$$

$$= 1 + 1 + (-1 + 1)(j) + \left(1 - 1 + (-1 + 1)(j)\right)(-w_8^3)$$

$$= 0$$

The total number of sum/product calculations is $8 \log_2 8 = 24$ in comparison with DFT where the total number of calculations equals $8^2 = 64$.

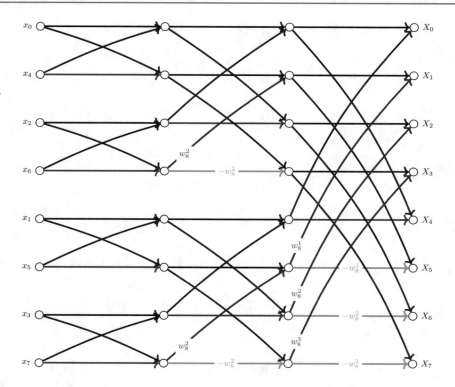

Fig. 20.17 Example 20.3-4(a)

References

[App17] W. Appel. *Mathématiques pour la physique et les physiciens*. Number 978–2–35141–339–5. H&K Éditions, 5th edn. (2017)

[Cou20a] MIT Open Courseware. *Multivariable Calculus* (2020), https://ocw.mit.edu/courses/mathematics/18-02sc-multivariable-calculus-fall-2010/. On-line July 2020

[Cou20b] MIT Open Courseware. *Signals and Systems* (2020), https://ocw.mit.edu/resources/res-6-007-signals-and-systems-spring-2011. On-line July 2020

[Cou20c] MIT Open Courseware. *Single Variable Calculus* (2020), https://ocw.mit.edu/courses/mathematics/18-01sc-single-variable-calculus-fall-2010/. On-line July 2020

[Dem63] B.P. Demidovic, *Collection of Problems and Exercises for Mathematical Analysis (in Russian)*. Number 517.2 D30. National Printing House for Literature in Mathematics and Physics, 5th edn. (1963)

[Ink20] Inkscape. *Drawings Generated by Inkscape* (2020), https://inkscape.org/. On-line July 2020

[LT09] Y. Leroyer, P. Tessen, *Mathématique pour l'ingénieur: exercices et problèmes*. Number 978–2–10–052186–9. Dunod, 1st edn. (2009)

[Mit67] D.S. Mitrinovic, *Matematika I: u obliku metodicke zbirke zadataka sa rešenjima*. Number 2043. Gradjevinska Knjiga, 3rd edn. (1967)

[Mit77] D.S. Mitrinovic, *Kompleksna Analiza*. Gradjevinska Knjiga, 4th edn. (1977)

[Mit78] D.S. Mitrinovic, *Matematika I: u obliku metodicke zbirke zadataka sa rešenjima*. Number 06–1785/1, Gradjevinska Knjiga, 5th edn. (1978)

[Mit79] D.S. Mitrinovic, *Kompleksna analiza: zbornik zadataka i problema*, vol. 3. Number 06–1431/1, Naucna Knjiga, 2nd edn. (1979)

[Mod64] P.S. Modenov, *Collection of problems in special program (in Russian)*. National Printing House for Literature in Mathematics and Physics, (1964)

[Mü20] D. Müller, *Mathématiques* (2020), http://www.apprendre-en-ligne.net/MADIMU2/INDEX.HTM. On-line July 2020

[Sob20] R. Sobot, *Private Collection of Course Notes in Mathematics*. (2020)

[Spi80] M.R. Spiegel, *Analyse de Fourier et Application aux Problémes de Valeurs aux Limites*. Number France 2–7042–1019–5. McGraw–Hill, 1st edn. (1980)

[Ste11a] Stewart, *Analyse Concepts et Contextes, Vol.1: Fonctions d'une Variable*. Number 978–2–8041–6306–8. De Boeck Supérieur, 3rd edn. (2011)

[Ste11b] Stewart, *Analyse Concepts et Contextes, Vol.2: Fonctions de plusieurs variables*. Number 978–2–8041–6327–3. De Boeck Supérieur, 3rd edn. (2011)

[Sym20] Symbolab, *Online Calculator* (2020), https://www.symbolab.com/. On-line July 2020

[Tik20] TikZ, *Graphs Generated by TikZ* (2020) , https://en.wikipedia.org/wiki/PGF/TikZ. On-line July 2020

[Ven96] T.B. Vene, *Zbirka rešenih zadataka iz matematike*, vol. 2. Number 86–17–09617–9. Zavod za Udžbenike i Nastavna Sredstva, 22nd edn. (1996)

[Ven01] T.B. Vene, *Zbirka rešenih zadataka iz matematike*, vol. 1. Number 86–17–09031–6. Zavod za Udžbenike i Nastavna Sredstva, 28th edn. (2001)

[Ven03] T.B. Vene, *Zbirka rešenih zadataka iz matematike*, vol. 3. Number 86–17–10786–3. Zavod za Udžbenike i Nastavna Sredstva, 27th edn. (2003)

© Springer Nature Switzerland AG 2021
R. Sobot, *Engineering Mathematics by Example*,
https://doi.org/10.1007/978-3-030-79545-0

[Ven04] T.B. Vene, *Zbirka rešenih zadataka iz matematike*, vol. 4. Number 86–17–11156–9. Zavod za Udžbenike i Nastavna Sredstva, 36th edn. (2004)

[Wik20a] Wikipedia, *Fourier Transform* (2020), https://en.wikipedia.org/wiki/Fourier_transform. On-line July 2020

[Wik20b] Wikipedia, *Linear Algebra* (2020), https://en.wikipedia.org/wiki/Linear_algebra. On-line July 2020

Index

© Springer Nature Switzerland AG 2021
R. Sobot, *Engineering Mathematics by Example*,
https://doi.org/10.1007/978-3-030-79545-0

Printed in the United States
by Baker & Taylor Publisher Services